NONLINEAR FUNCTIONAL ANALYSIS AND APPLICATIONS

VOLUME 1

NONLINEAR FUNCTIONAL ANALYSIS AND APPLICATIONS

Additional books in this series can be found on Nova's website
under the Series tab.

Additional E-books in this series can be found on Nova's website
under the E-book tab.

NONLINEAR FUNCTIONAL ANALYSIS AND APPLICATIONS

VOLUME 1

YEOL JE CHO
JONG KYU KIM
AND
SHIN MIN KANG
EDITORS

Nova Science Publishers, Inc.
New York

LIBRARY OF CONGRESS CATALOGING-IN-PUBLICATION DATA

Nonlinear functional analysis and applications / [edited by] Yoel Je Cho.
    p. cm.
  Includes index.
  ISBN 978-1-61942-059-5 (softcover)
  1. Nonlinear functional analysis. I. Cho, Yoel Je.
  QA321.5.N667 2009
  515'.7248--dc22
                                2010012210

Published by Nova Science Publishers, Inc. † New York

CONTENTS

PREFACE

On July 25~28, 2007, "the 9-th International Conference on Nonlinear Functional Analysis and Applications" was held at Gyeongsang National University (GNU) and Kyungnam University (KU) by the financial support of the Research Institute of Natural Sciences of GNU, the Institute of Basic Science Research of KU, the Korean Mathematical Society, the Pusan-Kyongnam Mathematical Society, Nova Science Publishers, Inc., New York, and others.

The aim of the conference which has been opened every two years is to introduce and exchange recent new topics on the areas of fixed point theory, convex and set-valued analysis, variational inequality and complementarity problem theory, nonlinear ergodic theory, difference, differential and integral equations, control and optimization theory, dynamic system theory, inequality theory, stochastic analysis and probability theory, and their applications.

We have published some books as the series in mathematical analysis and applications whenever the conference has been opened. The following books also consist of the reviewed papers by the referees, which were given at the 9-th Conference, and some invited papers of leading scholars in their fields as special contributors:

(∗) **Nonlinear Functional Analysis and Applications, Vol. 1 and 2**
   (Editors: Yeol Je Cho, Jong Kyu Kim, Shin Min Kang)

(∗) **Inequality Theory and Applications, Vol. 6**
   (Editors: Yeol Je Cho, Jong Kyu Kim, Sever S. Dragomir).

We are sure that these books will contribute and spur further research in the topics mentioned above. We are grateful and thankful to all the authors for their contribution to these books. These books would not have been published without their cooperation, support and help.

Finally, we would like to thank Dr. Frank Columbus, President of Nova Science Publishers, Inc., New York, and all the staffs for publishing the beautiful books.

August 1, 2009

YEOL JE CHO (Gyeongsang National University, Korea)
JONG KYU KIM (Kyungnam University, Korea)
SHIN MIN KANG (Gyeongsang National University, Korea)

Nonlinear Functional Analysis and Applications, Volume 1, 1–9

NEW ABSTRACT CONVEX SPACES
FOR THE KKM THEORY

SEHIE PARK[1]

ABSTRACT. We introduce some classes of abstract convex spaces, namely, abstract convex minimal spaces, minimal KKM spaces, and G-convex minimal spaces. Each of these subclasses is convenient for the KKM theory and contains G-convex spaces properly. The class of G-convex spaces contains the classes of L-spaces, FC-spaces, and others. A number of examples and related matters are also added.

1. Introduction

The KKM theory, originally called by the author [9], is nowadays the study of applications of various equivalent formulations of the Knaster-Kuratowski-Mazurkiewicz theorem (simply, the KKM principle) in 1929 [8]. In the last decade, the theory has been extensively studied for generalized convex spaces (simply, G-convex spaces) in a sequence of papers of the author; for details, see [10-14] and references therein.

Since the concept of G-convex spaces first appeared in 1993 [23], a number of modifications or imitations of the concept have followed. Such examples are L-spaces due to Ben-El-Mechaiekh et al. [2], spaces having property (H) due to Huang [7], FC-spaces due to Ding [3,4], and others. It is known that all of such examples are particular forms of G-convex spaces; see [20].

In our previous work [15], we introduced a new concept of abstract convex spaces and multimap classes $\mathfrak{K}$, $\mathfrak{KC}$, and $\mathfrak{KO}$ having certain KKM property. These new spaces and multimap classes are known to be adequate to establish the KKM theory; see [15-19]. Especially, in [19], we generalized and simplified known results of the theory on convex spaces, H-spaces, G-convex spaces, and others. It is noticed there that the class of abstract convex spaces $(E, D; \Gamma)$ satisfying the KKM principle play the major role in

Received August 10, 2007.

2000 *Mathematics Subject Classification.* 47H04, 47H10, 49J27, 49J35, 54H25, 91B50.

Key words and phrases. Abstract convex spaces, KKM spaces, minimal spaces, generalized (G-) convex space, L-spaces, FC-spaces, property (H).

[1] The National Academy of Sciences, Republic of Korea, and Department of Mathematical Sciences, Seoul National University, Seoul 151–747, Korea (*E-mail:* shpark@math.snu.ac.kr)

the KKM theory. Therefore, it seems to be quite natural to call such spaces the KKM spaces. In our work [21], we showed that a large number of well-known results in the KKM theory on G-convex spaces also hold on the KKM spaces.

Moreover, apparently motivated by the author's earlier works, Alimohammady et al. [1] introduced the notion of minimal G-convex spaces and obtained the open and closed versions of the KKM principle in this new setting. Their method is just replacing the topological structure in the relevant results by the more general minimal structure.

In our previous work [22], we introduced a new concept of abstract convex minimal spaces which can be also useful to establish some results in the KKM theory. With this new concept, we obtained generalizations of the KKM principle and some of their applications. In fact, the KKM type maps were used to obtain coincidence theorems, the Fan-Browder type fixed point theorems, the Fan intersection theorem, and the Nash equilibrium theorem on abstract convex minimal spaces. These generalize corresponding previously known results.

Now it is evident that the class of abstract convex spaces contains many subclasses on which it is convenient to establish the KKM theory. In the present paper, we introduce such new subclasses of the class of abstract convex spaces, namely, abstract convex minimal spaces, minimal KKM spaces, and generalized convex minimal spaces. Each of these contains G-convex spaces properly. Recall that the class of G-convex spaces contains the classes of L-spaces, FC-spaces, and others. Some related matters are also discussed.

2. Abstract convex spaces

In this section, we recall definitions of abstract convex spaces given in [15-19]:

Let $\langle D \rangle$ denote the set of all nonempty finite subsets of a set D.

Definitions. An *abstract convex space* $(E, D; \Gamma)$ consists of a nonempty set E, a nonempty set D, and a multimap $\Gamma : \langle D \rangle \multimap E$ with nonempty values. We may denote $\Gamma_A := \Gamma(A)$ for $A \in \langle D \rangle$.

For any $D' \subset D$, the Γ-*convex hull* of D' is denoted and defined by

$$\mathrm{co}_\Gamma D' := \bigcup \{ \Gamma_A \mid A \in \langle D' \rangle \}.$$

[co is reserved for the convex hull in topological vector spaces.] A subset X of E is called a Γ-*convex subset* of $(E, D; \Gamma)$ relative to D' if for any $N \in \langle D' \rangle$, we have $\Gamma_N \subset X$, that is, $\mathrm{co}_\Gamma D' \subset X$. This means that $(X, D'; \Gamma|_{\langle D' \rangle})$ itself is an abstract convex space called a *subspace* of $(E, D; \Gamma)$.

When $D \subset E$, the space is denoted by $(E \supset D; \Gamma)$. In such case, a subset X of E is said to be Γ-*convex* if $\mathrm{co}_\Gamma(X \cap D) \subset X$; in other words, X is Γ-convex relative to $D' := X \cap D$. In case $E = D$, let $(E; \Gamma) := (E, E; \Gamma)$.

If E is given a topology, then the abstract convex space $(E, D; \Gamma)$ is called *an abstract convex topological space*.

We already gave plenty of examples of abstract convex spaces in [15,19].

Definitions. ([1]) A family $\mathcal{M}$ of subsets of a set X is called a *minimal structure* on X if $\emptyset, X \in \mathcal{M}$. In this case, $(X, \mathcal{M})$ is called a *minimal space*. Any element of $\mathcal{M}$

is called an *m-open set* of X and a complement of an m-open set is called an *m-closed set* of X. For minimal spaces $(X, \mathcal{M})$ and $(Y, \mathcal{N})$, a function $f : X \to Y$ is said to be *continuous* (more precisely, *m-continuous* or $(\mathcal{M}, \mathcal{N})$-*continuous*) if $f^{-1}(V) \in \mathcal{M}$ for each $V \in \mathcal{N}$.

Definition. If E is given a minimal structure, then the abstract convex space $(E, D; \Gamma)$ is called *an abstract convex minimal space.*

Examples. 1. Any topological space is a minimal space and not conversely.

2. Any topological vector space is a minimal vector space. There is some linear minimal space which is not a topological vector space; see [1].

3. A triple $(X, D; \{\phi_A\}_{A \in \langle D \rangle})$ consisting of a topological [resp., minimal] space X, a nonempty set D, and a family of continuous functions $\phi_A : \Delta_n \to X$ for $A \in \langle D \rangle$ with the cardinality $|A| = n + 1$ and a standard n-simplex Δ_n, is an abstract convex topological [resp., minimal] space by putting $\Gamma_A := \phi_A(\Delta_n)$.

3. The KKM spaces

Recall the following in [15,19]:

Definitions. Let $(E, D; \Gamma)$ be an abstract convex space and Z a set. For a multimap $F : E \multimap Z$ with nonempty values, if a multimap $G : D \multimap Z$ satisfies

$$F(\Gamma_A) \subset G(A) := \bigcup_{y \in A} G(y) \qquad \text{for all } A \in \langle D \rangle,$$

then G is called a *KKM map* with respect to F. A *KKM map* $G : D \multimap E$ is a KKM map with respect to the identity map 1_E.

A multimap $F : E \multimap Z$ is said to have the *KKM property* and called *a $\mathfrak{K}$-map* if, for any KKM map $G : D \multimap Z$ with respect to F, the family $\{G(y)\}_{y \in D}$ has the finite intersection property. We denote

$$\mathfrak{K}(E, Z) := \{F : E \multimap Z \mid F \text{ is a } \mathfrak{K}\text{-map}\}.$$

Similarly, when Z is a topological space, a $\mathfrak{KC}$-map is defined for closed-valued maps G, and a $\mathfrak{KO}$-map for open-valued maps G. In this case, we have

$$\mathfrak{K}(E, Z) \subset \mathfrak{KC}(E, Z) \cap \mathfrak{KO}(E, Z).$$

Note that if Z is discrete then three classes $\mathfrak{K}$, $\mathfrak{KC}$, and $\mathfrak{KO}$ are identical.

Further, when $(Z, \mathcal{M})$ is a minimal space, an $m\mathfrak{KC}$-map is defined for m-closed-valued maps G, and an $m\mathfrak{KO}$-map for m-open-valued maps G. In this case, we have

$$\mathfrak{K}(E, Z) \subset m\mathfrak{KC}(E, Z) \cap m\mathfrak{KO}(E, Z).$$

Examples. 1. Every abstract convex space in our sense has a map $F \in \mathfrak{K}(E, Z)$ for any nonempty set Z and for any class of KKM maps $G : D \multimap Z$ with respect to F. In fact, for each $x \in E$, choose $F(x) := Z$ or $F(x)$ contains some $z_0 \in Z$.

2. Further examples were given in Section 5 of [15].

Definitions. For an abstract convex topological space $(E, D; \Gamma)$, the *KKM principle* is the statement $1_E \in \mathfrak{KC}(E, E) \cap \mathfrak{KO}(E, E)$. A *KKM space* is an abstract convex topological space satisfying the KKM principle [21].

For an abstract convex minimal space $(E, D; \Gamma)$, the *KKM principle* is the statement $1_E \in m\mathfrak{K}\mathfrak{C}(E, E) \cap m\mathfrak{K}\mathfrak{O}(E, E)$. A *minimal KKM space* (or simply, *mKKM space*) is an abstract convex minimal space satisfying the KKM principle.

Examples. We give examples of KKM spaces:

1. Every generalized convex space is a KKM space; see [11,12].

2. A connected ordered space $(X, \leq)$ can be made into an abstract convex topological space $(X \supset D; \Gamma)$ for any nonempty $D \subset X$ by defining $\Gamma_A := [\min A, \max A] = \{x \in X \mid \min A \leq x \leq \max A\}$ for each $A \in \langle D \rangle$. Further, it is a KKM space; see [18, Theorem 5(i)].

3. The extended long line L^* can be made into a KKM space $(L^* \supset D; \Gamma)$; see [18]. In fact, L^* is constructed from the ordinal space $D := [0, \Omega]$ consisting of all ordinal numbers less than or equal to the first uncountable ordinal Ω, together with the order topology. Recall that L^* is a generalized arc obtained from $[0, \Omega]$ by placing a copy of the interval $(0, 1)$ between each ordinal α and its successor $\alpha + 1$ and we give L^* the order topology. Now let $\Gamma : \langle D \rangle \multimap L^*$ be the one as in 2.

In our previous work [19], for G-convex spaces, there exist more than 15 equivalent formulations of the KKM principle such as Alexandroff-Pasynkoff theorem, Ky Fan type matching theorem, Tarafdar type intersection theorem, geometric or section properties, Fan-Browder type fixed point theorems, maximal element theorems, analytic alternatives, Ky Fan type minimax inequalities, variational inequalities, and others. This is also true for KKM spaces.

In our forthcoming paper [21], we show that some of well-known results in the KKM theory on G-convex spaces also hold on the KKM spaces. Examples of such results are theorems of Sperner and Alexandroff-Pasynkoff, the Horvath type fixed point theorem, the Fan-Browder type coincidence theorems, the Ky Fan type minimax inequalities, variational inequalities, the von Neumann type minimax theorem, and the Nash type equilibrium theorem.

4. Generalized convex spaces

Recall the following appeared in [10-14]:

Definition. A *generalized convex space* or a *G-convex space* $(X, D; \Gamma)$ consists of a topological space X, a nonempty set D, and a multimap $\Gamma : \langle D \rangle \multimap X$ such that for each $A \in \langle D \rangle$ with the cardinality $|A| = n + 1$, there exists a continuous function $\phi_A : \Delta_n \to \Gamma(A)$ such that $J \in \langle A \rangle$ implies $\phi_A(\Delta_J) \subset \Gamma(J)$.

Here, Δ_n is an n-simplex with vertices $\{e_i\}_{i=0}^n$, and Δ_J the face of Δ_n corresponding to $J \in \langle A \rangle$; that is, if $A = \{a_0, a_1, \ldots, a_n\}$ and $J = \{a_{i_0}, a_{i_1}, \ldots, a_{i_k}\} \subset A$, then $\Delta_J = \mathrm{co}\{e_{i_0}, e_{i_1}, \ldots, e_{i_k}\}$.

The original KKM principle [8] is for the triple $(\Delta_n, V; \mathrm{co})$, where V denotes the set of vertices and $\mathrm{co} : \langle V \rangle \multimap \Delta_n$ the convex hull operation, and Ky Fan's celebrated lemma [6] is for $(E, D; \mathrm{co})$, where D is a nonempty subset of a topological vector space E. These are the origins of our G-convex space $(X, D; \Gamma)$. Note that any KKM type theorem on $(X; \Gamma)$ can not generalize the KKM principle and the Ky Fan lemma.

Definition. A *generalized convex minimal space* or a *G-convex minimal space* $(X, D; \Gamma)$ consists of a minimal space X, a nonempty set D, and a multimap $\Gamma : \langle D \rangle \multimap X$ such that for each $A \in \langle D \rangle$ with the cardinality $|A| = n+1$, there exists a continuous function $\phi_A : \Delta_n \to \Gamma(A)$ such that $J \in \langle A \rangle$ implies $\phi_A(\Delta_J) \subset \Gamma(J)$. See [1].

Examples. 1. A G-convex space is a G-convex minimal space, and the converse does not hold; for example, see [1].

2. A G-convex space is a KKM space and the converse does not hold; for example, the extended long line L^* is a KKM space $(L^* \supset D; \Gamma)$, but not a G-convex space.

In fact, since $\Gamma\{0, \Omega\} = L^*$ is not path connected, for $A := \{0, \Omega\} \in \langle L^* \rangle$ and $\Delta_1 := [0, 1]$, there does not exist a continuous function $\phi_A : [0, 1] \to \Gamma_A$ such that $\phi_A\{0\} \subset \Gamma\{0\} = \{0\}$ and $\phi_A\{1\} \subset \Gamma\{\Omega\} = \{\Omega\}$. Therefore $(L^* \supset D; \Gamma)$ is not G-convex.

3. A G-convex minimal space $(X, D; \Gamma)$ is a minimal KKM space in view of the following:

Proposition 1. ([1]) *Let $(E, D; \Gamma)$ be a generalized convex minimal space and $F : D \multimap E$ a KKM map with m-closed values [resp., m-open values]. Then $\{F(z)\}_{z \in D}$ has the finite intersection property.*

Essentially, the proof of Proposition 1 [1, Theorems 3.2 and 3.5] is the one in [11,12] with slight modifications.

It is obvious that most facts on G-convex spaces (e.g. in [11]) can be extended to corresponding ones on G-convex minimal spaces.

In the category of topological vector spaces or C-spaces, the concepts of locally convex spaces, LC-spaces, Φ-spaces, subsets of the Zima-Hadžić type, admissible subsets, and Klee approximable sets are quite well-known. They were introduced in order to generalize known fixed point theorems.

In our previous work [14], we extended those concepts to G-convex uniform spaces and established the mutual relations among them as follows:

Proposition 2. *In the class of G-convex uniform spaces, the following hold:*

(1) *Any LG-space is of the Zima-Hadžić type.*

(2) *Every LG-space is locally G-convex whenever every singleton is Γ-convex.*

(3) *Any nonempty subset of a locally G-convex space is a Φ-set.*

(4) *Any Zima-Hadžić type subset of a G-convex uniform space such that every singleton is Γ-convex is a Φ-set.*

(5) *Every Φ-space is admissible. More generally, every nonempty compact Φ-subset is Klee approximable.*

Note that Proposition 2 can be extended to the KKM uniform spaces.

5. Spaces having a family $\{\phi_A\}_{A \in \langle D \rangle}$

In this section, we deal with particular subclasses or variants of G-convex spaces as follows:

Definition. A *space having a family* $\{\phi_A\}_{A \in \langle D \rangle}$ or simply a *ϕ_A-space*

$$(X, D; \{\phi_A\}_{A \in \langle D \rangle})$$

consists of a topological space X, a nonempty set D, and a family of continuous functions $\phi_A : \Delta_n \to X$ (that is, singular n-simplexes) for $A \in \langle D \rangle$ with the cardinality $|A| = n+1$.

Similarly, a *minimal ϕ_A-space* can be defined whenever X is a minimal space.

Examples. 1. [2] An *L-structure* on a topological space E is given by a nonempty set-valued map $\Gamma : \langle E \rangle \to E$ verifying

($*$) for each $A \in \langle E \rangle$, say $A = \{x_0, x_1, \ldots, x_n\}$, there exists a continuous function $f^A : \Delta_n \to \Gamma(A)$ such that for all $J \subset \{0, 1, \ldots, n\}$, $f^A(\Delta_J) \subset \Gamma(\{x_i \mid i \in J\})$.

The pair (E, Γ) is then called an *L-space*, and $X \subset E$ is said to be *L-convex* if $\forall A \in \langle X \rangle$, $\Gamma(A) \subset X$. Note that an L-space (E, Γ) is a particular form of a G-convex space $(E, D; \Gamma)$ with $E = D$.

2. [7] A topological space Y is said to have property (H) if, for each $N = \{y_0, \ldots, y_n\} \in \langle Y \rangle$, there exists a continuous mapping $\varphi_N : \Delta_n \to Y$.

Let X be a nonempty set and Y be a topological space with property (H). $T : X \to 2^Y$ is said to be a *generalized R-KKM mapping* if for each $\{x_0, \ldots, x_n\} \in \langle X \rangle$, there exists $N = \{y_0, \ldots, y_n\} \in \langle Y \rangle$ such that

$$\varphi_N(\Delta_k) \subset \bigcup_{j=0}^{k} Tx_{i_j},$$

for all $\{i_0, \ldots, i_k\} \subset \{0, \ldots, n\}$.

3. [3,4] $(Y, \{\varphi_N\})$ is said to be an *FC-space* if Y is a topological space and for each $N = \{y_0, \ldots, y_n\} \in \langle Y \rangle$ where some elements in N may be same, there exists a continuous mapping $\varphi_N : \Delta_n \to Y$. A subset D of $(Y, \{\varphi_N\})$ is said to be a *FC-subspace* of Y if for each $N = \{y_0, \ldots, y_n\} \in \langle Y \rangle$ and for each $\{y_{i_0}, \ldots, y_{i_k}\} \subset N \cap D$, $\varphi_N(\Delta_k) \subset D$ where $\Delta_k = \mathrm{co}\{e_{i_j} \mid j = 0, \ldots, k\}$.

4. For minimal spaces, we can also define L-spaces, spaces having property (H), and FC-spaces.

5. Any G-convex minimal space is a minimal ϕ_A-space. The converse also holds:

Proposition 3. *A minimal ϕ_A-space $(X, D; \{\phi_A\})$ can be made into a G-convex minimal space $(X, D; \Gamma)$.*

Proof. This can be done at least in three ways.

(1) For each $A \in \langle D \rangle$, by putting $\Gamma_A := X$, we obtain a trivial G-convex minimal space $(X, D; \Gamma)$.

(2) Let $\{\Gamma^\alpha\}_\alpha$ be the family of maps $\Gamma^\alpha : \langle D \rangle \multimap X$ giving a G-convex minimal space $(X, D; \Gamma^\alpha)$. Note that, by (1), this family is not empty. Then, for each α and each $A \in \langle D \rangle$ with $|A| = n + 1$, we have

$$\phi_A(\Delta_n) \subset \Gamma_A^\alpha \quad \text{and} \quad \phi_A(\Delta_J) \subset \Gamma_J^\alpha \quad \text{for} \quad J \subset A.$$

Let $\Gamma := \bigcap_\alpha \Gamma^\alpha$, that is, $\Gamma_A = \bigcap_\alpha \Gamma_A^\alpha$. Then

$$\phi_A(\Delta_n) \subset \Gamma_A \quad \text{and} \quad \phi_A(\Delta_J) \subset \Gamma_J \quad \text{for} \quad J \subset A.$$

Therefore, $(X, D; \Gamma)$ is a G-convex minimal space.

(3) Let $N \in \langle D \rangle$ with $|N| = n+1$. For each $M \in \langle D \rangle$ with $N \subset M$, $M = \{a_0, \ldots, a_m\}$ and $N = \{a_{i_0}, \ldots, a_{i_n}\}$, there exists a subset $\phi_M(\Delta_n^M)$ of X such that $\Delta_n^M := \mathrm{co}\{e_{i_j} \mid j = 0, \ldots, n\} \subset \Delta_m$. Now let

$$\Gamma_N = \Gamma(N) := \bigcup_{M \supset N} \phi_M(\Delta_n^M).$$

Then $\Gamma : \langle D \rangle \multimap X$ is well-defined and $(X, D; \Gamma)$ becomes a G-convex minimal space: For each $A \in \langle D \rangle$ with $|A| = n+1$, there exists a continuous map $\phi_A : \Delta_n \multimap \Gamma(A)$ such that $J \in \langle A \rangle$ implies $\phi_A(\Delta_J) \subset \Gamma(J)$. $\qquad \square$

Therefore, G-convex minimal spaces and minimal ϕ_A-spaces are essentially same.

Examples. Let $\Delta_3 = \mathrm{co}\, V$ where $V = \{e_0, e_1, e_2, e_3\}$.

1. We have a G-convex space $(\Delta_3, V; \mathrm{co})$ where $\mathrm{co} : \langle V \rangle \multimap \Delta_3$ is the convex hull operator.

2. Let $(\Delta_3, V; \Gamma)$ be another G-convex space given by $\Gamma\{e_0, e_1\} := \mathrm{co}\{e_0, e_1, e_2\}$ and $\Gamma(A) := \mathrm{co}\, A$ for all other $A \in \langle V \rangle$.

Let $(\Delta_3, V; \{\phi_A\})$ be a ϕ_A-space where $\phi_A(\Delta_n) = \Gamma(A)$. Then

$$\phi_{\{e_0, e_1\}}(\Delta_1) = \phi_{\{e_0, e_1\}}(\mathrm{co}\{e_0, e_1\}) = \Gamma\{e_0, e_1\} = \mathrm{co}\{e_0, e_1, e_2\},$$

where we may assume $\phi_{\{e_0, e_1\}}$ is a surjective space-filling curve such that $\phi_{\{e_0, e_1\}}(e_0) = \{e_0\}$ and $\phi_{\{e_0, e_1\}}(e_1) = \{e_1\}$. Then it is easily checked that Γ itself is the one in the proof (3) of Proposition 3 corresponding to $\{\phi_A\}$.

Example. The extended long line $(L^* \supset D; \Gamma)$ is not G-convex, and hence not a ϕ_A-space.

Recall that, in the recent study on abstract convex spaces in [15-19], many basic theorems on G-convex spaces are further generalized.

For a G-convex minimal space $(X, D; \Gamma)$, a multimap $G : D \multimap X$ is called a *KKM map* if $\Gamma_A \subset G(A)$ for each $A \in \langle D \rangle$.

Proposition 4. *For a minimal ϕ_A-space $(X, D; \{\phi_A\})$, any map $T : D \multimap X$ satisfying*

$$\phi_A(\Delta_J) \subset T(J) \quad \text{for each} \quad A \in \langle D \rangle \text{ and } J \in \langle A \rangle$$

is a KKM map on a G-convex minimal space $(X, D; \Gamma)$.

Proof. Define $\Gamma : \langle D \rangle \multimap X$ by $\Gamma_A := T(A)$ for each $A \in \langle D \rangle$. Then $(X, D; \Gamma)$ becomes a G-convex space. In fact, for each A with $|A| = n+1$, we have a continuous function $\phi_A : \Delta_n \to T(A) =: \Gamma(A)$ such that $J \in \langle A \rangle$ implies $\phi_A(\Delta_J) \subset T(J) =: \Gamma(J)$. Moreover, note that $\Gamma_A \subset T(A)$ for each $A \in \langle D \rangle$ and hence $T : D \multimap X$ is a KKM map on a G-convex minimal space $(X, D; \Gamma)$. $\qquad \square$

Similarly, we have the following:

Proposition 5. *A generalized R-KKM map $T : X \to 2^Y$ in [5,6] is simply a KKM map for some G-convex space $(Y, X; \Gamma)$.*

Contrary to Proposition 5, Ding in [5] claimed as follows: "The above class of generalized *R-KKM* mappings include those classes of *KKM* mappings, *H-KKM* mappings, *G-KKM* mappings, generalized *G-KKM* mappings, generalized *S-KKM* mappings, *GLKKM* mappings and *GMKKM* mappings defined in topological vector spaces, *H*-spaces, *G*-convex spaces, *G-H*-spaces, *L*-convex spaces and hyperconvex metric spaces, respectively, as true subclasses."

Therefore, all of the KKM type theorems on such variants are simple consequences of our *G*-convex space theory. Consequently, all results in [5] are artificial disguised forms of known ones having no proper examples.

REFERENCES

1. M. Alimohammady, M. R. Delavar, and M. Roohi, *Knaster-Kuratowski-Mazurkiewicz theorem in minimal generalized convex spaces*, Nonlinear Funct. Anal. Appl., to appear.
2. H. Ben-El-Mechaiekh, S. Chebbi, M. Florenzano, and J.-V. Llinares, *Abstract convexity and fixed points*, J. Math. Anal. Appl. **222** (1998), 138–150.
3. X.P. Ding, *Maximal element theorems in product FC-spaces and generalized games*, J. Math. Anal. Appl. **305** (2005), 29–42.
4. X.P. Ding, *Generalized KKM type theorems in FC-spaces with applications* (I), J. Glob. Optim. **36** (2006), 581–596.
5. X.P. Ding, *New generalized R-KKM type theorems in general topological spaces and applications*, Acta Math. Sinica, English Ser. (2006), DOI: 10.1007/s10114-005-0876-y.
6. Ky Fan, *A generalization of Tychonoff's fixed point theorem*, Math. Ann. **142** (1961), 305–310.
7. J. Huang, *The matching theorems and coincidence theorems for generalized R-KKM mapping in topological spaces*, J. Math. Anal. Appl. **312** (2005), 374–382.
8. B. Knaster, K. Kuratowski, S. Mazurkiewicz, *Ein Beweis des Fixpunktsatzes für n-Dimensionale Simplexe*, Fund. Math. **14** (1929), 132–137.
9. S. Park, *Some coincidence theorems on acyclic multifunctions and applications to KKM theory*, Fixed Point Theory and Applications (K.-K. Tan, ed.), 248–277, World Sci. Publ., River Edge, NJ, 1992.
10. S. Park, *Ninety years of the Brouwer fixed point theorem*, Vietnam J. Math. **27** (1999), 193–232.
11. S. Park, *Elements of the KKM theory for generalized convex spaces*, Korean J. Comput. & Appl. Math. **7** (2000), 1–28.
12. S. Park, *Remarks on topologies of generalized convex spaces*, Nonlinear Func. Anal. Appl. **5** (2000), 67–79.
13. S. Park, *New topological versions of the Fan-Browder fixed point theorem*, Nonlinear Anal. **47** (2001), 595–606.
14. S. Park, *A unified fixed point theory in generalized convex spaces*, Acta Math. Sinica, English Ser. **23(8)** (2007), 1509–1536.
15. S. Park, *On generalizations of the KKM principle on abstract convex spaces*, Nonlinear Anal. Forum **11** (2006), 67–77.
16. S. Park, *Fixed point theorems on ℜℭ-maps in abstract convex spaces*, Nonlinear Anal. Forum **11** (2006), 117–127.
17. S. Park, *Remarks on ℜℭ-maps and ℜ𝔒-maps on abstract convex spaces*, Nonlinear Anal. Forum, to appear.
18. S. Park, *Examples of ℜℭ-maps and ℜ𝔒-maps on abstract convex spaces*, Soochow J. Math., to appear.
19. S. Park, *Elements of the KKM theory on abstract convex spaces*, J. Korean Math. Soc., to appear.
20. S. Park, *Various subclasses of abstract convex spaces for the KKM theory*, Proc. Nat. Inst. Math. Sci. **2(4)** (2007), 35–47.
21. S. Park, *Equilibrium existence theorems in KKM spaces*, to appear.
22. S. Park, *On generalizations of the KKM principle on abstract convex minimal spaces*, to appear.

23. S. Park and H. Kim, *Admissible classes of multifunctions on generalized convex spaces*, Proc. Coll. Natur. Sci., Seoul Nat. Univ. **18** (1993), 1-21.

Nonlinear Functional Analysis and Applications, Volume 1, 11–20

STRONG CONVERGENCE THEOREMS OF COMMON FIXED POINTS FOR A FINITE FAMILY OF RELATIVELY NONEXPANSIVE MAPPINGS

Meijuan Shang[1,2] and Xiaolong Qin[3,*]

ABSTRACT. In this paper, we consider the problem of finding a common fixed point of Mann iteration and Halpern iteration for a finite family of relatively nonexpansive mappings and consider cyclic algorithms for solving this problem. Moreover, by applying the hybrid method in mathematical programming, we prove the two algorithms can be modified to have strong convergence. Our results extend and improve the recent ones announced by many others.

1. Introduction

Let E be a smooth Banach space, E^* the dual of E, C a nonempty closed convex subset of E, and $T : C \to C$ a mapping.

We denote by J the normalized duality mapping from E to E^* defined by

$$Jx = \{f^* \in E^* : \langle x, f^* \rangle = \|x\|^2 = \|f^*\|^2\}, \quad \forall x \in E,$$

where $\langle \cdot, \cdot \rangle$ denotes the generalized duality pairing.

Consider the function $\phi : E \times E \to \mathbb{R}$ defined by

$$\phi(x, y) = \|x\|^2 - 2\langle x, Jy \rangle + \|y\|^2$$

for all $x, y \in E$. It is obvious that

$$(\|x\| - \|y\|)^2 \le \phi(x, y) \le (\|x\| + \|y\|)^2$$

Received August 14, 2007.   * Corresponding author.

2000 *Mathematics Subject Classification.* 47H09, 47H10.

Key words and phrases. Relatively nonexpansive, asymptotic fixed point, generalized projection, cyclic algorithm; Banach space.

[1] Department of Mathematics, Tianjin Polytechnic University, Tianjin 300160, People's Republic of China (*E-mail:* meijuanshang@yahoo.com.cn)

[2] Department of Mathematics, Shijiazhuang University, Shijiazhuang 050035, People's Republic of China (*E-mail:* meijuanshang@yahoo.com.cn)

[3] Department of Mathematics, Gyeongsang National University, Chinju 660-701, Korea (*E-mail:* qxlxajh@163.com)

for all $x, y \in E$.

We use $F(T)$ to denote the set of fixed points of T, that is, $F(T) = \{x \in C : Tx = x\}$. A point p in C is said to be an asymptotic fixed point of T [15] if C contains a sequence $\{x_n\}$ which converges weakly to p such that the strong $\lim_{n\to\infty}(x_n - Tx_n) = 0$. The set of asymptotic fixed points of T will be denoted by $\hat{F}(T)$. A mapping T from C into itself is called nonexpansive if $\|Tx - Ty\| \leq \|x - y\|$ for all $x, y \in C$ and relatively nonexpansive [3-5] if $\hat{F}(T) = F(T)$ and $\phi(p, Tx) \leq \phi(p, x)$ for all $x \in C$ and $p \in F(T)$.

Two classical iteration processes are often used to approximate a fixed point of a nonexpansive mapping T. The first one is introduced by Halpern [6] and is defined as follows: Take an initial guess $x_0 \in C$ arbitrarily and define $\{x_n\}$ recursively by

$$(1.1) \qquad\qquad x_{n+1} = t_n x_0 + (1 - t_n)Tx_n, \quad \forall n \geq 0,$$

where $\{t_n\}_{n=0}^{\infty}$ is a sequence in the interval [0,1].

The second one is now known as Mann's iteration process [10] which is defined as

$$(1.2) \qquad\qquad x_{n+1} = \alpha_n x_n + (1 - \alpha_n)Tx_n, \quad \forall n \geq 0,$$

where the initial guess x_0 is taken in C arbitrarily and the sequence $\{\alpha_n\}_{n=0}^{\infty}$ is in the interval [0,1].

In general not much has been known regarding the convergence of the iteration process (1.1) and (1.2) unless the underlying space E has elegant properties which we briefly mention here.

Reich [13] shows that if E is uniformly convex and has a Fréchet differentiable norm and if the sequence $\{\alpha_n\}$ is such that $\sum_{n=0}^{\infty} \alpha_n(1 - \alpha_n) = \infty$, then the sequence $\{x_n\}$ generated by (1.2) converges weakly to a fixed point of T.

An advantage that process (1.2) is over process (1.1) though the former has only weak convergence in general is the use of the averaged mapping $\alpha I + (1 - \alpha)T$ in each iteration step. This averaged mapping behaves more regularly than the nonexpansive mapping T itself (see [2]). The weakness of process (1.2) is, however, its weak convergence.

In order to overcome this weakness, attempts have recently been made to modify process (1.2) so that strong convergence is guaranteed. Nakajo and Takahashi [11] proposed the following modification for process (1.2) in a Hilbert space H:

$$(1.3) \qquad \begin{cases} x_0 \in C \quad chosen\ arbitrarily, \\ y_n = \alpha_n x_n + (1 - \alpha_n)Tx_n, \\ C_n = \{z \in C : \|y_n - z\| \leq \|x_n - z\|\}, \\ Q_n = \{z \in C : \langle x_n - z, x_0 - x_n \rangle \geq 0\}, \\ x_{n+1} = P_{C_n \cap Q_n} x_0, \quad \forall n \geq 0, \end{cases}$$

where C is a closed convex subset of H and $T : C \to C$ is a nonexpansive mapping. They proved that if the sequence $\{\alpha_n\}$ is bounded above from one, then $\{x_n\}$ defined by (1.3) converges strongly to $P_{F(T)}x_0$.

On the other hand, regarding to the iteration sequence (1.1), it has been shown to be strongly convergent in Hilbert spaces [see [6], [14]] under the conditions:

(1) $t_n \to 0$;

(2) $\sum_{n=0}^{\infty} t_n = \infty$;

(3) either $\sum_{n=0}^{\infty} |t_{n+1} - t_n| < \infty$ or $\lim_{n \to \infty} \frac{t_n}{t_{n+1}} = 1$.

However, due to the restrictions of conditions (2) and (3), the convergence of $\{x_n\}$ is believed to be slow. Moreover, it is shown in [6] that conditions (1) and (2) are also necessary. So to improve the convergence rate of iteration (1.1), one has to perform some additional step of iteration.

Recently, Martinez-Yanes and Xu [9] modified the iteration (1.1) by using hybrid method in a Hilbert space H and proposed the following iteration algorithm:

$$(1.4) \quad \begin{cases} x_0 \in C \quad chosen\ arbitrarily, \\ y_n = t_n x_0 + (1 - t_n) T x_n, \\ C_n = \{v \in C : \|y_n - v\|^2 \le \|x_n - v\|^2 + t_n(\|x_0\|^2 + 2\langle x_n - x_0, v \rangle)\}, \\ Q_n = \{v \in C : \langle x_n - v, x_0 - x_n \rangle \ge 0\}, \\ x_{n+1} = P_{C_n \cap Q_n} x_0, \quad \forall n \ge 0, \end{cases}$$

where C is a closed convex subset of H and $T : C \to C$ is a nonexpansive mapping. They proved that the sequence $\{x_n\}$ defined by (1.4) has strong convergence under condition (1) only, moreover, converges strongly to $P_{F(T)} x_0$.

In recent years, the asymptotic behavior of relatively nonexpansive mapping has been extensively investigated [see 8,3-5]. Recently, Matsushita, Takahashi [8], Qin and Su [12] proposed the modification for process (1.2) and process (1.1) for a single relatively nonexpansive mapping T by applying the hybrid method, respectively, in a uniformly convex and uniformly smooth Banach space E. More precisely, they defined the iteration sequence $\{x_n\}$ recursively by the follows:

$$(1.5) \quad \begin{cases} x_0 = x \in C, \\ y_n = J^{-1}(\alpha_n J x_n + (1 - \alpha_n) J T x_n), \\ H_n = \{z \in C : \phi(z, y_n) \le \phi(z, x_n)\}, \\ W_n = \{z \in C : \langle x_n - z, Jx - J x_n \rangle \ge 0\}, \\ x_{n+1} = \Pi_{H_n \cap W_n} x, \quad \forall n \ge 0, \end{cases}$$

and

$$(1.6) \quad \begin{cases} x_0 \in C \quad chosen\ arbitrarily, \\ y_n = J^{-1}(\alpha_n J x_0 + (1 - \alpha_n) J T x_n), \\ C_n = \{v \in C : \phi(v, y_n) \le \alpha_n \phi(v, x_0) + (1 - \alpha_n) \phi(v, x_n)\}, \\ Q_n = \{v \in C : \langle x_n - v, J x_0 - J x_n \rangle \ge 0\}, \\ x_{n+1} = \Pi_{C_n \cap Q_n} x_0, \quad \forall n \ge 0, \end{cases}$$

where C is a closed convex subset of E and $T : C \to C$ is a relatively nonexpansive mapping, J is the single-valued duality mapping on E and Π_K is the generalized projection from C onto a nonempty closed convex subset K of C. They both proved the sequence $\{x_n\}$ converges in norm to $\Pi_{F(T)} x_0$.

This paper considers the following explicit cyclic algorithm:

$$x_1 = \alpha_0 x_0 + (1 - \alpha_0) T_0 x_0,$$
$$x_2 = \alpha_1 x_1 + (1 - \alpha_1) T_1 x_1,$$
$$\cdots$$
$$x_n = \alpha_{n-1} x_{n-1} + (1 - \alpha_{n-1}) T_{n-1} x_{n-1},$$
$$x_{n+1} = \alpha_n x_n + (1 - \alpha_n) T_0 x_n.$$

We can rewrite the above table into the compact form as follows

$$(1.7) \qquad\qquad x_{n+1} = \alpha_n x_n + (1 - \alpha_n) T_{[n]} x_n.$$

The purpose of this paper is to employ Matsushita and Takahashi's idea [8] to modify process (1.7) for a finite family of relatively nonexpasive mappings, accordingly, to modify Halpern iteration by cyclic algorithm. We obtain two convergence theorems in the framework of Banach spaces and prove the convergence is strong under no any compact on the operators or sets.

2. Preliminaries

A Banach space E is said to be strictly convex if $\|\frac{x+y}{2}\| < 1$ for all $x, y \in E$, with $\|x\| = \|y\| = 1$ and $x \neq y$. It is also said to be *uniformly convex* if $\lim_{n \to \infty} \|x_n - y_n\| = 0$ for any tow sequences $\{x_n\}$ and $\{y_n\}$ in E such that $\|x\| = \|y\| = 1$ and $\lim_{n \to \infty} \|\frac{x+y}{2}\| = 1$.

Let $U = \{x \in E : \|x\| = 1\}$ be the unit sphere of E. Then the Banach space E is said to be *smooth* provided $\lim_{t \to 0} \frac{\|x+ty\| - \|x\|}{t}$ exists for each $x, y \in U$. It is also said to be *uniformly smooth* if the limit is attained uniformly for $x, y \in E$.

It is well known that if E is smooth, then the duality mapping J is single valued. It is also known that E is uniformly smooth, then J is uniformly norm-to-norm continuous on each bounded subset of E; see for instance, [16].

Recall that a Banach space E is said to have the *Kadec-Klee* property if a sequence $\{x_n\}$ of E satisfying that $x_n \rightharpoonup x \in E$ and $\|x_n\| \to \|x\|$, then $x_n \to x$. It is known that if E is uniformly convex, then E has the *Kadec-Klee* property.

It is well known that if C is a nonempty closed convex subset of a Hilbert space H and $P_C : H \to C$ is the metric projection of H onto C, then P_C is nonexpansive. However, it is not available in more general Banach spaces. In this connection, Alber [1] recently introduced a generalized projection operator Π_C in a Banach space E. Recall that the generalized projection operator $\Pi_C : E \to C$ is a map that assigns to an arbitrary point $x \in E$ the minimum point of the function $\phi(x, y)$; that is, $\Pi_C = \overline{x}$, where $\overline{x}$ is the solution to the minimization problem $\phi(\overline{x}, x) = \min_{y \in C} \phi(y, x)$.

We need the following lemmas for the proof of our main results.

Lemma 2.1. (Kamimura and Takahashi [7]) *Let E be a uniformly convex and smooth Banach space and let $\{x_n\}$ and $\{y_n\}$ be two sequences of E. If $\phi(x_n, y_n) \to 0$ and either $\{x_n\}$ or $\{y_n\}$ is bounded, then $x_n - y_n \to 0$.*

Lemma 2.2. (Alber [1]) *Let C be a nonempty closed convex subset of a smooth Banach space E and $x \in E$. Then, $x_0 = \Pi_C x$ if and only if*

$$\langle x_0 - y, Jx - Jx_0 \rangle \geq 0, \quad \forall y \in C.$$

Lemma 2.3. (Alber [1]) *Let E be a reflexive, strictly convex and smooth Banach space, let C be a nonempty closed convex subset of E and let $x \in E$. Then*

$$\phi(y, \Pi_C x) + \phi(\Pi_c x, x) \leq \phi(y, x), \quad \forall y \in C.$$

Lemma 2.4. (Matsushita and Takahashi [8]) *Let E be a strictly convex and smooth Banach space, let C be a closed convex subset of E, and let T be a relatively nonexpansive mapping from C into itself. Then $F(T)$ is closed and convex.*

3. Results on Algorithmic Convergence Analysis

Now we will prove a strong convergence theorem of modified Mann iteration for a finite family of relatively nonexpansive mappings in a Banach space.

Theorem 3.1. *Let E be a uniformly convex and uniformly smooth Banach space and C a nonempty closed convex subset of E. Given an integer $N \geq 1$, let, for each i with $0 \leq i \leq N - 1$, T_i be a relatively nonexpansive mapping from C into itself and $\{\alpha_n\}$ a sequence of real numbers such that $0 \leq \alpha_n < 1$ and $\limsup_{n \to \infty} \alpha_n < 1$. Define a sequence $\{x_n\}_{n=0}^{\infty}$ by the following algorithm:*

(3.1)
$$\begin{cases} x_0 \in C \quad \text{chosen arbitrarily}, \\ y_n = J^{-1}(\alpha_n Jx_n + (1 - \alpha_n)JT_{[n]}x_n), \\ C_n = \{v \in C : \phi(v, y_n) \leq \phi(v, x_n)\}, \\ Q_n = \{v \in C : \langle x_n - v, Jx_0 - Jx_n \rangle \geq 0\}, \\ x_{n+1} = \Pi_{C_n \cap Q_n} x_0, \quad \forall n \geq 0, \end{cases}$$

where J is the duality mapping on E and $T_{[n]} = T_i$, $i = n(\mathrm{mod}\ N)$. If the common fixed point set $F := \bigcap_{i=0}^{N-1} F(T_i)$ of $\{T_i\}_{i=0}^{N-1}$ is nonempty, then $\{x_n\}$ converges strongly to $\Pi_F x_0$, where $\Pi_F x_0$ is the generalized projection from C onto F.

Proof. We first show that C_n and Q_n are closed and convex for each $n \geq 0$. From the definition of C_n and Q_n, it is obvious that C_n is closed and Q_n is closed and convex for each $n \geq 0$. Next, we show C_n is convex. Since $\phi(v, y_n) \leq \phi(v, x_n)$ is equivalent to

$$2\langle v, Jx_n - Jy_n \rangle + \|y_n\|^2 - \|x_n\|^2 \leq 0.$$

So, C_n is convex.

Next, we show that $F \subset C_n \bigcap Q_n$ for all n. Indeed, we have, for all $p \in F$,

$$
\begin{aligned}
\phi(p, y_n) &= \phi(p, J^{-1}(\alpha_n J x_n + (1 - \alpha_n) J T_{[n]} x_n)) \\
&= \|p\|^2 - 2\langle p, \alpha_n J x_n + (1 - \alpha_n) J T_{[n]} x_n \rangle \\
&\quad + \|\alpha_n J x_n + (1 - \alpha_n) J T_{[n]} x_n\|^2 \\
&\leq \|p\|^2 - 2\alpha_n \langle p, J x_n \rangle - 2(1 - \alpha_n)\langle p, J T_{[n]} x_n \rangle \\
&\quad + \alpha_n \|x\|^2 + (1 - \alpha_n)\|T_{[n]} x_n\|^2 \\
&\leq \alpha_n \phi(p, x_n) + (1 - \alpha_n)\phi(p, T_{[n]} x_n) \\
&\leq \alpha_n \phi(p, x_n) + (1 - \alpha_n)\phi(p, x_n) = \phi(p, x_n)
\end{aligned}
$$

So, $p \in C_n$ for all $n \geq 0$. Next we show that $F \subset Q_n$, for all $n \geq 0$. We prove this by induction. For $n = 0$, we have $F \subset C = Q_0$. Assume that $F \subset Q_n$. Since x_{n+1} is the generalized projection of x_0 onto $C_n \bigcap Q_n$, that is, $x_{n+1} = \Pi_{C_n \bigcap Q_n} x_0$, from Lemma 2.2, there holds $\langle x_{n+1} - z, J x_0 - J x_{n+1} \rangle \geq 0$, for each $z \in C_n \bigcap Q_n$. Since $F \subset C_n \bigcap Q_n$, we have $\langle x_{n+1} - p, J x_0 - J x_{n+1} \rangle \geq 0$ for every $p \in F$ and hence $F \subset Q_{n+1}$. Therefore we have $F \subset C_n \bigcap Q_n$ for all $n \geq 0$. This implies that $\{x_n\}$ is well defined.

It follows from the definition of Q_n and Lemma 2.2 that $x_n = \Pi_{Q_n} x_0$, which togethers with Lemma 2.3, we have

$$
\phi(x_n, x_0) = \phi(\Pi_{Q_n} x_0, x_0) \leq \phi(p, x_0) - \phi(p, \Pi_{Q_n} x_0) \leq \phi(p, x_0)
$$

So, $\phi(x_n, x_0)$ is bounded. Furthermore, from

$$
(\|x\| - \|y\|)^2 \leq \phi(x, y) \leq (\|x\| + \|y\|)^2, \quad \forall x, y \in E,
$$

we have that $\{x_n\}$ is bounded. Since $x_{n+1} = \Pi_{C_n \bigcap Q_n} x_0 \in Q_n$ and Lemma 2.3, we have $\phi(x_n, x_0) \leq \phi(x_{n+1}, x_0)$ for all $n \geq 0$. Therefore $\{\phi(x_n, x_0)\}$ is nondecreasing. So there exists the limit of $\phi(x_n, x_0)$.

From Lemma 2.3, we have

$$
\phi(x_{n+1}, x_n) = \phi(x_{n+1}, \Pi_{Q_n} x_0) \leq \phi(x_{n+1}, x_0) - \phi(x_n, x_0).
$$

This implies that

(3.2)
$$
\lim_{n \to \infty} \phi(x_{n+1}, x_n) = 0.
$$

Since $x_{n+1} = \Pi_{C_n \bigcap Q_n} x_0 \in C_n$, from the definition of C_n, we also have

$$
\phi(x_{n+1}, y_n) \leq \phi(x_{n+1}, x_n), \quad \forall n \geq 0.
$$

So we have $\lim_{n \to \infty} \phi(x_{n+1}, y_n) = 0$.

By using Lemma 2.1, we obtain

$$
\lim_{n \to \infty} \|x_{n+1} - y_n\| = \lim_{n \to \infty} \|x_{n+1} - x_n\| = 0.
$$

Since J is uniformly norm-to-norm continuous on bounded sets, we have

(3.3)
$$
\lim_{n \to \infty} \|J x_{n+1} - J y_n\| = \lim_{n \to \infty} \|J x_{n+1} - J x_n\| = 0.
$$

On the other hand, we have

$$\|Jx_{n+1} - Jy_n\| = \|Jx_{n+1} - (\alpha_n Jx_n + (1-\alpha_n)JT_{[n]}x_n)\|$$
$$= \|\alpha_n(Jx_{n+1} - Jx_n) + (1-\alpha_n)(Jx_{n+1} - JT_{[n]}x_n)\|$$
$$= \|(1-\alpha_n)(Jx_{n+1} - JT_{[n]}x_n) - \alpha_n(Jx_n - Jx_{n+1})\|$$
$$\geq (1-\alpha_n)\|Jx_{n+1} - JT_{[n]}x_n\| - \alpha_n\|Jx_n - Jx_{n+1}\|$$

and hence

$$\|Jx_{n+1} - JT_{[n]}x_n\| \leq \frac{1}{1-\alpha_n}(\|Jx_{n+1} - Jy_n\| + \alpha_n\|Jx_n - Jx_{n+1}\|)$$
$$\leq \frac{1}{1-\alpha_n}(\|Jx_{n+1} - Jy_n\| + \|Jx_n - Jx_{n+1}\|).$$

From (3.3) and $\limsup_{n\to\infty} \alpha_n < 1$, we obtain

$$\|Jx_{n+1} - JT_{[n]}x_n\| = 0.$$

Since J^{-1} is also uniformly norm-to-norm continuous on bounded sets, we obtain

$$\lim_{n\to\infty} \|x_{n+1} - T_{[n]}x_n\| = 0.$$

Noticing that

$$\|x_n - T_{[n]}x_n\| = \|x_n - x_{n+1} + x_{n+1} - T_{[n]}x_n\|$$
$$\leq \|x_n - x_{n+1}\| + \|x_{n+1} - T_{[n]}x_n\|,$$

we have

$$\lim_{n\to\infty} \|x_n - T_{[n]}x_n\| = 0.$$

Next we show that $\omega_\omega(x_n) \subset F$. Indeed, assume $\bar{x} \in \omega_\omega(x_n)$ and $x_{n_i} \rightharpoonup \bar{x}$ for some subsequence $\{x_{n_i}\}$ of $\{x_n\}$. We may further assume $n_i = l(mod\ N)$ for all i. By (3.2), we also have

$$\|x_{n_i+j} - T_{[l+j]}x_{n_i+j}\| = \|x_{n_i+j} - T_{[n_i+j]}x_{n_i+j}\| \to 0,$$

which implies $\bar{x} \in \widehat{F}(T_{[l+j]}) = F(T_{[l+j]})$ for all $j \geq 0$. Therefore, $\bar{x} \in \widehat{F} = F$, where $\widehat{F} := \bigcap_{n=0}^{N-1} \widehat{F}(T_i)$.

Finally, we show that $x_n \to \Pi_F x_0$. Let $\widetilde{x} = \Pi_F x_0$. From $x_{n+1} = \Pi_{C_n \bigcap Q_n} x_0$ and $\widetilde{x} \in F \subset C_n \bigcap Q_n$, we also have $\phi(x_{n+1}, x_0) \leq \phi(\widetilde{x}, x_0)$.

On the other hand, from weakly lower semicontinuity of the norm, we have

$$\phi(\bar{x}, x_0) = \|\bar{x}\|^2 - 2\langle \bar{x}, Jx_0 \rangle + \|x_0\|^2$$
$$\leq \liminf_{i\to\infty}(\|x_{n_i}\|^2 - 2\langle x_{n_i}, Jx_0 \rangle + \|x_0\|^2)$$
$$= \liminf_{i\to\infty} \phi(x_{n_i}, x_0)$$
$$\leq \limsup_{i\to\infty} \phi(x_{n_i}, x_0) \leq \phi(\widetilde{x}, x_0).$$

It follows from the definition of $\Pi_F x_0$ that we obtain $\bar{x} = \widetilde{x}$ and hence

$$\lim_{i\to\infty} \phi(x_{n_i}, x_0) = \phi(\widetilde{x}, x_0).$$

So, we have $\lim_{i\to\infty} \|x_{n_i}\| = \|\widetilde{x}\|$. By using the *Kadec-Klee* property of E, we obtain that $\{x_{n_i}\}$ converges strongly to $\Pi_F x_0$. Since $\{x_{n_i}\}$ is an arbitrarily weakly convergent sequence of $\{x_n\}$, we can conclude that $\{x_n\}$ converges strongly to $\Pi_F x_0$. This completes the proof. $\qquad\qquad\qquad\qquad\qquad\qquad\qquad\qquad\qquad\qquad\qquad\qquad\qquad\qquad\quad \square$

Now we will prove another strong convergence theorem of modified Halpern iteration for a finite family of relatively nonexpansive mappings in a Banach space.

Theorem 3.2. *Let E be a uniformly convex and uniformly smooth Banach space, let C be a nonempty closed convex subset of E. Given an integer $N \geq 1$, let, for each i with $0 \leq i \leq N-1$, T_i be a relatively nonexpansive mapping from C into itself. Assume the common fixed point set $F := \bigcap_{i=0}^{N-1} F(T_i)$ of $\{T_i\}_{i=0}^{N-1}$ is nonempty and let $\{\alpha_n\}$ be a sequence in (0,1) such that $\lim_{n\to\infty} \alpha_n = 0$. Define a sequence $\{x_n\}_{n=0}^{\infty}$ by the following algorithm:*

$$(3.4) \qquad \begin{cases} x_0 \in C \quad \text{chosen arbitrarily}, \\ y_n = J^{-1}(\alpha_n J x_0 + (1-\alpha_n) J T_{[n]} x_n), \\ C_n = \{v \in C : \phi(v, y_n) \leq \alpha_n \phi(v, x_0) + (1-\alpha_n)\phi(v, x_n)\}, \\ Q_n = \{v \in C : \langle x_n - v, J x_0 - J x_n \rangle \geq 0\}, \\ x_{n+1} = \Pi_{C_n \cap Q_n} x_0, \quad \forall n \geq 0, \end{cases}$$

where J is the duality mapping on E and $T_{[n]} = T_i$, $i = n(mod\ N)$. Then $\{x_n\}$ converges strongly to $\Pi_F x_0$, where $\Pi_F x_0$ is the generalized projection from C onto F.

Proof. We only conclude the difference. First, we show that C_n and Q_n are closed and convex for each $n \geq 0$. From the definition of C_n and Q_n, it is obvious that C_n is closed and Q_n is closed and convex, for each $n \geq 0$. By repeating the argument in the proof of Theorem 3.1, we can prove C_n is convex.

Next, we show that $F \subset C_n \bigcap Q_n$ for all $n \geq 0$. Indeed, we have, for all $p \in F$,

$$\phi(p, y_n)$$
$$= \phi(p, J^{-1}(\alpha_n J x_0 + (1-\alpha_n) J T_{[n]} x_n))$$
$$= \|p\|^2 - 2\langle p, \alpha_n J x_0 + (1-\alpha_n) J T_{[n]} x_n \rangle + \|\alpha_n J x_0 + (1-\alpha_n) J T_{[n]} x_n\|^2$$
$$\leq \|p\|^2 - 2\alpha_n \langle p, J x_0 \rangle - 2(1-\alpha_n)\langle p, J T_{[n]} x_n \rangle + \alpha_n \|x_0\|^2 + (1-\alpha_n)\|T_{[n]} x_n\|^2$$
$$\leq \alpha_n \phi(p, x_0) + (1-\alpha_n)\phi(p, T_{[n]} x_n)$$
$$\leq \alpha_n \phi(p, x_0) + (1-\alpha_n)\phi(p, x_n).$$

So $p \in C_n$ for all $n \geq 0$ and $F \subset C_n$. As in the proof of Theorem 3.1, we also obtain $F \subset Q_n$ for all n. Hence $\{x_n\}$ is well defined.

From the proof of Theorem 3.1, we also obtain $\{x_n\}$ is bounded and

$$(3.5) \qquad\qquad\qquad\qquad \lim_{n\to\infty} \phi(x_{n+1}, x_n) = 0.$$

Since $x_{n+1} = \Pi_{C_n \cap Q_n} x_0 \in C_n$, from the definition of C_n, we have

$$\phi(x_{n+1}, y_n) \leq \alpha_n \phi(x_{n+1}, x_0) + (1-\alpha_n)\phi(x_{n+1}, x_n).$$

It follows from $\lim_{n\to\infty} \alpha_n = 0$ and (3.5) that

$$(3.6) \qquad\qquad\qquad\qquad \phi(x_{n+1}, y_n) \to 0.$$

By using Lemma 2.1, we obtain

$$(3.7) \qquad \lim_{n \to \infty} \|x_{n+1} - y_n\| = \lim_{n \to \infty} \|x_{n+1} - x_n\| = 0.$$

Since J is uniformly norm-to-norm continuous on bounded sets, we have

$$(3.8) \qquad \lim_{n \to \infty} \|Jx_{n+1} - Jy_n\| = \lim_{n \to \infty} \|Jx_{n+1} - Jx_n\| = 0.$$

Note that

$$\|JT_{[n]}x_n - Jy_n\| = \|JT_{[n]}x_n - (\alpha_n Jx_0 + (1 - \alpha_n)JT_{[n]}x_n)\|$$
$$= \alpha_n \|Jx_0 - JT_{[n]}x_n\|.$$

Therefore, we have $\lim_{n \to \infty} \|JT_{[n]}x_n - Jy_n\| = 0$.
Since J^{-1} is also uniformly norm-to-norm continuous on bounded sets, we obtain

$$(3.9) \qquad \lim_{n \to \infty} \|T_{[n]}x_n - y_n\| = 0.$$

Notice that

$$\|x_n - T_{[n]}x_n\| \leq \|x_{n+1} - x_n\| + \|x_{n+1} - y_n\| + \|y_n - T_{[n]}x_n\|.$$

It follows from (3.5), (3.6) and (3.9) that $\lim_{n \to \infty} \|x_n - T_{[n]}x_n\| = 0$. By repeating the argument in the proof of Theorem 3.1, we can obtain $\omega_\omega(x_n) \subset F$. Furthermore, we can conclude that $\{x_n\}$ converges strongly to $\Pi_F x_0$. This completes the whole proof. $\qquad \square$

Remark 3.3. From the definition of relatively nonexpansive mappings, we know that $\widehat{F}(T) = F(T)$. However, this condition is very strict. So, it is very interesting to give Corollary 3.4 and Corollary 3.5, respectively in the framework of a Hilbert space.

Corollary 3.4. *Let C be a nonempty closed convex subset of a Hilbert space H. Given an integer $N \geq 1$, let, for each $0 \leq i \leq N-1$, T_i be a nonexpansive mapping from C into itself and $\{\alpha_n\}$ a sequence of real numbers such that $0 \leq \alpha_n < 1$ and $\limsup_{n \to \infty} \alpha_n < 1$. Define a sequence $\{x_n\}_{n=0}^{\infty}$ by the following algorithm:*

$$\begin{cases} x_0 \in C \quad \text{chosen arbitrarily,} \\ y_n = \alpha_n x_n + (1 - \alpha_n)T_{[n]}x_n, \\ C_n = \{v \in C : \|v - y_n\| \leq \|v - x_n\|\}, \\ Q_n = \{v \in C : \langle x_n - v, x_0 - x_n \rangle \geq 0\}, \\ x_{n+1} = P_{C_n \cap Q_n}x_0, \quad \forall n \geq 0. \end{cases}$$

If the common fixed point set $F := \bigcap_{i=0}^{N-1} F(T_i)$ of $\{T_i\}_{i=0}^{N-1}$ is nonempty, then $\{x_n\}$ converges strongly to $P_F x_0$, where $P_F x_0$ is the metric projection from C onto F.

Corollary 3.5. *Let C be a nonempty closed convex subset of a Hilbert space H. Given an integer $N \geq 1$, let, for each $0 \leq i \leq N - 1$, T_i be a nonexpansive mapping from C into itself and $\{\alpha_n\}$ a sequence in $(0, 1)$ such that $\lim_{n \to \infty} \alpha_n = 0$. Define a sequence*

$\{x_n\}_{n=0}^{\infty}$ *by the following algorithm:*

$$\begin{cases} x_0 \in C \quad \text{chosen arbitrarily}, \\ y_n = \alpha_n x_0 + (1 - \alpha_n) T_{[n]} x_n, \\ C_n = \{v \in C : \|v - y_n\| \leq \alpha_n \|v - x_0\| + (1 - \alpha_n)\|v - x_n\|\}, \\ Q_n = \{v \in C : \langle x_n - v, x_0 - x_n \rangle \geq 0\}, \\ x_{n+1} = P_{C_n \cap Q_n} x_0, \quad \forall n \geq 0. \end{cases}$$

If the common fixed point set $F := \bigcap_{i=0}^{N-1} F(T_i)$ *of* $\{T_i\}_{i=0}^{N-1}$ *is nonempty, then* $\{x_n\}$ *converges strongly to* $P_F x_0$, *where* $P_F x_0$ *is the metric projection from* C *onto* F.

REFERENCES

1. Ya. I. Alber, *Metric and generalized projection operators in Banach spaces: Properties and applications,* in: A. G. Kartsatos (Ed.), Theory and applications of Nonlinear Operators of Monotonic and Accretive Type, Marcel Dekker, New York, 1996, pp. 15–50.
2. C. Byrne, *A unified treatment of some iterative algorithms in signal processing and image reconstruction,* Inverse Problems **20** (2004), 103–120.
3. D. Butnariu, S. Reich and A. J. Zaslavski, *Asymptotic behavior of relatively nonexpansive operators in Banach spaces,* J. Appl. Anal. **7** (2001), 151–174.
4. D. Butnariu, S. Reich and A.J. Zaslavski, *Weak convergence of orbits of nonlinear operators in reflexive Banach spaces,* Numer. Funct. Anal. Optim. **24** (2003), 489–508.
5. Y. Censor and S. Reich, *Iterations of paracontractions and firmly nonexpansive operators with applications to feasibility and optimization,* 37 (1996), 323–339.
6. B. Halpern, *Fixed points of nonexpanding maps,* Bull. Amer. Math. Soc. **73** (1967), 957–961.
7. S. Kamimura and W. Takahashi, *Strong convergence of a proximal-type algorithm in a Banach space,* SIAM J. Optim. **13** (2003), 938–945.
8. S. Matsushita and W. Takahashi, *A strong convergence theorem for relatively nonexpansive mappings in a Banach space,* J. Approx. Theory **134** (2005), 257–266.
9. C. Martinez-Yanes and H.K. Xu, *Strong convergence of the CQ method for fixed point iteration processes,* Nonlinear Anal. **64** (2006), 2400–2411.
10. W. R. Mann, *Mean value methods in iteration,* Proc. Amer. Math. Soc. **4** (1953), 506–510.
11. K. Nakajo and W. Takahashi, *Mean value methods in iteration,* Proc. Amer. Math. Soc. **4** (1953), 506–510.
12. X. L. Qin and Y. F. Su, *Strong convergence theorems for relatively nonexpansive mappings in a Banach space,* Nonlinear Anal. **67** (2007), 1958–1965.
13. S. Reich, *Weak convergence theorems for nonexpansive mappings in Banach spaces,* J. Math. Anal. Appl. **67** (1979), 274–276.
14. S. Reich, *Strong convergence theorems for resolvents of accretive operators in Banach spaces,* J. Math. Anal. Appl. **75** (1980), 287–292.
15. S. Reich, *A weak convergence theorem for the alternating method with Bregman distance,* in: A.G.Kartsatos (Ed.), Theory and Applications of Nonlinear Operators of Accretive and Monotone Type, Marcel Dekker, New York, 1996, pp. 313–318.
16. W. Takahashi, *Nonlinear Functional Analysis,* Yokohama Publishers, Yokohama, 2000.

Nonlinear Functional Analysis and Applications, Volume 1, 21–32

CONVERGENCE ANALYSIS FOR STEFFENSEN-LIKE METHOD

IOANNIS K. ARGYROS[1]

ABSTRACT. Using more precise majorizing sequences than before [1] we provide a semilocal convergence analysis for a certain class of Steffensen-like methods with the following advantages: weaker convergence conditions, finer error bounds on the distances involved and an at least as precise information on the location of the solution for the associated nonlinear equation. The local convergence of the method not examined in [1] is also studied. Numerical example are provided.

1. Introduction

In this study we are concerned with the problem of approximating a locally unique solution x^* of the nonlinear equation

$$(1) \qquad\qquad F(x) = 0,$$

where F is a Fréchet-differentiable operator defined on a convex subset D of a Banach space X with values in X.

The most important aspects when studying iterative methods for generating a sequence approximating x^* are: convergence, and order of convergence.

There is an extensive literature on local as well as semilocal convergence theorems of Kantorovich-type which provide sufficient convergence conditions as long as the initial guess (pivot) is close enough to the solution. These theorems also provide estimates on the various distances involved (see, e.g., [1]-[7] and the references there). H. T. Kung and T. F. Traub [6] introduced a class of multipoint iterative methods without derivative whereas D. Chen [4] studied a particular case of these methods which contain the Steffensen method as a special case, but only in one dimension. S. Amat, S. Busquier

---

Received August 14, 2007.

2000 *Mathematics Subject Classification.* 65H10, 65G99.

Key words and phrases. Banach space, Steffensen-like method, local, semilocal convergence, majorizing sequence, Divided differences, Fréchet-derivative.

[1] Department of Mathematical Sciences, Cameron University, Lawton, OK 73505, USA (*E-mail:* iargyros@cameron.edu)

and V. Candela [1] introduced the Steffensen-like method

$$y_n = x_n + \alpha_n F(x_n), \quad (x_0 \in D, \, n \geq 0),$$

$$x_{n+1} = y_n - [x_n, y_n; F]^{-1} F(y_n),$$

where $\alpha_n \in (0, 1]$ are parameters keeping $\|\alpha_n F(x_n)\|$ small enough, and $[x, y; F]$ is a divided difference of order one in X [6]. Here we consider the following version of the above Steffensen-like method given by

(2) $$y_n = x_n + \alpha_n F'(x_0)^{-1} F(x_n),$$

and

(3) $$x_{n+1} = y_n - [x_n, y_n; F]^{-1} F(y_n)$$

The following assumptions were made in [1]: F is twice Fréchet-differentiable on D; $x_0 \in D$ is such that $F'(x_0)^{-1} \in L(X, X)$,

(4) $$\|F(x_0)\| \leq \overline{\eta},$$

(5) $$\left\|F'(x_0)^{-1}\right\| \leq 1,$$

(6) $$\|F''(x)\| \leq \ell \quad \text{for all } x \in D,$$

(7) $$\left\|\frac{I}{\alpha_n} + F'(x_0)\right\| \leq \frac{1}{\alpha_n} - 1,$$

(8) $$\overline{U}(x_0, \frac{1}{\ell}) = \{x \in X \mid \|x - x_0\| \leq \frac{1}{\ell}\} \subseteq D, \; \ell \neq 0,$$

and

(9) $$h = \ell \overline{\eta} \leq \frac{1}{2}.$$

It follows from (5) that the results in [1] cannot be used when

(10) $$\|F'(x_0)\| < 1.$$

We assume instead weaker conditions: F is Fréchet-differentiable on D, $F'(x_0)^{-1} \in L(X, X)$,

(11) $$\left\|F'(x_0)^{-1} F(x_0)\right\| \leq \eta,$$

(12) $$\left\|F'(x_0)^{-1}[F'(x) - F'(x_0)]\right\| \leq \ell_0 \|x - x_0\|,$$

(13) $$\left\|F'(x_0)^{-1}[F'(x) - F'(y)]\right\| \leq \ell \|x - y\|,$$

(14) $$\left\|\frac{F'(x_0)^{-1}}{\alpha_n} + I\right\| \leq \frac{1}{\alpha_n} - 1,$$

(15) $$\overline{U}(x_0, \frac{1}{\ell_0}) \subseteq D, \quad \ell_0, \neq 0$$

and

(16) $$h_A = \ell\eta \leq \frac{1}{2} \quad \text{(or the hypotheses of Lemma 1)}.$$

Note that

(17) $$\ell_0 \leq \ell$$

holds in general and $\frac{\ell}{\ell_0}$ can be arbitrarily large [2], [3]. Without loss of generality, we assume equality holds in (11). It then follows

(18) $$\eta \leq \left\|F'(x_0)^{-1}\right\| \cdot \|F(x_0)\| \leq \overline{\eta}.$$

That is,

(19) $$h \leq \frac{1}{2} \implies h_A \leq \frac{1}{2},$$

but not necessarily vice versa. We show in the semilocal case that under the same or weaker hypotheses than in [1] we can provide finer estimates on the distances $\|x_n - y_n\|$, $\|x_{n+1} - y_n\|$, $\|x_n - x^*\|$, $\|y_n - x^*\|$ and an at least as precise information on the location of the solution x^*.

The local convergence not examined in [1] is also studied. Numerical examples are also provided. We shall assume throughout the study that divided difference $[x, y; F]$ satisfies

(20) $$[x, y; F](x - y) = F(x) - F(y)$$

and

(21) $$[x, y; F] = \int_0^1 F'[y + \theta(x - y)]dt$$

for all $x, y \in D$.

Note that, if F is Lipschitz on an interval D containing x and y, then F is differentiable almost everywhere according to Rademacher's theorem and, by using the Lebesque integral, we obtain (21) as the integral representation of divided difference $[x, y; F]$.

2. Majorizing Sequences

Let us define the scalar iterations $\{s_n\}$, $\{t_n\}$ $(n \geq 0)$ by

(22) $$t_0 = 0, \ s_0 = \alpha_0\eta, \quad s_n = t_n + \frac{\alpha_n\ell}{2}(t_n - t_{n-1})(t_n - s_{n-1}) \quad (n \geq 1)$$

and for $\varepsilon_n > 0$

(23) $$t_{n+1} = s_n + \frac{[\ell(s_n - t_n) + 2\ell_0 t_n + \beta_n]}{2[-\ell_0 s_n + 1 + \varepsilon_n]}(s_n - t_n) \quad (n \geq 0),$$

where,

(24) $$\beta_n = 2\left(\frac{1}{\alpha_n} - 1\right) \quad (n \geq 0).$$

Then we can show the following result on the convergence of majorizing sequences $\{s_n\}$, $\{t_n\}$:

Lemma 1. *Assume there exist parameters* $\delta_1 \in [0,2)$, $\delta_2 \in [0,2)$, $\alpha \geq 0$, $\beta \geq 0$, *and* $\varepsilon \geq 0$ *such that for* $\delta = \frac{\delta_1 \delta_2}{4}$, $t = \frac{t_1}{1-\delta}$ *the following hold true for all* $n \geq 0$:

$$(25) \qquad\qquad\qquad \alpha_n \leq \alpha,$$

$$(26) \qquad\qquad\qquad \varepsilon_n \leq \varepsilon,$$

$$(27) \qquad\qquad\qquad \beta_n \leq \beta,$$

$$(28) \qquad\qquad\qquad \ell \alpha t_1 \leq \delta_1,$$

$$(29) \qquad\qquad\qquad \ell_0 t < 1 + \varepsilon$$

and

$$(30) \qquad\qquad \frac{\ell(s_0 - t_0) + 2\ell_0 t + \beta}{1 - \ell_0 t + \varepsilon} \leq \delta_2.$$

Then sequences $\{t_n\}$, $\{s_n\}$ *are monotonically increasing and converge to their common limit* t^* *such that, for all* $n \geq 0$,

$$(31) \qquad\qquad \delta_n \leq t_{n+1} \leq t^* \leq t,$$

$$(32) \qquad\qquad 0 \leq s_n - t_n \leq \delta^n (s_0 - t_0)$$

and

$$(33) \qquad\qquad 0 \leq t_{n+1} - s_n \leq \delta^n (t_1 - s_0).$$

Proof. We shall show that, for all $k \geq 0$,

$$(34) \qquad\qquad a_k \ell(t_k - t_{n-1}) \leq \delta_1,$$

$$(35) \qquad\qquad \frac{\ell(s_k - t_k) + 2\ell_0 t_k + b_k}{1 - \ell_0 s_k + \varepsilon_k} \leq \delta_2,$$

$$(36) \qquad\qquad \ell_0 s_k < 1 + \varepsilon_k,$$

$$(37) \qquad\qquad s_k \leq t_{k+1},$$

$$(38) \qquad\qquad 0 \leq s_k - t_k \leq \delta^k (s_0 - t_0)$$

and

$$(39) \qquad\qquad 0 \leq t_{k+1} - s_k \leq \delta^k (t_1 - s_0).$$

For $k = 0$ estimates (34)-(39) hold true by the initial conditions. Let us assume that they hold true for all $k \leq n - 1$. It follows from the induction hypotheses that:

$$
\begin{aligned}
t_{k+1} &= (t_{k+1} - t_k) + t_k = (t_{k+1} - t_k) + (t_k - t_{k-1}) + t_{k-1} \\
&= (t_{k+1} - t_k) + (t_k - t_{k-1}) + \dots + (t_1 - t_0) + t_0 \\
&\leq (\delta^k + \delta^{k-1} + \dots + 1)(t_1 - t_0) + t_0 \\
&\leq \frac{1 - \delta^{k+1}}{1 - \delta}(t_1 - t_0) + t_0 \leq t,
\end{aligned}
$$
(40)

$$
s_k - t_k \leq \frac{\delta_1}{2}(t_k - s_{k-1}) \leq \frac{\delta_1}{2}\frac{\delta_2}{2}(s_{k-1} - t_{k-1}) = \delta(s_{k-1} - t_{k-1})
$$
(41)

and, similarly,

$$
t_{k+1} - s_k \leq \delta^k(t_1 - s_0).
$$
(42)

It follows from (19)-(42) that (34)-(39) hold true by (25)-(30). The induction is completed. It follows that sequences $\{s_n\}$, $\{t_n\}$ are monotonically increasing and bounded above by t and as such they converge to their common limit satisfying (31). That completes the proof of the Lemma. $\qquad\square$

Remark 2. (1) The proof goes through if $\delta_1 = 2$ or $\delta_2 = 2$ but not at the same time.

(2) Let us define scalar sequences $\{\bar{t}_n\}$, $\{\bar{s}_n\}$, $\{\bar{\bar{t}}_n\}$, $\{\bar{\bar{s}}_n\}$ by

$$
\bar{s}_n = \bar{t}_n + \alpha_n f(\bar{t}_n), \bar{t}_0 = 0,
$$
(43)

$$
\bar{t}_{n+1} = \bar{s}_n - \frac{f(\bar{s}_n)}{f'(\bar{s}_n) - \varepsilon_n},
$$
(44)

$$
\bar{\bar{s}}_n = \bar{\bar{t}}_n + \alpha_n f(\bar{\bar{t}}_n), \bar{\bar{t}}_0 = 0,
$$
(45)

$$
\bar{\bar{t}}_{n+1} = \bar{\bar{s}}_n - \frac{f(\bar{\bar{s}}_n)}{f'(\bar{\bar{s}}_n) - \varepsilon_n},
$$
(46)

where

$$
f(t) = \frac{l}{2}t^2 - t + \eta
$$
(47)

and

$$
f_0(t) = \frac{l_0}{2}t^2 - t + \eta.
$$
(48)

Moreover, define the scalar sequence $\{\bar{\bar{\bar{t}}}\}$, $\{\bar{\bar{\bar{s}}}_n\}$, used in [1], by

$$
\bar{\bar{\bar{s}}}_n = \bar{\bar{\bar{t}}}_n + \alpha_n f(\bar{\bar{\bar{t}}}_n), \quad \bar{\bar{\bar{t}}}_n = 0
$$
(49)

$$
\bar{\bar{\bar{t}}}_{n+1} = \bar{\bar{\bar{s}}}_n - \frac{f_1(\bar{\bar{\bar{s}}}_n)}{f_1^1(\bar{\bar{\bar{s}}}_n) - \varepsilon_n},
$$
(50)

where

$$(51) \qquad f_1(t) = \frac{l}{2}t^2 - t + \overline{\eta}.$$

Under hypothesis (9) it can easily be seen using induction on $n \geq 0$

$$(52) \qquad \overline{t}_{n+1} \leq \overline{s}_n \leq \overline{t}_n,$$

$$(53) \qquad \overline{\overline{t}}_{n+1} \leq \overline{\overline{s}}_n \leq \overline{\overline{t}}_n,$$

$$(54) \qquad \overline{\overline{\overline{t}}}_{n+1} \leq \overline{\overline{\overline{s}}}_n \leq \overline{\overline{\overline{t}}}_n,$$

$$(55) \qquad \overline{s}_n - \overline{t}_n \leq \overline{\overline{s}}_n - \overline{\overline{t}}_n \leq \overline{\overline{\overline{s}}}_n - \overline{\overline{\overline{t}}}_n,$$

$$(56) \qquad \overline{t}_{n+1} - \overline{s}_n \leq \overline{\overline{t}}_{n+1} - \overline{\overline{s}}_n \leq \overline{\overline{\overline{t}}}_{n+1} - \overline{\overline{\overline{s}}}_n,$$

$$(57) \qquad \overline{t}_n \leq \overline{\overline{t}}_n \leq \overline{\overline{\overline{t}}}_n,$$

$$(58) \qquad \overline{s}_n \leq \overline{\overline{s}}_n \leq \overline{\overline{\overline{s}}}_n$$

and

$$(59) \qquad \overline{t^*} \leq \overline{\overline{t^*}} \leq \overline{\overline{\overline{t^*}}}_n,$$

where

$$(60) \qquad \lim_{n \to \infty} \overline{t}_n = \overline{t^*},$$

$$(61) \qquad \lim_{n \to \infty} \overline{\overline{t}}_n = \overline{\overline{t^*}}$$

and

$$(62) \qquad \lim_{n \to \infty} \overline{\overline{\overline{t}}}_n = \overline{\overline{\overline{t^*}}}.$$

Finally, under hypotheses (25)-(30) and (9)

$$(63) \qquad t_n \leq \overline{t}_n,$$

$$(64) \qquad s_n \leq \overline{s}_n,$$

$$(65) \qquad s_n - t_n \leq \overline{s}_n - t_n,$$

$$(66) \qquad t_{n+1} - s_n \leq \overline{t}_{n+1} - \overline{s}_n,$$

and

$$(67) \qquad t^* \leq \overline{t}^*.$$

for all $n \geq 0$. In case (17) or (18) hold as strict inequalities so do (63)-(66) for all $n > 1$. Hence we conclude that $\{t_n\}$ and $\{s_n\}$ are finer sequences than all the rest defined above. The quadratic convergence of $\{\overline{\overline{t}}_n\}$, $\{\overline{\overline{s}}_n\}$ was established in [1, pp. 401]. It follows from the above that the rest of the t, s sequences are also at least of guadratic order.

3. Semilocal Convergence Analysis of Method (2)-(3).

We can show the following semilocal convergence theorem for method (2)-(3):

Theorem 3. *Assume* (11)-(15) *and the hypotheses of Lemma 1 hold. Let us also assume:*

$$(68) \qquad\qquad t_0^* \ell < 1.$$

Then we have the following:
(a) *For all* $\varepsilon_n > 0$ *there are* $\alpha_n \in (0,1]$ *such that*

$$(69) \qquad\qquad \|x_{n+1} - x_n\| \le t_{n+1} - t_n;$$

(b) *The sequence* $\{x_n\}$ *generated by Steffensen-like method* (2)-(3) *is well defined remains in* $\overline{U}(x_0, t^*)$ *for all* $n \ge 0$ *and converges to a unique solution* $x^* \in \overline{U}(x_0, t^*)$ *of the equation* $F(x) = 0$;
(c) *Moreover, the following estimates hold:*

$$(70) \qquad\qquad \|x^* - x_n\| \le t^* - t_n \quad (n \ge 0);$$

(d) *Furthermore, if there exists* $R \in (t^*, \frac{1}{\ell_0})$ *such that*

$$(71) \qquad\qquad \ell_0 (t^* + R) \le 2,$$

then the solution x^* *is unique in* $U(x_0, R)$.

Proof. We shall show (69) by induction on $n \ge 0$. If $n = 0$, we have $x_0 \in \overline{U}(x_0, \frac{1}{\ell_0})$. By (68), there exists $\alpha_0 \in (0,1]$ such that $y_0 \in \overline{U}(x_0, \frac{1}{\ell_0})$. In view of (12), (7) and the identity

$$(72) \qquad \begin{aligned} F'(x_0)^{-1} F(y_0) &= F'(x_0)^{-1} \int_0^1 \{F'[x_0 + t(y_0 - x_0)] - F'(x_0)\}(y_0 - x_0) dt \\ &+ \left(\frac{F'(x_0)^{-1}}{\alpha_0} + I \right)(y_0 - x_0), \end{aligned}$$

we get

$$(73) \qquad \begin{aligned} \|F'(x_0)^{-1} F(y_0)\| &\le \frac{\ell_0}{2}(s_0 - t_0)^2 + \left(\frac{1}{\alpha_0} - 1 \right)(s_0 - t_0) \\ &\le \frac{\ell}{2}(s_0 - t_0)^2 + \left(\frac{1}{\alpha_0} - 1 \right)(s_0 - t_0). \end{aligned}$$

We also have, by (12) and (68), for $x, y \in U(x_0, t)$,

$$(74) \qquad \begin{aligned} &\|F'(x_0)^{-1}(F'(x_0) - [x, y; f])\| \\ &\int_0^1 \|F'(x_0)^{-1}(F'(x + \theta(y - x)) - F'(x_0))\| \, d\theta \\ &\le \ell_0 \int_0^1 [(1 - \theta)\|x - x_0\| + \theta\|y - x_0\|] d\theta \\ &\le \ell_0 t < 1. \end{aligned}$$

It follows from (74) and the Banach Lemma on invertible operators [3] that $[x, y; F]^{-1}$ is invertible. In particular, for $x = x_0$, and $y = y_0$, we obtain in turn

$$\left\| [x_0, y_0; F]^{-1} F'(x_0) \right\| \leq \frac{1}{1 - \|F'(x_0)^{-1}(F'(x_0) - [x_0, y_0; F])\|}$$

(75)
$$\leq \frac{1}{1 - \|F'(x_0)^{-1}(F'(x_0) - F'(y_0) + F'(x_0)^{-1}(F'(y_0) - [x_0, y_0; F])\|}$$

$$\leq \frac{1}{1 + \varepsilon_0 - l_0 s_0}.$$

In view of (2)-(3), (73) and (75), we get

$$\|x_1 - x_0\| = \left\| \alpha_0 F'(x_0)^{-1} F(x_0) - [x_0, y_0; F]^{-1} F(y_0) \right\|$$

(76)
$$\leq \alpha_0 n + \frac{l(s_0 - t_0) + 2l_0 t_0 + b_0}{2(1 + \varepsilon_0 - l_0 s_0)} (s_0 - t_0)$$

$$= t_1 - t_0.$$

Let us assume $\|x_k - x_{k-1}\| \leq t_k - t_{k-1})$ for all $k = 1, 2, ..., n$. In particular, we have

$$\|x_k - x_0\| \leq \|x_k - x_{k-1}\| + \|x_{k-1} - x_{k-2}\| + ... + \|x_1 - x_0\|$$
$$\leq (t_k - t_{k-1}) + (t_{k-1} - t_{k-2}) + ... + (t_1 - t_0)$$
$$= t_k \leq t < \frac{1}{\ell_0}.$$

That is, $x_k \in U\left(x_0, \frac{1}{l_0}\right)$. In view of (2) and (68), we can choose α_k such that $y_k \in U\left(x_0, \frac{1}{l_0}\right)$. Using the identity

$$F'(x_0)^{-1} F(y_k) = F'(x_0)^{-1} \int_0^1 \{(F'[x_k + \theta(y_k - x_k)] - F'(x_k))(y_k - x_k) d\theta$$

(77)
$$+ F'(x_0)^{-1} (F'(x_k) - F'(x_0))(y_k - x_k)$$

$$+ \left(\frac{F'(x_0)^{-1}}{\alpha_k} + I \right) (y_k - x_k),$$

(12), (13) and the induction hypotheses, we obtain

$$\left\| F'(x_0)^{-1} F(y_k) \right\| \leq \frac{\ell}{2} (s_k - t_k)^2 + \ell_0 (s_k - t_k)(t_k - t_0)$$

(78)
$$+ \left(\frac{1}{\alpha_k} - 1 \right) (s_k - t_k).$$

Moreover, from the identity

$$F'(x_0)^{-1} F(x_k) = F'(x_0)^{-1} \{F(x_k) - F(y_{k-1}) - [x_{k-1}, y_{k-1}; F](x_k - y_{k-1})\}$$

(79)
$$= F'(x_0)^{-1} \int_0^1 [F'(y_{k-1} + \theta(x_k - y_{k-1}))$$

$$- F'(y_{k-1} + \theta(x_{k-1} - y_{k-1})](x_k - y_{k-1}) d\theta,$$

the induction hypotheses and (13), we get

$$(80) \qquad \left\| F'(x_0)^{-1} F(x_k) \right\| \leq \frac{\ell}{2} \left\| x_k - x_{k-1} \right\| \left\| x_n - y_{k-1} \right\|.$$

In view of (2), (22) and (80), we obtain

$$(81) \qquad \left\| y_k - x_k \right\| \leq \frac{\alpha_k}{2} \ell (t_k - t_{k-1})(t_k - s_{k-1}) = s_k - t_k.$$

As in (75), for x_0, y_0 being replaced by x_k, y_k, we get

$$(82) \qquad \left\| [x_k, y_k; F]^{-1} F'(x_0) \right\| \leq \frac{1}{1 + \varepsilon_k - \ell_0 s_k}.$$

Using (3), (23), (78) and (82), we get

$$(83) \qquad \begin{aligned} \left\| x_{k+1} - y_k \right\| &\leq \left\| [x_k, y_k; F]^{-1} F'(x_0) \right\| \cdot \left\| F'(x_0)^{-1} F(y_k) \right\| \\ &\leq t_{k+1} - s_k. \end{aligned}$$

Hence we obtain

$$\begin{aligned} \left\| x_{k+1} - x_k \right\| &\leq \left\| \alpha_k F'(x_0)^{-1} F(x_k) - ([x_k, y_k; F]^{-1} F'(x_0)) F'(x_0)^{-1} F(y_k) \right\| \\ &\leq \frac{\alpha_k \ell}{2}(t_k - t_{k-1})(t_k - s_{k-1}) - \frac{[\ell(s_k - t_k) + 2\ell_0 t_k + \beta_k]}{2[1 + \varepsilon_k - \ell_0 s_k]}(s_k - t_k) \\ &= t_{k+1} - t_k, \end{aligned}$$

which completes the induction for (69). It follows from Lemma 1 and (69) that sequence $\{x_n\}$ is Cauchy in a Banach space X and as such it converges to a point $x^* \in \overline{U}(x_0, t^*)$ (since $\overline{U}(x_0, t^*)$ is a closed set). By letting $k \to \infty$ in (80), we get $F(x^*) = 0$. Estimate (70) follows from (69) by using standard majorization techniques [3], [6]. To show uniqueness in $\overline{U}(x_0, t^*)$, let y^* be a solution of equation $F(x) = 0$ in $\overline{U}(x_0, t^*)$. As in (74) we get

$$(84) \qquad \begin{aligned} \left\| [x^*, y^*; F]^{-1} F'(x_0) \right\| &\leq \ell_0 \int_0^1 [\theta \left\| x^* - x_0 \right\| + (1 - \theta) \left\| y^* - x_0 \right\|] d\theta \\ &\leq \frac{\ell_0}{2}(t^* + t^*) < 1 \end{aligned}$$

by (68). That is $[x^*, y^*]$ is invertible. Moreover by the identity

$$(85) \qquad F(x^*) - F(y^*) = [x^*, y^*; F](x^* - y^*),$$

we get

$$(86) \qquad x^* = y^*.$$

Furthermore, to show uniqueness in $U(x_0, R)$, let y^* be a solution of equation $F(x) = 0$ in $U(x_0, R)$. This time the left hand side of (84) becomes strictly smaller than

$$\frac{\ell_0}{2}(t^* + R) \leq 1$$

by (71). Hence again we deduce

$$x^* = y^*.$$

That completes the proof. □

Remark 4. Condition (65) can be replaced with

$$(87) \qquad\qquad\qquad t\ell_0 < 1.$$

In this case t^* is also replaced by t throughout Theorem 3.

Exampl 5. Let $X = \mathbf{R}$, $x_0 = 1$, $D = [q, 2 - q]$, $q \in [0, \frac{1}{2})$ and define F on D by

$$(88) \qquad\qquad\qquad F(x) = x^3 - q.$$

Using (88), (4), (12) and (13) we obtain $\bar{n} = 1 - q$, $n = \frac{1}{3}(1 - q)$, $l_0 = 3 - q$ and $l = 2(2 - q)$. However, hypothesis (9) is violated since

$$h = 2(2 - q)(1 - q) > \frac{1}{2} \quad \text{for all } q \in \left[0, \frac{1}{2}\right).$$

That is there is no guarantee that Steffensen-like method (2)-(3) starting at $x_0 = 1$ converges to $x^* = \sqrt[3]{q}$. Let us choose $\alpha_n = \alpha = \frac{1}{4}$, $\varepsilon_n = \varepsilon = 10^{-8}$ and $s_1 = s_2 = 1.9$. Then we obtain $b_n = 6$. Let also $q = .48$. Then it can easily be seen that all hypotheses of Theorem 3 are satisfied with the above choices. Hence the conclusions of it follow.

Example 6. Let $X = \mathbf{R}^2$ be any neighborhood of $x_0 = (0,0)$. Let us define F on D by

$$(89) \qquad\qquad F(x, y) = \begin{bmatrix} \lambda_1 x^2 + \lambda_2 y^2 - \lambda_3 x + m_1 \\ \lambda_4 x^2 + \lambda_5 y^2 - \lambda_6 y + m_2 \end{bmatrix},$$

where λ_i, $i = 1, ..., 6$, m_1, m_2 are given real parameters. It follows by (89) that $\|F'(x_0)\| = v = \max\{|\lambda_3|, |\lambda_6|\}$. If $v < 1$, then (5) cannot hold and therefore the results in [1] cannot be applied to solve equation $F(x, y) = 0$ no matter how the rest of the parameters are chosen. However, for suitable choices of the parameters involved our Theorem 3 can be applied.

4. Local Convergence of Method (2)-(3)

Let us assume there exists a simple zero x^* of the equation $F(x) = 0$, a nonnegative sequence $\{\gamma_n\}$ and constants γ, L_0, L such that

$$(90) \qquad\qquad \left\| I + \alpha_n F'(x_0)^{-1}[x, y; F] \right\| \leq \gamma_n \leq \gamma,$$

$$(91) \qquad\qquad \left\| F'(x^*)^{-1}[F'(x) - F'(y)] \right\| \leq L \|x - y\|,$$

$$(92) \qquad\qquad \left\| F'(x^*)^{-1}[F'(x) - F'(x^*)] \right\| \leq L_0 \|x - x^*\|$$

for all $x, y \in D$, and

$$(93) \qquad\qquad\qquad \overline{U}(x^*, p) \subseteq D,$$

where

$$(94) \qquad\qquad\qquad p = \frac{2}{\gamma(1 + 2\gamma)L + 2(1 + \gamma)L_0}$$

with γ, L_0 and L not all zero at the same time. Then we can show the following local convergence theorem for method (2)-(3):

Theorem 7. *Under the above stated hypotheses and provided that $x_0 \in U(x^*, p)$ and the sequence $\{x_n\}$ generated by Steffensen-Like method (2)-(3) converges to x^* so that, for all $n \geq 0$,*

$$(95) \qquad \|y_n - x^*\| \leq \gamma \|x_n - x^*\|$$

and

$$(96) \qquad \|x_{n+1} - x^*\| \leq \frac{\gamma(1 + 2\gamma)L \|x_n - x^*\|^2}{2[1 - L_0(\|x_n - x^*\| + \|y_n - x^*\|)]}$$

Proof. By hypothesis $x_0 \in U(x^*, p)$. Using induction on $k \geq 0$, (90)-(94) and the identities:

$$
\begin{aligned}
(97) \qquad x_{k+1} - x^* &= y_k - x^* - [x_k, y_k; F]^{-1} F(y_k) \\
&= [x_k, y_k; F]^{-1} F'(x^*) F'(x^*)^{-1} \{([x_k, y_k; F] - [y_k, x^*])(y_k - x^*)\} \\
&= [x_k, y_k; F]^{-1} F'(x^*) F'(x^*)^{-1} \\
&\quad \int_0^1 \{[F'(y_k + \theta(x_k - y_k)) - F'(x^* + \theta(y_k - x^*))] d\theta(y_k - x^*)\}
\end{aligned}
$$

$$
\begin{aligned}
(98) \qquad y_k - x^* &= x_k - x^* + \alpha_k F'(x_0)^{-1} F(x_k) \\
&= x_k - x^* + \alpha_k F'(x_0)^{-1} [x_k, x^*; F](x_k - x^*) \\
&= \{I + \alpha_k F'(x_0)^{-1} [x_k, x^*; F]\}(x_k - x^*),
\end{aligned}
$$

we can easily see that (95) and (96) hold true. It also follows from (95), (96) and the choice of p that

$$(99) \qquad \|x_{k+1} - x^*\| < \|x_k - x^*\| < p,$$

which shows $x_n \in U(x^*, p)$ for all $n \geq 0$ and $\lim_{n \to \infty} x_n = x^*$. $\qquad \square$

Remark 8. The results obtained here can be extended immediately for divided differences $[x, y; F]$ satisfying conditions more general than (21). Indeed, simply replace conditions (12), (13), (91) and (92) by

$$\left\| F'(x_0)^{-1}([x, y; F] - [x_0, x_0; F]) \right\| \leq \bar{l}_0(\|x - x_0\| + \|y - x_0\|),$$

$$\left\| F'(x_0)^{-1}([x, y; F] - [v, w; F]) \right\| \leq \bar{l}(\|x - v\| + \|y - w\|),$$

$$\left\| F'(x^*)^{-1}([x, y; F] - [v, w; F]) \right\| \leq \bar{L}(\|x - v\| + \|y - w\|)$$

and

$$\left\| F'(x^*)^{-1}([x, y; F] - F'(x^*)) \right\| \leq \bar{L}_0 \|x - x^*\| + \|y - x^*\|)$$

for all $x, y, v, w \in D$, respectively. Parameters l_0, l, L_0 and L should also be replaced by $2\bar{l}_0$, $2\bar{l}$, $2\bar{L}_0$ and $2\bar{L}$, respectively.

References

1. S. Amat, S. Busqier and V. Candela, *A class of quasi-Newton generalized Steffensen methods an Banach spaces,* J. Comput. Appl. Math. **149** (2002), 397–406.
2. I. K. Argyros, *A unifying local-semilocal convergence analysis and applications for two point Newton-Like methods in Banach space,* J. Math. Anal. Appl. **298** (2004), 374–397.
3. I. K. Argyros, *Computational theory of iterative methods,* Series: Computational Mathematics 15, Editors C.K.Chui, and L.Wyutack, Elsevier Publ. Comp. 2007, New York, U.S.A.
4. D. Chen, *On the convergence of a class of generalized Steffensen's iterative procedures and error analysis,* Int. J. Comput. Math. **31** (1989), 195–203.
5. T. M. Gutiercez, M. A. Hernandez and M. A. Salanova, *A family of the Chebyshev-Halley type methods in Banach spaces,* Bull. Austral. Math. Soc. **55** (1997), 113–130.
6. T. M. Ortega and W. C. Rheinboldt, *Iterative solution of nonlinear equations in Banach spaces,* Academic Press, New York, 1970.
7. T. T. Ypma, *Local convergence of inexact Nweton methods,* SIAM J. Numer. Anal. **21** (1984), 583–590.

Nonlinear Functional Analysis and Applications, Volume 1, 33–39

GENERALIZED VARIATIONAL INEQUALITIES INVOLVING RELAXED MONOTONE MAPPINGS AND NONEXPANSIVE MAPPINGS IN HILBERT SPACES

YONGFU SU[1],[*] AND XIAOLONG QIN[2]

ABSTRACT. In this paper, we consider the solvability of generalized variational inequalities involving multivalued relaxed monotone operators and singlevalued nonexpansive mappings in the framework of Hilbert spaces. We also study the convergence criteria of iterative methods under some mild conditions. Our results include the previous results as special cases and may be considered as an improvement and refinement of the previously known results.

1. Introduction

Variational inequalities [1, 4] and hemi-variational inequalities [2] have significant applications in various fields of mathematics, physics, economics, engineering sciences and others. The associated operator equations are equally essential in the sense that these turn out to be powerful tools to the solvability of variational inequalities. Relaxed monotone operators have applications to constrained hemi-variational inequalities. Since in the study of constrained problems in reflexive Banach spaces E the set of all admissible elements is non-convex but star-shaped, corresponding variational formulations are no longer variational inequalities. Using hemi-variational inequalities, one can prove the existence of solutions to the following type of non-convex constrained problems (P): find u in C such that

$$(1.1) \qquad \langle Au - g, v \rangle \geq 0, \quad \forall v \in T_C(u),$$

Received August 17, 2007.   * Corresponding author.

2000 *Mathematics Subject Classification.* 47J05, 47J20, 47J25, 47H09.

Key words and phrases. Variational inequality; Hilbert space; Nonexpansive mapping; Relaxed cocoercive mapping.

[1] Department of Mathematics, Tianjin Polytechnic University, Tianjin 300160, People's Republic of China (*E-mail*: suyongfu@gmail.com)

[2] Department of Mathematics, Gyeongsang National University, Chinju 660-701, Korea (*E-mail*: qxlxljh@163.com)

where the admissible set $C \subset E$ is a star-shaped set with respect to a certain ball $B_E(u_0, \rho)$ and $T_C(u)$ denotes Clarke's tangent cone of C at u in C. It is easily seen that when C is convex, (1.1) reduces to the variational inequality of finding u in C such that

$$(1.2) \qquad \langle Au - g, v \rangle \geq 0, \quad \forall v \in C.$$

Example. ([2]) Let $A : E \to E^*$ be a maximal monotone operator from a reflexive Banach space E into E^* with strong monotonicity and let $C \subset E$ be star-shaped with respect to a bail $B_E(u_0, \rho)$. Suppose that $Au_0 - g \neq 0$ and that distance function d_C satisfies the condition of relaxed monotonicity

$$\langle u^* - v^*, u - v \rangle \geq -c\|u - v\|^2, \quad \forall u, v \in E,$$

and for any $u^* \in \partial d_C(u)$ and v^* in $\partial d_C(v)$ with c satisfying $0 < c < \frac{4a^2\rho}{\|Au_0 - g\|^2}$, where a is the constant for strong monotonicity of A. Here ∂d_C is a relaxed monotone operator. Then the problem (P) has at least one solution.

2. Preliminaries

Let H be a separable real Hilbert space, whose inner product and norm are denoted by $\langle \cdot, \cdot \rangle$ and $\| \cdot \|$, respectively. Let C be a nonempty clsoed convex set in H. Given the nonlinear continuous operator $A, g : H \to H$, we consider the problem of finding $u \in H$ such that $g(u) \in C$ and

$$(2.1) \qquad \langle Au + w, g(v) - g(u) \rangle \geq 0, \quad \forall g(v) \in C, w \in Tu,$$

where T is a multi-valued nonlinear mapping. We call the inequality of the type (2.1) generalized variational inequality involving three nonlinear operators. We shall denote the solution of variational inequality (2.1) by $GVI(C, A)$. Many mathematical problems either arise or can be formulated in terms of (2.1).

Next, we will consider some special cases of generalized variational inequality (2.1).

(I) If $w \equiv 0$, then the problem (2.1) is equivalent to finding $u \in C$ such that

$$(2.2) \qquad \langle Au, g(v) - g(u) \rangle \geq 0, \quad \forall g(v) \in C,$$

which is known as general (Noor) variational inequality which was introduced by Noor [3] in 1988.

(II) If $g \equiv I$, the identity operator, then the problem (2.1) equivalent to find $u \in C$ such that

$$(2.3) \qquad \langle Au + w, v - u \rangle \geq 0, \quad \forall v \in C, w \in Tu,$$

which is originally introduced and studied by Verma [5].

(III) If $g \equiv I$, the identity operator and $w \equiv 0$, then problem (2.1) equivalent to find $u \in C$ such that

$$(2.4) \qquad \langle Au, v - u \rangle \geq 0, \quad \forall v \in C,$$

which is known as the variational inequality problem, originally introduced and studied by Stampacchia [4].

(IV) If $w \equiv 0$ and $C^* = \{u \in H : \langle u, v \rangle \geq 0, \text{for } \forall v \in C\}$ is a polar cone of the convex cone C in H and $C \subset g(C)$, then problem (2.1) is equivalent to

(2.5) $$g(u) \in C, \quad Au \in C \quad \text{and} \quad \langle Au, g(u) \rangle = 0.$$

(2.5) is known as the general nonlinear complementarity problem, which includes many previously known complementarity problems as special cases.

(V) If $w \equiv 0$ and $C = H$, then (2.1) is equivalent to finding $u \in H$ such that

$$\langle Au, g(v) \rangle = 0, \quad \forall g(v) \in H,$$

which is known as the weak formulation of the odd-order boundary-value problems.

Recall the following definitions:
(1) A is called *v-strongly monotone* if there exists a constant $v > 0$ such that

$$\langle Ax - Ay, x - y \rangle \geq v \|x - y\|^2, \quad \forall x, y \in C.$$

This implies that

$$\|Ax - Ay\| \geq v \|x - y\|, \quad \forall x, y \in C,$$

that is, A is v-expansive and, when $v = 1$, it is expansive.

(2) A is said to be *r-cocoercive* if there exists a constant $r > 0$ such that

$$\langle Ax - Ay, x - y \rangle \geq r \|Ax - Ay\|^2, \quad \forall x, y \in C.$$

Clearly, every r-cocoercive mapping A is $\frac{1}{r}$-Lipschitz continuous.

(3) A is called *relaxed γ-cocoerceive* if there exists a constant $\gamma > 0$ such that

$$\langle Ax - Ay, x - y \rangle \geq (-\gamma) \|Ax - Ay\|^2, \quad \forall x, y \in C.$$

(4) A is said to be *relaxed (γ, r)-cocoercive* if there exist two constants $u, v > 0$ such that

$$\langle Ax - Ay, x - y \rangle \geq (-\gamma) \|Ax - Ay\|^2 + r \|x - y\|^2, \quad \forall x, y \in C.$$

For $\gamma = 0$, A is r-strongly monotone. This class of mappings is more general that the class of strongly monotone mappings. It is easy to see that we have the following implication:

r-strongly monotonicity $\Rightarrow$ relaxed (γ, r)-cocoercivity.

(5) $T : H \to 2^H$ is said to be a *relaxed monotone operator* if there exists a constant $k > 0$ such that

$$\langle w_1 - w_2, u - v \rangle \geq -k \|u - v\|^2,$$

where $w_1 \in Tu$ and $w_2 \in Tv$.

(6) A multivalued operator T is *Lipschitz continuous* if there exists a constant $\lambda > 0$ such that

$$\|w_1 - w_2\| \leq \lambda \|u - v\|,$$

where $w_1 \in Tu$ and $w_2 \in Tv$.

(7) $S : C \to C$ is said to be *nonexpansive* if

$$\|Sx - Sy\| \le \|x - y\|, \quad \forall x, y \in C.$$

In order to prove our main results, we need the following lemmas and definitions.

Lemma 2.1. ([6]) *Assume that $\{a_n\}$ is a sequence of nonnegative real numbers such that*

$$a_{n+1} \le (1 - \lambda_n)a_n + b_n, \quad \forall n \ge n_0,$$

where n_0 is some nonnegative integer, $\{\lambda_n\}$ is a sequence in $(0, 1)$ with $\sum_{n=1}^{\infty} \lambda_n = \infty$, $b_n = \circ(\lambda_n)$, then $\lim_{n \to \infty} a_n = 0$.

Lemma 2.2. *For any $z \in H$, $u \in C$ satisfies the inequality:*

$$\langle u - z, v - u \rangle \ge 0, \quad \forall v \in C,$$

if and only if $u = P_C z$.

From Lemma 2.2, one can easily get the following results.

Lemma 2.3. *The function $u \in H$ such that $g(u) \in C$ is a solution of the $GVI(C, A)$ if and only if u and $g(u)$ satisfy the relation*

$$(2.6) \qquad g(u) = P_C[g(u) - \rho(Au + w)],$$

where w is in Tu and $\rho > 0$ is a constant. If $g(u)$ is a fixed point of some nonexpansive mapping S_1, we have

$$(2.7) \qquad g(u) = S_1 P_C[g(u) - \rho(Au + w)],$$

by rewriting (2.7), we have

$$u = u - g(u) + S_1 P_C[g(u) - \rho(Au + w)].$$

If u is also a fixed point of some nonexpansive mapping S_2, one can easily see

$$(2.2) \qquad \begin{aligned} u = S_2 u &= u - g(u) + S_1 P_C[g(u) - \rho(Au + w)] \\ &= S_2\{u - g(u) + S_1 P_C[g(u) - \rho(Au + w)]\}. \end{aligned}$$

where $\rho > 0$ is a constant and $w \in Tu$.

This formulation is used to suggest the following one-step iterative methods for finding a common element of two different sets of fixed points of a nonexpansive mapping and solutions of the general variational inequalities involving a multivalued nonlinear mapping.

3. Algorithms

Algorithm 3.1. For any $x_0 \in C$ and $w_0 \in Tx_0$, compute the sequence $\{u_n\}$ by the iterative processes:

$$(3.1) \qquad u_{n+1} = (1 - a_n)u_n + a_n S_2\{u_n - g(u_n) + S_1 P_C[g(u_n) - \rho(Au_n + w_n)]\},$$

where $\{\alpha_n\}$ is a sequence in $[0, 1]$ for all $n \ge 0$ and S_1, S_2 are two nonexpansive mappings.

(I) If $S_1 = g = I$ in Algorithm 3.1, then we have the following algorithm:

Algorithm 3.2. For any $x_0 \in C$ and $w_0 \in Tx_0$, compute the sequence $\{u_n\}$ by the iterative processes:

$$(3.2) \qquad u_{n+1} = (1 - a_n)u_n + a_n S_2 P_C[u_n - \rho(Au_n + w_n)],$$

where $\{\alpha_n\}$ is a sequence in $[0, 1]$ for all $n \geq 0$ and S_2 is a nonexpansive mapping.

(II) If $S_1 = S_2 = g = I$ and $\{\alpha_n\} = 1$ in Algorithm 3.1, then we have the following algorithm:

Algorithm 3.3. For any $x_0 \in C$ and $w_0 \in Tx_0$, compute the sequence $\{u_n\}$ by the iterative processes which is basicall due to Verma [5]:

$$(3.3) \qquad u_{n+1} = P_C[u_n - \rho(Au_n + w_n)],$$

where $\{\alpha_n\}$ is a sequence in $[0, 1]$ for all $n \geq 0$.

4. Main Results

Theorem 4.1. *Let C be a closed convex subset of a separable real Hilbert space H. Let $A : C \to H$ be a relaxed (γ_1, r_1)-cocoerceive and μ_1-Lipschitz continuous mapping, $g : C \to H$ be a relaxed (γ_2, r_2)-cocoerceive and μ_2-Lipschitz continuous mapping and let S_1, S_2 be two nonexpansive mappings from C into itself such that $F(S_2) \cap VI(C, A) \neq \emptyset$ and $F(S_1) \neq \emptyset$, respectively. Let $T : H \to 2^H$ be a multivalued relaxed monotone and Lipschitz continuous operator with corresponding constant $k > 0$ and $m > 0$. Let $\{u_n\}$ be a sequence generated by Algorithm 3.1. $\{\alpha_n\}$ is a sequence in $[0, 1]$ satisfying the following conditions:*

(a) $\sum_{n=0}^{\infty} \alpha_n = \infty$;
(b) $\theta_1 + 2\theta_2 < 1$, where

$$\theta_2 = \sqrt{1 + \mu_2^2 - 2r_2 + 2\gamma_2 \mu_2^2},$$

$$\theta_1 = \sqrt{1 + 2\rho(\gamma_1 \mu_1 - r_1 + k) + \rho^2(\mu_1 + m)^2}.$$

Then the sequence $\{u_n\}$ and $\{g(u_n)\}$ converge strongly to $u^* \in F(S_2) \cap VI(C, A)$ and $g(u^*) \in F(S_1)$, respectively.

Proof. Let $u \in C$ is the common elements of $F(S_2) \cap GVI(C, A)$, we have

$$u^* = (1 - \alpha)u^* + \alpha S_2\{u^* - g(u^*) + S_1 P_C[g(u^*) - \rho(Au^* + w^*)]\},$$

where $w^* \in Tu^*$ and $g(u^*) \in F(S_1)$. Observing (3.1), we obtain

$$
\begin{aligned}
&\|u_{n+1} - u^*\| \\
&= \|(1 - a_n)u_n + a_n S_2\{u_n - g(u_n) + S_1 P_C[g(u_n) - \rho(Au_n + w_n)]\} - u^*\| \\
&= \|(1 - a_n)u_n + a_n S_2\{u_n - g(u_n) + S_1 P_C[g(u_n) - \rho(Au_n + w_n)]\} \\
&\qquad - (1 - \alpha)u^* + \alpha S_2\{u^* - g(u^*) + S_1 P_C[g(u^*) - \rho(Au^* + w^*)]\}\| \\
&\leq (1 - a_n)\|u_n - u^*\| + a_n\|\{u_n - g(u_n) + S_1 P_C[g(u_n) - \rho(Au_n + w_n)]\} \\
&\qquad - \{u^* - g(u^*) + S_1 P_C[g(u^*) - \rho(Au^* + w^*)]\}\| \\
&\leq (1 - a_n)\|u_n - u^*\| + 2a_n\|u_n - g(u_n) - (u^* - g(u^*))\| \\
&\qquad + a_n\|u_n - u^* - \rho[(Au_n + w_n) - (Au^* + w^*)]\|
\end{aligned}
$$

(4.1)

Now, we consider the second term of rightside of (4.1). By the assumption that g is relaxed (γ_2, r_2)-cocoercive, μ_2-Lipschitz continuous, we obtain

$$
\begin{aligned}
(4.2) \quad & \|u_n - u^* - g(u_n) - g(u^*)\|^2 \\
&= \|u_n - u^*\|^2 - 2\langle g(u_n) - g(u^*), u_n - u^* \rangle + \|g(u_n) - g(u^*)\|^2 \\
&\leq \|u_n - u^*\|^2 - 2[-\gamma_2 \|g(u_n) - g(u^*)\|^2 + r_2 \|u_n - u^*\|^2] + \|g(u_n) - g(u^*)\|^2 \\
&\leq \|u_n - u^*\|^2 + 2\mu_2^2 \gamma_2 \|u_n - u^*\|^2 - 2r_2 \|u_n - u^*\|^2 + \mu_2^2 \|u_n - u^*\|^2 \\
&= (1 + 2\mu_2^2 \gamma_2 - 2r_2 + \mu_2^2) \|u_n - u^*\|^2 \\
&= \theta_2^2 \|u_n - u^*\|^2,
\end{aligned}
$$

where $\theta_2 = \sqrt{1 + \mu_2^2 - 2r_2 + 2\gamma_2 \mu_2^2}$.

Next, we consider the third term of rightside of (4.1). By the assumption that A is relaxed (γ_1, r_1)-cocoercive, μ_1-Lipschitz continuous and T is relaxed monotone, m-Lipschitz continuous, we obtain

$$
\begin{aligned}
(4.3) \quad & \|u_n - u^* - \rho[(Au_n + w_n) - (Au^* + w^*)]\|^2 \\
&= \|u_n - u^*\|^2 - 2\rho \langle (Au_n + w_n) - (Au^* + w^*), u_n - u^* \rangle \\
&\quad + \rho^2 \|(Au_n + w_n) - (Au^* + w^*)\|^2 \\
&= \|u_n - u^*\|^2 - 2\rho \langle Au_n - Au^*, u_n - u^* \rangle - 2\rho \langle w_n - w^*, u_n - u^* \rangle \\
&\quad + \rho^2 \|(Au_n + w_n) - (Au^* + w^*)\|^2 \\
&\leq \|u_n - u^*\|^2 - 2\rho(-\gamma_1 \|Au_n - Au^*\| + r_1 \|u_n - u^*\|) + 2\rho k \|u_n - u^*\| \\
&\quad + \rho^2 \|(Au_n + w_n) - (Au^* + w^*)\|^2 \\
&\leq \|u_n - u^*\|^2 + 2\rho(\gamma_1 \mu_1 - r_1 + k) \|u_n - u^*\| + \rho^2 \|(Au_n + w_n) - (Au^* + w^*)\|^2
\end{aligned}
$$

Next, we consider the second term of rightside of (4.3).

$$
\begin{aligned}
(4.4) \quad & \|(Au_n + w_n) - (Au^* + w^*)\| = \|(Au_n - Au^*) + (w_n - w^*)\| \\
&\leq \|Au_n - Au^*\| + \|w_n - w^*\| \leq (\mu_1 + m) \|u_n - u^*\|.
\end{aligned}
$$

Substituting (4.4) into (4.3) yields that

$$
\begin{aligned}
(4.5) \quad & \|u_n - u^* - \rho[(Au_n + w_n) - (Au^* + w^*)]\|^2 \\
&\leq \|u_n - u^*\|^2 + 2\rho(\gamma_1 \mu_1 - r_1 + k) \|u_n - u^*\| + \rho^2 (\mu_1 + m)^2 \|u_n - u^*\|^2 \\
&= [1 + 2\rho(\gamma_1 \mu_1 - r_1 + k) + \rho^2 (\mu_1 + m)^2] \|u_n - u^*\|^2 \\
&= \theta_1^2 \|u_n - u^*\|^2,
\end{aligned}
$$

where $\theta_1 = \sqrt{1 + 2\rho(\gamma_1 \mu_1 - r_1 + k) + \rho^2 (\mu_1 + m)^2}$. From the condition (b), we have $\theta < 1$. Substituting (4.2) and (4.5) into (4.1), we have

$$
\begin{aligned}
(4.6) \quad & \|u_{n+1} - u^*\| \leq (1 - a_n) \|u_n - u^*\| + 2a_n \theta_1 \|u_n - u^*\| + a_n \theta_2 \|u_n - u^*\| \\
&\leq [1 - a_n(1 - 2\theta_2 - \theta_1)] \|u_n - u^*\|
\end{aligned}
$$

Observing the condition (a) and Applying Lemma 2.1 into (4.6), we can get

$$\lim_{n \to \infty} \|u_n - u^*\| = 0.$$

On the other hand, observing that $\|g(u_n) - g(u^*)\| \leq \mu_2 \|u_n - u^*\|$, we have

$$\lim_{n \to \infty} \|g(u_n) - g(u^*)\| = 0.$$

This completes the proof.   □

From Theorem 4.1, we have the following theorems immediately.

Theorem 4.2. *Let C be a closed convex subset of a separable real Hilbert space H. Let $A : C \to H$ be a relaxed (γ_1, r_1)-cocoerceive and μ_1-Lipschitz continuous mapping and let S_2 be a nonexpansive mapping from C into itself such that $F(S_2) \cap VI(C, A) \neq \emptyset$. Let $T : H \to 2^H$ be a multivalued relaxed monotone and Lipschitz continuous operator with corresponding constant $k > 0$ and $m > 0$. Let $\{u_n\}$ be a sequence generated by Algorithm 3.2. $\{\alpha_n\}$ is a sequence in $[0, 1]$ satisfying the following conditions:*

(a) $\sum_{n=0}^{\infty} \alpha_n = \infty$;

(b) $0 < \rho < \frac{2(r_1 - \gamma_1 \mu_1 - k)}{(\mu_1 + m)^2}$,

then the sequence $\{u_n\}$ converges strongly to $u^* \in F(S_2) \cap VI(C, A)$.

Theorem 4.3. *Let C be a closed convex subset of a separable real Hilbert space H. Let $A : C \to H$ be a relaxed (γ_1, r_1)-cocoerceive and μ_1-Lipschitz continuous mapping such that $VI(C, A) \neq \emptyset$. Let $T : H \to 2^H$ be a multivalued relaxed monotone and Lipschitz continuous operator with corresponding constant $k > 0$ and $m > 0$. Let $\{u_n\}$ be a sequence generated by Algorithm 3.3. Assume that the following condition is satisfied:*

$$0 < \rho < \frac{2(r_1 - \gamma_1 \mu_1 - k)}{(\mu_1 + m)^2},$$

then the sequence $\{u_n\}$ converges strongly to $u^* \in VI(C, A)$.

Remark. Theorem 4.1 includes Verma [5] as a special case when A is a strong monotone mapping.

REFERENCES

1. D. Kinderlehrer and G. Starnpacchia, *An Introduction to Variational Inequalities and Their Applications,* Academic Press, New York, 1980.
2. Z. Naniewicz and P. D. Panagiotopoulos, *Mathematical Theory of Hemivariational Inequalities and Applications,* Marcel Dekker, New York, 1995.
3. M. Aslam Noor, *General variational inequalities,* Appl. Math. Lett. **1** (1988), 119,-121.
4. G. Stampacchia, *Formes bilineaires coercivites sur les ensembles convexes,* Comptes Rendus de lAcademie des Sciences, Paris, **258** (1964),,4413–4416.
5. R. U. Verma, *Generalized variational inequalities involving multivalued relaxed monotone operators,* Appl. Math. Lett. **10** (1997), 107–109.
6. X. L. Weng, *Fixed point iteration for local strictly pseudocontractive mappings,* Proc. Amer. Math. Soc. **113** (1991), 727–731.

Nonlinear Functional Analysis and Applications, Volume 1, 41–51

ITERATIVE SCHEMES FOR A FAMILY OF FINITE ASYMPTOTICALLY PSEUDOCONTRACTIVE MAPPINGS IN BANACH SPACES

FENG GU[1]

ABSTRACT. Let E be a real Banach space and K be a nonempty closed convex and bounded subset of E. Let $T_i : K \to K$, $i = 1, 2, \cdots, N$, be N uniformly L-Lipschitzian, uniformly asymptotically regular with the sequences $\{\varepsilon_n^{(i)}\}$ and asymptotically pseudocontractive mappings with sequences $\{k_n^{(i)}\}$, where $\{k_n^{(i)}\}$ and $\{\varepsilon_n^{(i)}\}$ for $i = 1, 2, \cdots, N$ satisfy certain mild conditions. In this paper, we proved that, if a sequence $\{x_n\}$ be generated from $x_1 \in K$ by

$$\begin{cases} z_n := \sigma_n \tau_n x_1 + [1 - \sigma_n(1 + \tau_n)]x_n + \sigma_n T_n^n x_n, \\ x_{n+1} := \lambda_n \theta_n x_1 + [1 - \lambda_n(1 + \theta_n)]x_n + \lambda_n T_n^n z_n, \quad \forall n \geq 1, \end{cases}$$

where $T_n = T_{n(mod\ N)}$, and $\{\lambda_n\}$, $\{\theta_n\}$, $\{\sigma_n\}$, $\{\tau_n\}$ be four real sequences in $[0, 1]$ satisfying appropriate conditions, then $\|x_n - T_l x_n\| \to 0$ as $n \to \infty$ for each $l \in \{1, 2, \cdots, N\}$. The results presented in this paper generalize and improve the corresponding results of Chidume and Zegeye [1], Reinermann [10], Rhoades [11] and Schu [13].

1. Introduction

Let E be a real normed linear space and E^* its dual space. Let $J : E \to 2^{E^*}$ is the normalized duality mapping defined by

$$J(x) = \{f \in E^* : \langle x, f \rangle = \|x\|^2, \|x\| = \|f\|\}, \quad \forall x \in E,$$

Received August 17, 2007.

2000 *Mathematics Subject Classification.* 47H06, 47H09, 47J05.

Key words and phrases. Approximated fixed point sequence, Uniformly asymptotically regular mapping, Asymptotically pseudocontractive mapping.

This paper was supported by the Natural Science Foundation of Zhejiang Province (Y605191), the Natural Science Foundation of Heilongjiang Province (A0211) and the Scientific Research Foundation from Zhejiang Province Education Committee (20051897).

[1] Institute of Applied Mathematics and Department of Mathematics, Hangzhou Normal University, Hangzhou, Zhejiang 310036, China (*E-mail:* mathgufeng@yahoo.com.cn)

where $\langle \cdot, \cdot \rangle$ denotes the generalized duality pairing. It is well known that if E^* is strictly convex, then J is single-valued. In the sequel, we shall denote the single-valued normalized duality mapping by j.

Let E be a normed linear space, $\emptyset \neq K \subset E$. A mapping $T : K \to K$ is said to be *nonexpansive* if for all $x, y \in K$ we have $\|Tx - Ty\| \leq \|x - y\|$. It is said to be *uniformly L-Lipschitzian* if there exists $L > 0$ such that $\|T^n x - T^n y\| \leq L\|x - y\|$ for all integers $n \geq 1$ and all $x, y \in K$. It is said to be *asymptotically nonexpansive* if there exists a sequence $\{k_n\}$ with $k_n \geq 1$ and $\lim_{n \to \infty} k_n = 1$ such that $\|T^n x - T^n y\| \leq k_n \|x - y\|$ for all integers $n \geq 1$ and all $x, y \in K$. Clearly, every nonexpansive mapping is asymptotically nonexpansive with sequence $k_n \equiv 1, \forall n \geq 1$. There are however, asymptotically nonexpansive mappings which are not nonexpansive (see, e.g., [4]).

The class of asymptotically nonexpansive mappings were introduced by Goebel and Kirk [3] in 1972 and has been studied by severval authors (see, e.g., [5, 11-13, 15]).

An important class of nonlinear mappings generalizing the class of asymptotically nonexpansive mapping has been introduced and studied. Let K be a subset of real Banach space E and $T : K \to E$ any mapping. T is said to be *asymptotically pseudocontractive* if there exists a sequence $\{k_n\} \subset [1, +\infty)$ such that $\lim_{n \to \infty} k_n = 1$, and there exists $j(x - y) \in J(x - y)$ such that the inequality

$$(1.1) \qquad \langle T^n x - T^n y, j(x - y) \rangle \leq k_n \|x - y\|^2$$

holds for all integers $n \geq 1$ and all $x, y \in K$. It is easy to know that every asymptotically nonexpansive mapping is asymptotically pseudocontracive mapping.

The class of asymptotically pseudocontractive mappings was introduced by Schu [14] and has been studied by various authors.

The mapping T is called *uniformly asymptotically regular* if for each $\varepsilon > 0$ there exists integer $n_0 \in \mathbb{N}$, such that $\|T^{n+1} x - T^n x\| \leq \varepsilon$ for all $n \geq n_0$ and all $x \in K$ and it is called *uniformly asymptotically regular with sequence* $\{\varepsilon_n\}$ if $\|T^{n+1} x - T^n x\| \leq \varepsilon_n$ for all integers $n \geq 1$ and all $x \in K$, where $\varepsilon_n \to 0$ as $n \to \infty$.

A family mappings $\{T_i\}_{i=1}^N$ is called *uniformly asymptotically regular* if for each $\varepsilon > 0$ there exists integer $n_0 \in \mathbb{N}$, such that

$$\max_{1 \leq i,j \leq N} \|T_i^{n+1} x - T_j^n x\| \leq \varepsilon$$

for all $n \geq n_0$ and all $x \in K$ and the mapping family $\{T_i\}_{i=1}^N$ is called *uniformly asymptotically regular with sequence* $\{\varepsilon_n\}$ if

$$\max_{1 \leq i,j \leq N} \{\|T_i^{n+1} x - T_j^n x\|\} \leq \varepsilon_n$$

for all integers $n \geq 1$ and all $x \in K$, where $\varepsilon_n \to 0$ as $n \to \infty$.

Let K be a nonempty closed convex and bounded subset of a real Banach space E. A mapping $T : K \to K$ is called *pseudocontractive* if there exists $j(x - y) \in J(x - y)$ such that

$$(1.2) \qquad \langle Tx - Ty, j(x - y) \rangle \leq \|x - y\|^2$$

for all $x, y \in K$. It follows from a result of Kato [6] that the inequality(1.2) is equivalent to

$$(1.3) \qquad \|x - y\| \leq \|x - y + t((I - T)x - (I - T)y)\|$$

for all $x, y \in K$ and all $t > 0$, where I denotes the identity mapping.

A mapping T is called *strongly pseudocontractive* if, for each $x, y \in D(T)$, there exists $j(x - y) \in J(x - y)$ and $k \in (0, 1)$ such that

$$\langle Tx - Ty, j(x - y) \rangle \leq k \|x - y\|^2.$$

Any sequence $\{x_n\}$ satisfying that $\|x_n - T_l x_n\| \to 0$ as $n \to \infty$ for each $l \in \{1, 2, \cdots, N\}$ is called an *approximate fixed point sequence* for a family mappings $\{T_i\}_{i=1}^N$.

The importance of approximate fixed point sequences is that once a sequence has been constructed and proved to be an appropriate fixed point sequence for a continuous mapping T, convergence of that sequence to a fixed point of T is then generally achieved.

For an asymptotically pseudocontractive self-mapping T of a nonempty closed convex and bounded subset of a Hilbert space H, Schu [13] proved the following theorem:

Theorem S. ([13]) *Let H be a Hilbert space, $K \subset E$ be nonempty closed convex and bounded. Let T be a uniformly L-Lipschitzian and asymptotically pseudocontractive self-mapping of K with $\{k_n\} \subset [1, \infty)$; $\sum (q_n^2 - 1) < \infty$, where $q_n = (2k_n - 1)$ for all $n \geq 1$, $\alpha_n, \beta_n \in [0, 1]$, $\varepsilon \leq \alpha_n \leq \beta_n \leq b$ for all integers $n \geq 1$ and some $\varepsilon > 0$; and some $b \in (0, L^{-1}[(1 + L^2)^{1/2} - 1])$; pick $x_0 \in K$; and define $x_{n+1} := \alpha_n T^n z_n + (1 - \alpha_n) x_n$; $z_n = \beta_n T^n x_n + (1 - \beta_n) x_n$ for all $n \geq 0$. Then $\lim_{n \to \infty} \|x_n - T x_n\| = 0$.*

In 2003, Chidume and Zegeye [1] constructed an approximate fixed point sequence for the class of asymptotically pseudocontractive mappings in Banach spaces and proved the following theorem:

Theorem CZ. ([1]) *Let K be a nonempty closed convex and bounded subset of a real Banach space E. Let $T : K \to K$ be a uniformly L-Lipschitzian, uniformly asymptotically regular with sequence $\{\varepsilon_n\}$ and asymptotically pseudocontractive with sequence $\{k_n\}$ such that for $\lambda_n, \theta_n \in (0, 1)$, $\forall n \geq 0$, the following conditions are satisfied:*

(a) $\lambda_n (1 + \theta_n) \leq 1$, $\sum_{n=1}^{\infty} \lambda_n \theta_n = \infty$;

(b) $\theta_n \to 0$, $\frac{\lambda_n}{\theta_n} \to 0$, $\frac{\left(\frac{\theta_{n-1}}{\theta_n} - 1 \right)}{\lambda_n \theta_n} \to 0$, $\frac{\varepsilon_{n-1}}{\lambda_n \theta_n^2} \to 0$;

(c) $k_{n-1} - k_n = o(\lambda_n \theta_n^2)$;

(d) $k_n - 1 = o(\theta_n)$.

Let a sequence $\{x_n\}$ be iteratively generated from $x_1 \in K$ by

(1.4) $$x_{n+1} := \lambda_n \theta_n x_1 + [1 - \lambda_n (1 + \theta_n)] x_n + \lambda_n T^n x_n, \quad \forall n \geq 1.$$

Then $\|x_n - T x_n\| \to 0$ as $n \to \infty$.

In this paper, we introduce a new two-step iteration process as follows:

(1.5) $$\begin{cases} x_{n+1} := \lambda_n \theta_n x_1 + [1 - \lambda_n (1 + \theta_n)] x_n + \lambda_n T_n^n z_n, \\ z_n := \sigma_n \tau_n x_1 + [1 - \sigma_n (1 + \tau_n)] x_n + \sigma_n T_n^n x_n, \quad \forall n \geq 1, \end{cases}$$

where $\{T_i\}_{i=1}^N : K \to K$ be N asymptotically pseudocontractive mappings, $T_n = T_{n(\bmod N)}$, $\{\lambda_n\}$, $\{\theta_n\}$, $\{\sigma_n\}$ and $\{\tau_n\}$ are four real sequences in $[0, 1]$ satisfying $\lambda_n (1 + \theta_n) \leq 1$, $\sigma_n (1 + \tau_n) \leq 1$ for all $n \geq 1$ and x_0 is a given point in K.

Especially, if $\{\lambda_n\}$, $\{\theta_n\}$ be two sequences in $[0, 1]$ satisfying $\lambda_n(1 + \theta_n) \leq 1$ for all $n \geq 1$ and x_0 is a given point in K, then the sequence $\{x_n\}$ defined by

(1.6) $x_{n+1} := \lambda_n \theta_n x_1 + [1 - \lambda_n(1 + \theta_n)]x_n + \lambda_n T_n^n z_n, \quad \forall n \geq 1.$

Remark 1.1. If $T_1 = T_2 = \cdots = T_N = T$ or $N = 1$, then (1.6) reduces to (1.4).

The purpose of this paper is to construct an approximate fixed point sequence for a finite family of asymptotically pseudocontractive mappings $\{T_i\}_{i=1}^N$ in Banach spaces. The results presented in this paper generalize and improve the corresponding results of Chidume and Zegeye [1], Reinermann [10], Rhoades [11] and Schu [13].

2. Preliminaries

In order to prove the main result of this paper, we need the following Lemmas:

Lemma 2.1. ([2,8]) *Let E be a real normed linear space, then, for any $x, y \in E$ and $j(x + y) \in J(x + y)$, we have*

$$\|x + y\|^2 \leq \|x\|^2 + 2\langle y, j(x + y)\rangle,$$

Lemma 2.2. ([7]) *Let $\{\rho_n\}$, $\{\sigma_n\}$ and $\{\alpha_n\}$ be three sequences of nonnegative numbers satisfying the conditions $\lim_{n\to\infty} \alpha_n = 0$, $\Sigma_{n=1}^\infty \alpha_n = \infty$ and $\frac{\sigma_n}{\alpha_n} \to 0$ as $n \to \infty$. Let the recursive inequality*

(2.1) $\rho_{n+1}^2 \leq \rho_n^2 - \alpha_n \psi(\rho_{n+1}) + \sigma_n, \quad \forall n \geq 1,$

be given, where $\psi : [0, +\infty) \to [0, +\infty)$ is a strictly increasing function such that it is positive on $(0, +\infty)$ and $\psi(0) = 0$. Then $\rho_n \to 0$ as $n \to \infty$.

3. Main Results

Lemma 3.1. *Let E be a real Banach space, K is a nonempty closed convex and bounded subset of E. Let $\{T_i\}_{i=1}^N : K \to K$ be N uniformly asymptotically regular, uniformly L-Lipschitzian and asymptotically pseudocontractive mappings with sequences $\{k_n^{(i)}\}$, $i = 1, 2, \cdots, N$. Then for $u \in K$ and $t_n \in (0, 1)$ such that $t_n \to 1$ as $n \to \infty$, there exists a sequence $\{y_n\} \subset K$ satisfies the following condition:*

(3.1) $$y_n = \frac{t_n}{k_n} T_n^n y_n + \left(1 - \frac{t_n}{k_n}\right) u,$$

where $k_n = \max\{k_n^{(1)}, k_n^{(2)}, \cdots, k_n^{(N)}\}$ and $T_n = T_{n(mod\ N)}$.
Furthermore, we have $\|y_n - T_n y_n\| \to 0$ as $n \to \infty$.

Proof. Since $T_i : K \to K$, $i = 1, 2, \cdots, N$, is uniformly L-Lpschitzian, there exists $L_i > 0$, $i = 1, 2, \cdots, N$, such that

(3.2) $$\|T_i^n x - T_i^n y\| \leq L_i \|x - y\| \leq L \|x - y\|$$

for all $n \geq 1$ and all $x, y \in K$, where $L = \max\{L_1, L_2, \cdots, L_N\}$.
For each $n \geq 1$, define the mapping $S_n : K \to K$ by

$$S_n(y) := \frac{t_n}{k_n} T_n^n y + \left(1 - \frac{t_n}{k_n}\right) u, \quad \forall y \in K.$$

Then $S_n : K \to K$ is continuous and strongly pseudocontrative. Therefore, by Theorem 5 of Reich [9], S_n has a unique fixed point (say) $y_n \in K$. This means that the equation

$$y_n = \frac{t_n}{k_n} T_n^n y_n + \left(1 - \frac{t_n}{k_n}\right) u$$

has a unique solution for each $t_n \in (0,1)$. Moreover, since K is bounded, we have

(3.3)
$$\begin{aligned}
\|y_n - T_n^n y_n\| &= \left\|\left(1 - \frac{t_n}{k_n}\right) u + \left(\frac{t_n}{k_n} - 1\right) T_n^n y_n\right\| \\
&= \left(1 - \frac{t_n}{k_n}\right) \|u - T_n^n y_n\| \to 0 \quad \text{as } n \to \infty.
\end{aligned}$$

Thus it follows that

(3.4)
$$\begin{aligned}
&\|y_n - T_n y_n\| \\
&= \left\|\left(1 - \frac{t_n}{k_n}\right)(u - T_n y_n) + \frac{t_n}{k_n}(T_n^n y_n - T_n y_n)\right\| \\
&\leq \left(1 - \frac{t_n}{k_n}\right)\|u - T_n y_n\| + \frac{t_n}{k_n}\|T_n^n y_n - T_n^{n+1} y_n\| + \frac{t_n}{k_n}\|T_n^{n+1} y_n - T_n y_n\| \\
&\leq \left(1 - \frac{t_n}{k_n}\right)\|u - T_n y_n\| + \frac{t_n}{k_n}\|T_n^n y_n - T_n^{n+1} y_n\| + \frac{t_n}{k_n} L\|T_n^n y_n - y_n\|.
\end{aligned}$$

In view of the uniformly asymptotic regularity of $\{T_i\}_{i=1}^N$, it follows from (3.3) and (3.4) that $\|y_n - T_n y_n\| \to 0$ as $n \to \infty$. This completes the proof. $\square$

Theorem 3.2. *Let K be a nonempty closed convex and bounded subset of a real Banach space E. Let $\{T_i\}_{i=1}^N : K \to K$ be N uniformly L-Lipschitzian, asymptotically pseudocontractive with the sequenceS $\{k_n^{(i)}\}$, $i = 1, 2, \cdots, N$, and uniformly asymptotically regular with A sequence $\{\varepsilon_n\}$. Let $\{\lambda_n\}$, $\{\theta_n\}$, $\{\sigma_n\}$ and $\{\tau_n\}$ be four real sequences in $[0,1]$ satisfying the following conditions:*

(a) $\lambda_n(1 + \theta_n) \leq 1$, $\sigma_n(1 + \tau_n) \leq 1$, $\sum_{n=1}^\infty \lambda_n \theta_n = \infty$;

(b) $\theta_n \to 0$, $\frac{\lambda_n}{\theta_n} \to 0$, $\frac{\sigma_n}{\theta_n} \to 0$, $\frac{\left|\frac{\theta_{n-1}}{\theta_n} - 1\right|}{\lambda_n \theta_n} \to 0$, $\frac{\varepsilon_{n-1}}{\lambda_n \theta_n^2} \to 0$;

(c) $|k_{n-1} - k_n| = o(\lambda_n \theta_n^2)$;

(d) $k_n - 1 = o(\theta_n)$,

Where $k_n = \max\{k_n^{(1)}, k_n^{(2)}, \cdots, k_n^{(N)}\}$. Suppose, further, that $x_1 \in K$ be any given point and $\{x_n\}$ is the iterative sequence defined by (1.5). Then $\|x_n - T_l x_n\| \to 0$ as $n \to \infty$ for each $l \in \{1, 2, \cdots, N\}$.

Proof. Let $\{y_n\}$ denote the sequence defined as in (3.1) with $t_n = \frac{1}{1+\theta_n}$ and $u = x_1$. Then, form (1.5) and Lemma 2.1, we get the following estimates:

$$
\begin{aligned}
&\|x_{n+1} - y_n\|^2 \\
&= \|x_n - y_n - \lambda_n((x_n - T_n^n z_n) + \theta_n(x_n - x_1))\|^2 \\
&\leq \|x_n - y_n\|^2 - 2\lambda_n\langle (x_n - T_n^n z_n) + \theta_n(x_n - x_1), j(x_{n+1} - y_n)\rangle \\
&= \|x_n - y_n\|^2 - 2\lambda_n\langle \theta_n(x_{n+1} - y_n), j(x_{n+1} - y_n)\rangle \\
&\quad + 2\lambda_n\langle \theta_n(x_{n+1} - y_n) - (x_n - T_n^n z_n) - \theta_n(x_n - x_1), j(x_{n+1} - y_n)\rangle \\
&= \|x_n - y_n\|^2 - 2\lambda_n\theta_n\|x_{n+1} - y_n\|^2 \\
&\quad + 2\lambda_n\langle \theta_n(x_{n+1} - x_n) - (x_n - T_n^n z_n) + \theta_n(x_1 - y_n), j(x_{n+1} - y_n)\rangle \\
&\leq \|x_n - y_n\|^2 - 2\lambda_n\theta_n\|x_{n+1} - y_n\|^2 \\
&\quad + 2\lambda_n\Big\langle \theta_n(x_{n+1} - x_n) - (x_n - T_n^n z_n) + \theta_n(x_1 - y_n) - \Big(y_n - \frac{1}{k_n}T_n^n y_n\Big) \\
&\quad + \Big(y_n - \frac{1}{k_n}T_n^n y_n\Big) - \Big(x_{n+1} - \frac{1}{k_n}T_n^n x_{n+1}\Big) \\
&\quad + \Big(x_{n+1} - \frac{1}{k_n}T_n^n x_{n+1}\Big), j(x_{n+1} - y_n)\Big\rangle \\
&= \|x_n - y_n\|^2 - 2\lambda_n\theta_n\|x_{n+1} - y_n\|^2 \\
&\quad + 2\lambda_n\Big\langle \theta_n(x_{n+1} - x_n) + \Big[\theta_n(x_1 - y_n) - \Big(y_n - \frac{1}{k_n}T_n^n y_n\Big)\Big] \\
&\quad - \Big[\Big(x_{n+1} - \frac{1}{k_n}T_n^n x_{n+1}\Big) - \Big(y_n - \frac{1}{k_n}T_n^n y_n\Big)\Big] \\
&\quad + \Big[\Big(x_{n+1} - \frac{1}{k_n}T_n^n x_{n+1}\Big) - (x_n - T_n^n z_n)\Big], j(x_{n+1} - y_n)\Big\rangle
\end{aligned}
$$

(3.5)

Observe that from the properties of y_n and the asymptotically pseudocontractivity of T_n, we have

$$
\theta_n(x_1 - y_n) - \Big(y_n - \frac{1}{k_n}T_n^n y_n\Big) + \Big(1 - \frac{1}{k_n}\Big)x_1 = 0 \tag{3.6}
$$

and

$$
\Big\langle \Big(x_{n+1} - \frac{1}{k_n}T_n^n x_{n+1}\Big) - \Big(y_n - \frac{1}{k_n}T_n^n y_n\Big), j(x_{n+1} - y_n)\Big\rangle \geq 0. \tag{3.7}
$$

Combining (3.6), (3.7) and (3.5), we have

$$\|x_{n+1} - y_n\|^2$$
$$\leq \|x_n - y_n\|^2 - 2\lambda_n\theta_n\|x_{n+1} - y_n\|^2$$
$$+ 2\lambda_n\Big\langle (\theta_n + 1)(x_{n+1} - x_n) - \frac{1}{k_n}(T_n^n x_{n+1} - T_n^n z_n)$$
$$+ \frac{k_n - 1}{k_n}(T_n^n z_n - x_1), j(x_{n+1} - y_n)\Big\rangle$$
$$- 2\lambda_n\Big\langle \Big(x_{n+1} - \frac{1}{k_n}T_n^n x_{n+1}\Big) - \Big(y_n - \frac{1}{k_n}T_n^n y_n\Big), j(x_{n+1} - y_n)\Big\rangle$$
$$\leq \|x_n - y_n\|^2 - 2\lambda_n\theta_n\|x_{n+1} - y_n\|^2$$

(3.8)
$$+ 2\lambda_n\Big[(\theta_n + 1)\|x_{n+1} - x_n\| + \frac{1}{k_n}\|T_n^n z_n - T_n^n x_{n+1}\|$$
$$+ \frac{k_n - 1}{k_n}\|T_n^n z_n - x_1\|\Big]\|x_{n+1} - y_n\|$$
$$\leq \|x_n - y_n\|^2 - 2\lambda_n\theta_n\|x_{n+1} - y_n\|^2$$
$$+ 2\lambda_n\Big[2\|x_{n+1} - x_n\| + L\|z_n - x_{n+1}\| + \frac{k_n - 1}{k_n}(\|T_n^n z_n\| + \|x_1\|)\Big]\|x_{n+1} - y_n\|$$
$$\leq \|x_n - y_n\|^2 - 2\lambda_n\theta_n\|x_{n+1} - y_n\|^2$$
$$+ 2\lambda_n\Big[(2 + L)\|x_{n+1} - x_n\| + L\|z_n - x_n\|$$
$$+ \frac{k_n - 1}{k_n}(\|T_n^n z_n\| + \|x_1\|)\Big]\|x_{n+1} - y_n\|$$

Notice the fact that

$$x_{n+1} - x_n = \lambda_n(\theta_n x_1 - (1 + \theta_n)x_n + T_n^n z_n) = \lambda_n u_n$$

and

$$z_n - x_n = \sigma_n(\tau_n x_1 - (1 + \tau_n)x_n + T_n^n x_n) = \sigma_n v_n,$$

where $u_n = \theta_n x_1 - (1 + \theta_n)x_n + T_n^n z_n$ and $v_n = \tau_n x_1 - (1 + \tau_n)x_n + T_n^n x_n$. Since K is bounded, $\{x_n\}$, $\{y_n\}$, $\{z_n\}$, $\{T_n^n x_n\}$ and $\{T_n^n z_n\}$ are also all bounded. Thus there exists $M_1 > 0$ such that

(3.9)
$$\max\{\|x_{n+1} - y_n\|, \|u_n\|, \|v_n\|, \|T_n^n z_n\| + \|x_1\|\} \leq M_1,$$

and so

(3.10)
$$\|x_{n+1} - x_n\| = \lambda_n\|u_n\| \leq \lambda_n M_1, \quad \|z_n - x_n\| = \sigma_n\|v_n\| \leq \sigma_n M_1.$$

Substituting (3.9) and (3.10) into (3.8), we have

(3.11)
$$\|x_{n+1} - y_n\|^2 \leq \|x_n - y_n\|^2 - 2\lambda_n\theta_n\|x_{n+1} - y_n\|^2$$
$$+ 2(2 + L)\lambda_n^2 M_1^2 + 2\lambda_n L\sigma_n M_1^2 + 2\lambda_n\frac{k_n - 1}{k_n}M_1^2.$$

Moreover, observe that $\overline{T} := \frac{1}{k_n} T_n^n$ is a pseudocntractive mapping. Thus it follows from (1.3) that

$$
\begin{aligned}
&\|y_{n-1} - y_n\| \\
&\leq \left\| y_{n-1} - y_n + \frac{1}{\theta_n} \left[(I - \overline{T}) y_{n-1} - (I - \overline{T}) y_n \right] \right\| \\
&= \left\| y_{n-1} - y_n + \frac{1}{\theta_n} \left[\left(y_{n-1} - \frac{1}{k_n} T_n^n y_{n-1} \right) - \left(y_n - \frac{1}{k_n} T_n^n y_n \right) \right] \right\| \\
&= \left\| y_{n-1} - y_n + \frac{1}{\theta_n} \left[\left(y_{n-1} - \frac{1}{k_{n-1}} T_{n-1}^{n-1} y_{n-1} \right) \right. \right. \\
&\qquad \left. \left. + \left(\frac{1}{k_{n-1}} T_{n-1}^{n-1} y_{n-1} - \frac{1}{k_n} T_n^n y_{n-1} \right) - \left(y_n - \frac{1}{k_n} T_n^n y_n \right) \right] \right\| \\
&= \left\| y_{n-1} - y_n + \frac{1}{\theta_n} \left\{ \left[\theta_{n-1}(x_1 - y_{n-1}) + \left(1 - \frac{1}{k_{n-1}} \right) x_1 \right] \right. \right. \\
&\qquad \left. \left. + \left(\frac{1}{k_{n-1}} T_{n-1}^{n-1} y_{n-1} - \frac{1}{k_n} T_n^n y_{n-1} \right) - \left[\theta_n(x_1 - y_n) + \left(1 - \frac{1}{k_n} \right) x_1 \right] \right\} \right\| \\
&= \left\| \left(\frac{\theta_{n-1}}{\theta_n} - 1 \right)(x_1 - y_{n-1}) + \frac{1}{\theta_n k_{n-1}} (T_{n-1}^{n-1} y_{n-1} - T_n^n y_{n-1}) \right. \\
&\qquad \left. + \frac{1}{\theta_n} \left(\frac{1}{k_{n-1}} - \frac{1}{k_n} \right)(T_n^n y_{n-1} - x_1) \right\| \\
&\leq \left| \frac{\theta_{n-1}}{\theta_n} - 1 \right| (\|x_1\| + \|y_{n-1}\|) + \frac{1}{\theta_n k_{n-1}} \|T_{n-1}^{n-1} y_{n-1} - T_n^n y_{n-1}\| \\
&\qquad + \frac{1}{\theta_n} \left| \frac{1}{k_{n-1}} - \frac{1}{k_n} \right| (\|T_n^n y_{n-1}\| + \|x_1\|) \\
&\leq \left| \frac{\theta_{n-1}}{\theta_n} - 1 \right| (\|x_1\| + \|y_{n-1}\|) + \frac{\varepsilon_{n-1}}{\theta_n k_{n-1}} + \frac{1}{\theta_n} \frac{|k_n - k_{n-1}|}{k_n k_{n-1}} (\|T_n^n y_{n-1}\| + \|x_1\|).
\end{aligned}
$$
(3.12)

Since $\{x_n\}$, $\{y_n\}$, $\{T_n^n x_n\}$, $\{T_n^n y_n\}$ and $\{T_n^n y_{n-1}\}$ are bounded, thus there exists $M_2 > 0$ such that

$$
\max\{2(\|x_n - y_{n-1}\| + \|y_{n-1} - y_n\|), \|x_1\| + \|y_{n-1}\|, \|T_n^n y_{n-1}\| + \|x_1\|\} \leq M_2.
$$

Notice that

$$
\begin{aligned}
\|x_n - y_n\|^2 &\leq (\|x_n - y_{n-1}\| + \|y_{n-1} - y_n\|)^2 \\
&= \|x_n - y_{n-1}\|^2 + 2\|y_{n-1} - y_n\|(\|x_n - y_{n-1}\| + \|y_{n-1} - y_n\|) \\
&\leq \|x_n - y_{n-1}\|^2 + \|y_{n-1} - y_n\| \cdot M_2.
\end{aligned}
$$
(3.13)

Combing (3.12), (3.13) and (3.11), we have

$$
\begin{aligned}
\|x_{n+1} - y_n\|^2 &\leq \|x_n - y_{n-1}\|^2 - 2\lambda_n \theta_n \|x_{n+1} - y_n\|^2 + 2\lambda_n L \sigma_n M_1^2 \\
&\quad + 2(2 + L)\lambda_n^2 M_1^2 + 2\lambda_n (k_n - 1) M_1^2 + \left| \frac{\theta_{n-1}}{\theta_n} - 1 \right| M_2^2 \\
&\quad + \frac{\varepsilon_{n-1}}{\theta_n k_{n-1}} M_2 + \frac{1}{\theta_n} \frac{|k_n - k_{n-1}|}{k_n k_{n-1}} M_2^2.
\end{aligned}
$$
(3.14)

Thus, by Lemma 2.2 and the conditions (a)-(d) on $\{\lambda_n\}$, $\{\theta_n\}$, $\{\sigma_n\}$, $\{\tau_n\}$, $\{k_n\}$ and $\{\varepsilon_n\}$, we get $\|x_{n+1} - y_n\| \to 0$ as $n \to \infty$. Consequently, $\|x_n - y_n\| \to 0$ as $n \to \infty$.

Next, we prove that $\|x_n - T_l x_n\| \to 0$ as $n \to \infty$ for each $l \in \{1, 2, \cdots, N\}$. Indeed, by Lemma 3.1, we have that $\|y_n - T_n y_n\| \to 0$ as $n \to \infty$. Thus

(3.15)
$$\|x_n - T_n x_n\| \leq \|x_n - y_n\| + \|y_n - T_n y_n\| + \|T_n y_n - T_n x_n\|$$
$$\leq L(1 + L)\|x_n - y_n\| + \|y_n - T_n y_n\| \to 0 \quad \text{as } n \to \infty$$

From the condition $\lambda_n \to 0$ as $n \to \infty$ and (3.10), we have $\|x_{n+1} - x_n\| \leq \lambda_n M_1 \to 0$ as $n \to \infty$ and so

(3.16)
$$\|x_n - x_{n+l}\| \to 0 \quad \text{as } n \to \infty \quad \text{for each } 1 \in \{1, 2, \cdots, N\}.$$

Thus, for each $l \in \{1, 2, \cdots, N\}$, from (3.15) and (3.16), we have
$$\|x_n - T_{n+l} x_n\| \leq \|x_n - x_{n+l}\| + \|x_{n+l} - T_{n+l} x_{n+l}\| + \|T_{n+l} x_{n+l} - T_{n+l} x_n\|$$
$$\leq (1 + L)\|x_n - x_{n+l}\| + \|x_{n+l} - T_{n+l} x_{n+l}\| \to 0 \quad \text{as } n \to \infty,$$

which implies that the sequence
$$\bigcup_{l=1}^{N} \{\|x_n - T_{n+l} x_n\|\}_{n=1}^{\infty} \to 0 \quad \text{as } n \to \infty.$$

For each $l \in \{1, 2, \cdots, N\}$, observe that
$$\{\|x_n - T_l x_n\|\}_{n=1}^{\infty} = \{\|x_n - T_{n+(l-n)} x_n\|\}_{n=1}^{\infty}$$
$$= \{\|x_n - T_{n+l_n} x_n\|\}_{n=1}^{\infty}$$
$$\subset \bigcup_{l=1}^{N} \{\|x_n - T_{n+l} x_n\|\}_{n=1}^{\infty},$$

where $l - n = l_n (mod N)$ for $l_n \in \{1, 2, \cdots, N\}$. Therefore, we have $\|x_n - T_l x_n\| \to 0$ as $n \to \infty$. This completes the proof.   □

Theorem 3.3. *Let K be a nonempty closed convex and bounded subset of a real Banach space E, Let $\{T_i\}_{i=1}^{N} : K \to K$ be N uniformly L-Lipschitzian, asymptotically pseudocontractive with the sequences $\{k_n^{(i)}\}$, $i = 1, 2, \cdots, N$, and uniformly asymptotically regular with a sequence $\{\varepsilon_n\}$. Let $\{\lambda_n\}$ and $\{\theta_n\}$ be two real sequences in $[0, 1]$ satisfying the following conditions:*

(a) $\lambda_n(1 + \theta_n) \leq 1$, $\sum_{n=1}^{\infty} \lambda_n \theta_n = \infty$;
(b) $\theta_n \to 0$, $\frac{\lambda_n}{\theta_n} \to 0$, $\frac{\left|\frac{\theta_{n-1}}{\theta_n} - 1\right|}{\lambda_n \theta_n} \to 0$, $\frac{\varepsilon_{n-1}}{\lambda_n \theta_n^2} \to 0$;
(c) $|k_{n-1} - k_n| = o(\lambda_n \theta_n^2)$;
(d) $k_n - 1 = o(\theta_n)$,

where $k_n = \max\{k_n^{(1)}, k_n^{(2)}, \cdots, k_n^{(N)}\}$. Suppose, further, that $x_1 \in K$ be any given point and $\{x_n\}$ is the iterative sequence defined by (1.6). Then $\|x_n - T_l x_n\| \to 0$ as $n \to \infty$ for each $l \in \{1, 2, \cdots, N\}$.

Proof. Taking $\sigma_n \equiv 0$ in Theorem 3.2, then the conclusion of Theorem 3.3 can be obtained from Theorem 3.2 immediately. This completes the proof.   □

Remark 3.1. Theorem 3.3 is a generalization of Theorem CZ, that is, if $T_1 = T_2 = \cdots = T_N = T$ or $N = 1$, then Theorem 3.3 will reduce to the Theorem CZ.

For $T_1 = T_2 = \cdots = T_N = T$ or $N = 1$ in Theorem 3.2, we can obtain the following result.

Theorem 3.4. *Let K be a nonempty closed convex and bounded subset of a real Banach space E, Let $T : K \to K$ be an uniformly L-Lipschitzian, asymptotically pseudocontractive with a sequence $\{k_n\}$ and uniformly asymptotically regular with a sequence $\{\varepsilon_n\}$. Let $\{\lambda_n\}$, $\{\theta_n\}$, $\{\sigma_n\}$ and $\{\tau_n\}$ be four real sequences in $[0,1]$ satisfying the following conditions:*

(a) $\lambda_n(1 + \theta_n) \leq 1$, $\sigma_n(1 + \tau_n) \leq 1$, $\sum_{n=1}^{\infty} \lambda_n \theta_n = \infty$;

(b) $\theta_n \to 0$, $\frac{\lambda_n}{\theta_n} \to 0$, $\frac{\sigma_n}{\theta_n} \to 0$, $\frac{\left| \frac{\theta_{n-1}}{\theta_n} - 1 \right|}{\lambda_n \theta_n} \to 0$, $\frac{\varepsilon_{n-1}}{\lambda_n \theta_n^2} \to 0$;

(c) $|k_{n-1} - k_n| = o(\lambda_n \theta_n^2)$;

(d) $k_n - 1 = o(\theta_n)$.

Suppose further that $x_1 \in K$ be any given point and $\{x_n\}$ is the iterative sequence defined by

$$(3.17) \qquad \begin{cases} x_{n+1} := \lambda_n \theta_n x_1 + [1 - \lambda_n(1 + \theta_n)]x_n + \lambda_n T^n z_n, \\ z_n := \sigma_n \tau_n x_1 + [1 - \sigma_n(1 + \tau_n)]x_n + \sigma_n T^n x_n, \quad \forall n \geq 1. \end{cases}$$

Then $\|x_n - Tx_n\| \to 0$ as $n \to \infty$.

Remark 3.2. Theorems 3.2, 3.3 and 3.4 also improve and extend the corresponding results of Reinermann [10], Rhoades [11] and Schu [13].

References

1. C. E. Chidume and H. Zegeye, *Approximate fixed point sequences and convergence theorems for asymptotically pseudocontractive mappings,* J. Math. Anal. Apple. **278** (2003), 354–366.

2. C. E. Chidume, H. Zegeye and B. Ntatin, *A generalized steepest descent approximation for the zerors of m-accretive operators,* J. Math. Anal. Apple. **236** (1999), 48–73.

3. K. Goebel and Kirk, *A fixed point theorem for asymptotically nonexpansive mappings,* Proc. Amer. Math. Soc. **35**(1972), 171–174.

4. K. Goebel and Kirk, *Topics in Metric Fixed Point Theory,* in: Cambridge Studies in Advanced Mathematics, Vol. 28, Cambridge University Press, Cambridge, 1990.

5. J. Gornicki, *Weak convergence theorems for asymptotically nonexpanssive mappings in uniformly convex Banach space,* Comment. Math. Univ. Carolin. (1989), 249–252

6. T. Kato, *Nonlinear semi-groups and evolution equations,* J. Math. Soc. Japan **19**(1967), 508–520.

7. C. Moore and B. V. C. Nnoli, *Iteration solution of nonlinear equations involving set-valued uniformly accretive operators,* Comput. Math. Anal. **42**(2001), 131–140.

8. C. H. Morales and J. S. Jung, *Convergence of paths for pseudo-contractive mappings in Banach spaces,* Proc. Amer. Math. Soc. **128**(2000), 3411–3419.

9. S. Reich, *Iterative methods for accretive sets,* in: Nonlinear Equations in Abstract Spaces, Academic Press, New York, 1978, pp. 317–326.

10. J. Reinermann, *Über Fipunkte kontrahierender Abbildungen und schwach konvergente Toeplitz-Verfahren,* Arch. Math. **20**(1969), 59–64.

11. B. E. Rhoades, *Comments on two fixed point iteration methods,* J. Math. Anal. Appl. **56**(1976), 741–750.

12. B. E. Rhoades, *Fixed point iteration methods for certain nonlinear mappings,* J. Math. Anal. Appl. **183**(1994), 118–120.

13. J. Schu, *Iteration construction of fixed points of asymptotically nonexpansive mappings,* J. Math. Anal. Apple. **158**(1991), 407–413.

14. J. Schu, *Approximating of fixed points of asymptotically nonexpansive mappings,* Proc. Amer. Math. Soc. **112**(1991), 143–151.

15. J. Schu, *Weak and strong convergence of fixed points of asymptotically nonexpansive maps,* Bull. Austral. Math. Soc. **43**(1991), 153–159.

Nonlinear Functional Analysis and Applications, Volume 1, 53–60

COMMON FIXED POINT THEOREMS FOR
A CLASS MAPS IN $\mathcal{L}$-FUZZY METRIC SPACES

Shaban Sedghi Ghadikolaei[1]

ABSTRACT. In this paper, we first prove a common fixed point theorem in $\mathcal{L}$-fuzzy metric space. Then we prove fixed point theorems for various compatible maps in $\mathcal{L}$-fuzzy metric spaces.

1. Introduction and Preliminaries

The notion of fuzzy sets was introduced by Zadeh [27]. Various concepts of fuzzy metric spaces were considered in [8, 9, 14, 15]. Many authors have studied fixed theory in fuzzy metric spaces; see, for example, [3-5, 12, 13, 17, 18, 23-25]. In the sequel, we shall adopt the usual terminology, notation and conventions of $\mathcal{L}$-fuzzy metric spaces introduced by Saadati et al. [20], which are a generalization of fuzzy metric sapces [11] and intuitionistic fuzzy metric spaces [19, 21].

Definition 1.1. ([12]) Let $\mathcal{L} = (L, \leq_L)$ be a complete lattice and U a non-empty set called a universe. An $\mathcal{L}$-fuzzy set $\mathcal{A}$ on U is defined as a mapping $\mathcal{A} : U \longrightarrow L$. For each u in U, $\mathcal{A}(u)$ represents the degree (in L) to which u satisfies $\mathcal{A}$.

Lemma 1.2. ([6, 7]) *Consider the set L^* and the operation $\leq_{L^*}$ defined by:*

$$L^* = \{(x_1, x_2) : (x_1, x_2) \in [0, 1]^2 \text{ and } x_1 + x_2 \leq 1\},$$

$(x_1, x_2) \leq_{L^*} (y_1, y_2) \iff x_1 \leq y_1 \text{ and } x_2 \geq y_2 \text{ for every } (x_1, x_2), (y_1, y_2) \in L^*.$ *Then* $(L^*, \leq_{L^*})$ *is a complete lattice.*

Classically, a triangular norm T on $([0, 1], \leq)$ is defined as an increasing, commutative, associative mapping $T : [0, 1]^2 \to [0, 1]$ satisfying $T(1, x) = x$ for all $x \in [0, 1]$. These

---

Received August 17, 2007.

2000 *Mathematics Subject Classification*. 54E40, 54E35, 54H25.

Key words and phrases. $\mathcal{L}$-fuzzy contractive mapping, complete $\mathcal{L}$-fuzzy metric space, fixed point theorem.
[1] Department of Mathematics, Islamic Azad University-Ghaemshahr Branch Ghaemshahr P.O. Box 163, Iran (*E-mail*: sedghi_gh@yahoo.com)

definitions can be straightforwardly extended to any lattice $\mathcal{L} = (L, \leq_L)$. Define first $0_\mathcal{L} = \inf L$ and $1_\mathcal{L} = \sup L$.

Definition 1.3. A triangular norm (t-norm) on $\mathcal{L}$ is a mapping $\mathcal{T} : L^2 \to L$ satisfying the following conditions:
  (1) $(\forall x \in L)(\mathcal{T}(x, 1_\mathcal{L}) = x)$;   (boundary condition)
  (2) $(\forall (x, y) \in L^2)(\mathcal{T}(x, y) = \mathcal{T}(y, x))$;   (commutativity)
  (3) $(\forall (x, y, z) \in L^3)(\mathcal{T}(x, \mathcal{T}(y, z)) = \mathcal{T}(\mathcal{T}(x, y), z))$;   (associativity)
  (4) $(\forall (x, x', y, y') \in L^4)(x \leq_L x'$ and $y \leq_L y' \Rightarrow \mathcal{T}(x, y) \leq_L \mathcal{T}(x', y'))$.
  (monotonicity)

A t-norm $\mathcal{T}$ on $\mathcal{L}$ is said to be *continuous* if, for any $x, y \in \mathcal{L}$ and any sequences $\{x_n\}$ and $\{y_n\}$ which converge to x and y, we have
$$\lim_n \mathcal{T}(x_n, y_n) = \mathcal{T}(x, y).$$

For example, $\mathcal{T}(x, y) = \min(x, y)$ and $\mathcal{T}(x, y) = xy$ are two continuous t-norms on $[0, 1]$. A t-norm can also be defined recursively as an $(n+1)$-ary operation $(n \in \mathbf{N})$ by $\mathcal{T}^1 = \mathcal{T}$, and
$$\mathcal{T}^n(x_1, \cdots, x_{n+1}) = \mathcal{T}(\mathcal{T}^{n-1}(x_1, \cdots, x_n), x_{n+1})$$
for all $n \geq 2$ and $x_i \in L$.

Definition 1.4. A negation on $\mathcal{L}$ is any decreasing mapping $\mathcal{N} : L \to L$ satisfying $\mathcal{N}(0_\mathcal{L}) = 1_\mathcal{L}$ and $\mathcal{N}(1_\mathcal{L}) = 0_\mathcal{L}$. If $\mathcal{N}(\mathcal{N}(x)) = x$ for all $x \in L$, then $\mathcal{N}$ is called an *involutive negation*.

Definition 1.5. The 3-tuple $(X, \mathcal{M}, \mathcal{T})$ is said to be an $\mathcal{L}$-*fuzzy metric space* if X is an arbitrary (non-empty) set, $\mathcal{T}$ is a continuous t–norm on $\mathcal{L}$ and $\mathcal{M}$ is an $\mathcal{L}$-fuzzy set on $X^2 \times]0, +\infty[$ satisfying the following conditions: for every x, y, z in X and t, s in $]0, +\infty[$,
  (1) $\mathcal{M}(x, y, t) >_L 0_\mathcal{L}$;
  (2) $\mathcal{M}(x, y, t) = 1_\mathcal{L}$ for all $t > 0$ if and only if $x = y$;
  (3) $\mathcal{M}(x, y, t) = \mathcal{M}(y, x, t)$;
  (4) $\mathcal{T}(\mathcal{M}(x, y, t), \mathcal{M}(y, z, s)) \leq_L \mathcal{M}(x, z, t + s)$;
  (5) $\mathcal{M}(x, y, \cdot) :]0, \infty[\to L$ is continuous.

Let $(X, \mathcal{M}, \mathcal{T})$ be an $\mathcal{L}$-fuzzy metric space. For any $t \in]0, +\infty[$, we define the *open ball* $B(x, r, t)$ with center $x \in X$ and radius $r \in L \setminus \{0_\mathcal{L}, 1_\mathcal{L}\}$ as
$$B(x, r, t) = \{y \in X : \mathcal{M}(x, y, t) >_L \mathcal{N}(r)\}.$$

A subset $A \subseteq X$ is called *open* if, for each $x \in A$, there exist $t > 0$ and $r \in L \setminus \{0_\mathcal{L}, 1_\mathcal{L}\}$ such that $B(x, r, t) \subseteq A$. Let $\tau_\mathcal{M}$ denote the family of all open subsets of X. Then $\tau_\mathcal{M}$ is called the *topology induced by the $\mathcal{L}$-fuzzy metric $\mathcal{M}$*.

Example 1.6. ([22]) Let (X, d) be a metric space. Denote $\mathcal{T}(a, b) = (a_1 b_1, \min(a_2 + b_2, 1))$ for all $a = (a_1, a_2)$, $b = (b_1, b_2)$ in L^* and let M, N be fuzzy sets on $X^2 \times (0, \infty)$ be defined as follows:
$$\mathcal{M}_{M,N}(x, y, t) = (M(x, y, t), N(x, y, t)) = \left(\frac{t}{t + d(x, y)}, \frac{d(x, y)}{t + d(x, y)} \right).$$

Then $(X, \mathcal{M}_{M,N}, \mathcal{T})$ is an intuitionistic fuzzy metric space.

Example 1.7. Let $X = \underline{\mathbb{N}}$. Define $\mathcal{T}(a,b) = (\max(0, a_1 + b_1 - 1), a_2 + b_2 - a_2 b_2)$ for all $a = (a_1, a_2)$, $b = (b_1, b_2)$ in L^* and let $\mathcal{M}(x, y, t)$ on $X^2 \times (0, \infty)$ be defined as follows:

$$\mathcal{M}(x, y, t) = \begin{cases} (\frac{x}{y}, \frac{y-x}{y}) & \text{if } x \leq y, \\ (\frac{y}{x}, \frac{x-y}{x}) & \text{if } y \leq x \end{cases}$$

for all $x, y \in X$ and $t > 0$. Then $(X, \mathcal{M}, \mathcal{T})$ is an $\mathcal{L}$-fuzzy metric space.

Lemma 1.8. ([11]) *Let $(X, \mathcal{M}, \mathcal{T})$ be an $\mathcal{L}$-fuzzy metric space. Then $\mathcal{M}(x, y, t)$ is nondecreasing with respect to t for all x, y in X.*

Definition 1.9. A sequence $\{x_n\}_{n \in \mathbb{N}}$ in an $\mathcal{L}$-fuzzy metric space $(X, \mathcal{M}, \mathcal{T})$ is called a *Cauchy sequence* if for each $\varepsilon \in L \setminus \{0_\mathcal{L}\}$ and $t > 0$, there exists $n_0 \in \mathbb{N}$ such that, for all $m \geq n \geq n_0$ $(n \geq m \geq n_0)$,

$$\mathcal{M}(x_m, x_n, t) >_L \mathcal{N}(\varepsilon).$$

The sequence $\{x_n\}_{n \in \mathbb{N}}$ is said to be *convergent* to a point $x \in X$ in the $\mathcal{L}$-fuzzy metric space $(X, \mathcal{M}, \mathcal{T})$ (denoted by $x_n \xrightarrow{\mathcal{M}} x$) if $\mathcal{M}(x_n, x, t) = \mathcal{M}(x, x_n, t) \to 1_\mathcal{L}$ whenever $n \to +\infty$ for every $t > 0$. A $\mathcal{L}$-fuzzy metric space is said to be *complete* if and only if every Cauchy sequence is convergent.

Henceforth, we assume that $\mathcal{T}$ is a continuous t-norm on the lattice $\mathcal{L}$ such that, for every $\mu \in L \setminus \{0_\mathcal{L}, 1_\mathcal{L}\}$, there is a $\lambda \in L \setminus \{0_\mathcal{L}, 1_\mathcal{L}\}$ such that

$$\mathcal{T}^{n-1}(\mathcal{N}(\lambda), ..., \mathcal{N}(\lambda)) >_L \mathcal{N}(\mu).$$

For more information see [20].

Definition 1.10. Let $(X, \mathcal{M}, \mathcal{T})$ be an $\mathcal{L}$-fuzzy metric space. $\mathcal{M}$ is said to be *continuous* on $X \times X \times]0, \infty[$ if

$$\lim_{n \to \infty} \mathcal{M}(x_n, y_n, t_n) = \mathcal{M}(x, y, t)$$

whenever a sequence $\{(x_n, y_n, t_n)\}$ in $X \times X \times]0, \infty[$ converges to a point $(x, y, t) \in X \times X \times]0, \infty[$, i.e., $\lim_n \mathcal{M}(x_n, x, t) = \lim_n \mathcal{M}(y_n, y, t) = 1_\mathcal{L}$ and $\lim_n \mathcal{M}(x, y, t_n) = \mathcal{M}(x, y, t)$.

Lemma 1.11. *Let $(X, \mathcal{M}, \mathcal{T})$ be an $\mathcal{L}$-fuzzy metric space. Then $\mathcal{M}$ is continuous function on $X \times X \times]0, \infty[$.*

Proof. The proof is the same as that for fuzzy spaces (see Proposition 1 of [16]).

Definition 1.12. Let A and S be mappings from an $\mathcal{L}$-fuzzy metric space $(X, \mathcal{M}, \mathcal{T})$ into itself. Then the mappings are said to be *weak compatible* if they commute at their coincidence point, that is, $Ax = Sx$ implies that $ASx = SAx$.

Definition 1.13. Let A and S be mappings from an $\mathcal{L}$-fuzzy metric space $(X, \mathcal{M}, \mathcal{T})$ into itself. Then the mappings are said to be *compatible* if

$$\lim_{n \to \infty} \mathcal{M}(ASx_n, SAx_n, t) = 1_\mathcal{L}, \quad \forall t > 0$$

whenever $\{x_n\}$ is a sequence in X such that

$$\lim_{n \to \infty} Ax_n = \lim_{n \to \infty} Sx_n = x \in X.$$

Proposition 1.14. ([26]) *If self-mappings A and S of an $\mathcal{L}$-fuzzy metric space $(X, \mathcal{M}, \mathcal{T})$ are compatible, then they are weak compatible.*

Lemma 1.15. ([1, 20]) *Let $(X, \mathcal{M}, \mathcal{T})$ be an $\mathcal{L}$-fuzzy metric space. Define $E_{\lambda, \mathcal{M}} : X^2 \longrightarrow \mathbf{R}^+ \cup \{0\}$ by*

$$E_{\lambda, \mathcal{M}}(x, y) = \inf\{t > 0 : \mathcal{M}(x, y, t) >_L \mathcal{N}(\lambda)\}$$

for each $\lambda \in L \setminus \{0_{\mathcal{L}}, 1_{\mathcal{L}}\}$ and $x, y \in X$. Then we have rhe following:
  (a) *For any $\mu \in L \setminus \{0_{\mathcal{L}}, 1_{\mathcal{L}}\}$ there exists $\lambda \in L \setminus \{0_{\mathcal{L}}, 1_{\mathcal{L}}\}$ such that*

$$E_{\mu, \mathcal{M}}(x_1, x_n) \leq E_{\lambda, \mathcal{M}}(x_1, x_2) + E_{\lambda, \mathcal{M}}(x_2, x_3) + \cdots + E_{\lambda, \mathcal{M}}(x_{n-1}, x_n)$$

for any $x_1, \cdots, x_n \in X$;
  (b) *The sequence $\{x_n\}_{n \in \mathbf{N}}$ is convergent with respect to $\mathcal{L}$-fuzzy metric $\mathcal{M}$ if and only if $E_{\lambda, \mathcal{M}}(x_n, x) \to 0$. Also the sequence $\{x_n\}_{n \in \mathbf{N}}$ is Cauchy sequence with respect to $\mathcal{L}$-fuzzy metric $\mathcal{M}$ if and only if it is a Cauchy sequence with $E_{\lambda, \mathcal{M}}$.*

Lemma 1.16. *Let $(X, \mathcal{M}, \mathcal{T})$ be an $\mathcal{L}$-fuzzy metric space. If*

$$\mathcal{M}(x_n, x_{n+1}, t) \geq_L \mathcal{M}(x_0, x_1, k^n t)$$

for some $k > 1$ and $n \in \mathbf{N}$. Then $\{x_n\}$ is a Cauchy sequence.

Proof. For every $\lambda \in L \setminus \{0_{\mathcal{L}}, 1_{\mathcal{L}}\}$ and $x_n \in X$, we have

$$
\begin{aligned}
E_{\lambda, \mathcal{M}}(x_{n+1}, x_n) &= \inf\{t > 0 : \mathcal{M}(x_{n+1}, x_n, t) >_L \mathcal{N}(\lambda)\} \\
&\leq \inf\{t > 0 : \mathcal{M}(x_0, x_1, k^n t) >_L \mathcal{N}(\lambda)\} \\
&= \inf\left\{\frac{t}{k^n} : \mathcal{M}(x_0, x_1, t) >_L \mathcal{N}(\lambda)\right\} \\
&= \frac{1}{k^n} \inf\{t > 0 : \mathcal{M}(x_0, x_1, t) >_L \mathcal{N}(\lambda)\} \\
&= \frac{1}{k^n} E_{\lambda, \mathcal{M}}(x_0, x_1).
\end{aligned}
$$

From Lemma 1.15, for every $\mu \in L \setminus \{0_{\mathcal{L}}, 1_{\mathcal{L}}\}$ there exists $\lambda \in L \setminus \{0_{\mathcal{L}}, 1_{\mathcal{L}}\}$, such that

$$
\begin{aligned}
E_{\mu, \mathcal{M}}&(x_n, x_m) \\
&\leq E_{\lambda, \mathcal{M}}(x_n, x_{n+1}) + E_{\lambda, \mathcal{M}}(x_{n+1}, x_{n+2}) + \cdots + E_{\lambda, \mathcal{M}}(x_{m-1}, x_m) \\
&\leq \frac{1}{k^n} E_{\lambda, \mathcal{M}}(x_0, x_1) + \frac{1}{k^{n+1}} E_{\lambda, \mathcal{M}}(x_0, x_1) + \cdots + \frac{1}{k^{m-1}} E_{\lambda, \mathcal{M}}(x_0, x_1) \\
&= E_{\lambda, \mathcal{M}}(x_0, x_1) \sum_{j=n}^{m-1} \frac{1}{k^j} \longrightarrow 0.
\end{aligned}
$$

Hence sequence $\{x_n\}$ is a Cauchy sequence.

2. The Main Results

Now, we give some main results in this paper.

Let Φ be the set of all continuous functions $\phi : L \longrightarrow L$ such that $\phi(t) > t$ for every $t \in L \setminus \{1_L\}$.

Theorem 2.1. *Let* $(X, \mathcal{M}, \mathcal{T})$ *be a complete $\mathcal{L}$-fuzzy metric space and assume that* $S, T, I, J : X \longrightarrow X$ *are four mappings,such that*

$$(*) \qquad\qquad TX \subseteq JX, \quad SX \subseteq IX,$$

and

$$\mathcal{M}(Tx, Sy, t)) \geq_L \phi\left(\frac{\min\{\mathcal{M}(Ix, Tx, kt), \mathcal{M}(Jy, Sy, kt), \mathcal{M}(Ix, Jy, kt)\}}{\max\{\mathcal{M}(Ix, Sy, kt), \mathcal{M}(Jy, Tx, kt)\}} \right)$$

for every $x, y \in X$*, some* $k > 1$ *and* $T(X)$ *or* $S(X)$ *is a closed subset of* X*.*
Suppose in addition that either
(a) T, I *are compatible,* I *is continuous and* S, J *are weak compatible*
or
(b) S, J *are compatible,* J *is continuous and* T, I *are weak compatible.*
Then I, J, T *and* S *have a unique common fixed point.*

Proof. Let $x_0 \in X$ be given. By $(*)$, one can choose a point $x_1 \in X$ such that $Tx_0 = Jx_1 = y_1$ and a point $x_2 \in X$ such that $Sx_1 = Tx_2 = y_2$. Continuing this way, we define by induction a sequence $\{x_n\}$ in X such that, for all $n \geq 0$,

$$Ix_{2n+2} = Sx_{2n+1} = y_{2n+2},$$
$$Jx_{2n+1} = Tx_{2n} = y_{2n+1}.$$

For simplicity, we set
$$d_n(t) = \mathcal{M}(y_n, y_{n+1}, t), \quad \forall n \geq 0.$$
It follows from assume that, for all $n \geq 0$,

$d_{2n+1}(t))$

$= \mathcal{M}(y_{2n+1}, y_{2n+2}, t) = \mathcal{M}(Tx_{2n}, Sx_{2n+1}, t)$

$\geq_L \phi\left(\frac{\min\{\mathcal{M}(Ix_{2n}, Tx_{2n}, kt), \mathcal{M}(Jx_{2n+1}, Sx_{2n+1}, kt), \mathcal{M}(Ix_{2n}, Jx_{2n+1}, kt)\}}{\max\{\mathcal{M}(Ix_{2n}, Sx_{2n+1}, kt), \mathcal{M}(Jx_{2n+1}, Tx_{2n}, kt)\}} \right)$

$\geq_L \phi\left(\min\{d_{2n}(kt), d_{2n+1}(kt), d_{2n}(kt)\}, \max\{1_L, 1_L\} \right)$

Now, if $d_{2n+1}(kt) <_L d_{2n}(kt)$, then
$$d_{2n+1}(t) \geq_L \phi(d_{2n+1}(kt)) >_L d_{2n+1}(kt).$$

Hence $d_{2n+1}(t) >_L d_{2n+1}(kt)$, which is a contradiction. Therefore, $d_{2n+1}(t) \geq_L d_{2n}(kt)$. That is $\mathcal{M}(y_{2n+1}, y_{2n+2}, t) \geq_L \mathcal{M}(y_{2n}, y_{2n+1}, kt)$. So we have

$$\mathcal{M}(y_n, y_{n+1}, t) \geq_L \mathcal{M}(y_{n-1}, y_n, kt) \geq_L \cdots \geq_L \mathcal{M}(y_0, y_1, k^n t).$$

By Lemma 1.16, sequence $\{y_n\}$ is Cauchy sequence and so it is converges to a point $a \in X$. That is,

$$\lim_{n\to\infty} y_n = a = \lim_{n\to\infty} Jx_{2n+1} = \lim_{n\to\infty} Sx_{2n+1} = \lim_{n\to\infty} Ix_{2n+2} = \lim_{n\to\infty} Tx_{2n}.$$

Now suppose that (a) is satisfied. Then $I^2 x_{2n} \longrightarrow Ia$ and $ITx_{2n} \longrightarrow Ia$ since T and I are compatible, implies that $TIx_{2n} \longrightarrow Ia$.

Now we wish to show that a is common fixed point of I, J, T and S.

(a) a is fixed point of I. Indeed, if $Ia \neq a$, we have

$$\mathcal{M}(TIx_{2n}, Sx_{2n+1}, t)$$
$$\geq_L \phi\left(\frac{\min\{\mathcal{M}(I^2x_{2n}, TIx_{2n}, kt), \mathcal{M}(Jx_{2n+1}, Sx_{2n+1}, kt), \mathcal{M}(I^2x_{2n}, Jx_{2n+1}, kt)\}}{\max\{\mathcal{M}(I^2x_{2n}, Sx_{2n+1}, kt), \mathcal{M}(Jx_{2n+1}, TIx_{2n}, kt)\}} \right).$$

Letting $n \to \infty$ (since $Ia \neq a$) yields

$$\mathcal{M}(Ia, a, t)) \geq_L \phi\left(\frac{\min\{\mathcal{M}(Ia, Ia, kt), \mathcal{M}(a, a, kt), \mathcal{M}(Ia, a, kt)\}}{\max\{\mathcal{M}(Ia, a, kt), \mathcal{M}(a, Ia, kt)\}} \right)$$
$$= \phi(\mathcal{M}(Ia, a, kt)) >_L \mathcal{M}(Ia, a, kt),$$

which is a contradiction and hence $Ia = a$.

(b) a is fixed point of T. Indeed, if $Ta \neq a$, we have

$$\mathcal{M}(Ta, Sx_{2n+1}, t)$$
$$\geq_L \phi\left(\frac{\min\{\mathcal{M}(Ia, Ta, kt), \mathcal{M}(Jx_{2n+1}, Sx_{2n+1}, kt), \mathcal{M}(Ia, Jx_{2n+1}, kt)\}}{\max\{\mathcal{M}(Ia, Sx_{2n+1}, kt), \mathcal{M}(Jx_{2n+1}, Ta, kt)\}} \right).$$

Letting $n \to \infty$ yields

$$\mathcal{M}(Ta, a, t)) \geq_L \phi\left(\frac{\min\{\mathcal{M}(Ia, Ta, kt), \mathcal{M}(a, a, kt), \mathcal{M}(Ia, a, kt)\}}{\max\{\mathcal{M}(Ia, a, kt), \mathcal{M}(a, Ta, kt)\}} \right)$$
$$= \phi(\mathcal{M}(Ta, a, kt)) >_L \mathcal{M}(Ta, a, kt),$$

which is a contradiction. Hence $Ta = a$.

Since $TX \subseteq JX$ for all $x \in X$, there is a point $b \in X$ such that

$$Ta = a = Jb.$$

We show that b is coincidence point for J and S. Indeed, if $Jb \neq Sb$, we have

$$\mathcal{M}(Jb, Sb, t) = \mathcal{M}(a, Sb, t) = \mathcal{M}(Ta, Sb, t)$$
$$\geq \phi\left(\frac{\min\{\mathcal{M}(Ia, Ta, kt), \mathcal{M}(Jb, Sb, kt), \mathcal{M}(Ia, Jb, kt)\}}{\max\{\mathcal{M}(Ia, Sb, kt), \mathcal{M}(Jb, Ta, kt)\}} \right)$$
$$>_L \mathcal{M}(Jb, Sb, kt),$$

which is a contradiction. Thus $Ta = Sb = Jb = a$. Since J and S are weak compatible, we deduce that $SJb = JSb \implies Sa = Ja$. We show that $Ta = Sa$. Indeed, if $Ta \neq Sa$, we have

$$\mathcal{M}(Ta, Sa, t) \geq_L \phi\left(\frac{\min\{\mathcal{M}(Ia, Ta, kt), \mathcal{M}(Ja, Sa, kt), \mathcal{M}(Ia, Ja, kt)\}}{\max\{\mathcal{M}(Ia, Sa, kt), \mathcal{M}(Ja, Ta, kt)\}} \right)$$
$$>_L \mathcal{M}(Ta, Sa, kt),$$

which is a contradiction, that is, $Ta = Sa$. Therefore, $Sa = Ta = Ia = Ja = a$.

To show the uniqueness of fixed point a, let $b \neq a$ be another fixed point of I, J, T and S. Then we have

$$\mathcal{M}(Ta, Sb, t))$$
$$\geq_L \phi\left(\frac{\min\{\mathcal{M}(Ia, Ia, kt), \mathcal{M}(a, a, kt), \mathcal{M}(Ia, a, kt)\}}{\max\{\mathcal{M}(Ia, a, kt), \mathcal{M}(a, Ia, kt)\}} \right)$$
$$>_L \mathcal{M}(a, b, kt) = \mathcal{M}(Ta, Sb, t).$$

This is a contradiction. That is, a is unique common fixed point. This completes the proof. ☐

Next, let $\{S_\alpha\}_{\alpha \in A}$ and $\{T_\beta\}_{\beta \in B}$ be the set of all self-mappings of a complete $\mathcal{L}$-fuzzy metric space $(X, \mathcal{M}, \mathcal{T})$.

Theorem 2.2. *Let I, J and $\{T_\alpha\}_{\alpha \in A}, \{S_\beta\}_{\beta \in B}$ be self-mappings of a complete $\mathcal{L}$-fuzzy metric space $(X, \mathcal{M}, \mathcal{T})$ satisfying:*
(a) there exist $\alpha_0 \in A$ and $\beta_0 \in B$ such that $T_{\alpha_0}(X) \subseteq J(X)$, $S_{\beta_0}(X) \subseteq I(X)$ and $T_{\alpha_0}(X)$ or $S_{\beta_0}(X)$ is a closed subset of X,
(b)

$$\mathcal{M}(T_{\alpha_0}x, S_{\beta_0}y, t))$$
$$\geq_L \phi\left(\frac{\min\{\mathcal{M}(Ix, T_{\alpha_0}x, kt), \mathcal{M}(Jy, S_{\beta_0}y, kt), \mathcal{M}(Ix, Jy, kt)\}}{\max\{\mathcal{M}(Ix, S_{\beta_0}y, kt), \mathcal{M}(Jy, T_{\alpha_0}x, kt)\}} \right)$$

for every $x, y \in X$, some $k > 1$ and $\phi \in \Phi$. Suppose in addition that either
(1) T_{α_0}, I are compatible, I is continuous and S_{β_0}, J are weak compatible,
or
(2) S_{β_0}, J are compatible, J is continuous and T_{α_0}, I are weak compatible.
Then I, J, T_α and S_β have a unique common fixed point.

Proof. By Theorem 2.1 I, J, S_{α_0} and T_{β_0} for some $\alpha_0 \in A, \beta_0 \in B$ have a unique common fixed point in X. That is there exist a unique $a \in X$ such that $I(a) = J(a) = S_{\alpha_0}(a) = T_{\beta_0}(a) = a$. Let there exist $\lambda \in B$ such that $\lambda \neq \beta_0$ and $\mathcal{M}(T_\lambda a, a, t) < 1_L$ then we have

$$\mathcal{M}(a, S_\lambda a, t)$$
$$= \mathcal{M}(T_{\alpha_0}a, S_\lambda a, t)$$
$$\geq_L \phi\left(\frac{\min\{\mathcal{M}(Ia, T_{\alpha_0}a, kt), \mathcal{M}(Ja, S_\lambda a, kt), \mathcal{M}(Ia, Ja, kt)\}}{\max\{\mathcal{M}(Ia, S_\lambda a, kt), \mathcal{M}(Ja, T_{\alpha_0}a, kt)\}} \right)$$
$$= \phi(\mathcal{M}(a, S_\lambda a, kt)) >_L \mathcal{M}(a, S_\lambda a, kt),$$

is a contradiction. Hence for every $\lambda \in B$ we have $S_\lambda(a) = a = I(a) = J(a)$. Similarly for every $\gamma \in A$ we get $T_\gamma(a) = a$. Therefore for every $\gamma \in A, \lambda \in B$ we have

$$T_\gamma(a) = S_\lambda(a) = I(a) = J(a) = a.$$

☐

REFERENCES

1. H. Adibi, Y. J. Cho, D. O'Regan and R. Saadati, *Common fixed point theorems in L-fuzzy metric spaces*, Appl. Math. Comput. (doi:10.1016/j.amc.2006.04.045)
2. A. T. Atanassov, *Intuitionistic fuzzy sets,* Fuzzy Sets Syst. **20** (1986), 87–96.
3. S. S. Chang, Y. J. Cho, B. S. Lee, J. S. Jung and S. M. Kang, *Coincidence point and minimization theorems in fuzzy metric spaces*, Fuzzy Sets Syst. **88** (1997), 119–128.
4. Y. J. Cho, H. K. Pathak, S. M. Kang and J. S. Jung, *Common fixed points of compatible maps of type (β) on fuzzy metric spaces*, Fuzzy Sets Syst. **93** (1998), 99–111.
5. Y. J. Cho, S. Sedghi and N. Shobe, *Generalized fixed point theorems for compatible mappings with some types in fuzzy metric spaces*, to appear in Chaos Solitons and Fractals.
6. G. Deschrijver, C. Cornelis and E. E. Kerre, *On the representation of intuitionistic fuzzy t-norms and t-conorms*, IEEE Transactions on Fuzzy Sys. **12** (2004), 45–61.
7. G. Deschrijver and E. E Kerre, *On the relationship between some extensions of fuzzy set theory*, Fuzzy Sets Syst. **33** (2003), 227–35.
8. Z. K. Deng, *Fuzzy pseduo-metric spaces*, J. Math. Anal. Appl. **86** (1982), 74–95.
9. M. A. Erceg, *Metric spaces in fuzzy set theory*, J. Math. Anal. Appl. **69** (1979), 205–230.
10. J. X. Fang, *On fixed point theorems in fuzzy metric spaces*, Fuzzy Sets Syst. **46** (1992), 107–113.
11. A. George and P. Veeramani, *On some results in fuzzy metric spaces*, Fuzzy Sets Syst. **64** (1994), 395–399.
12. J. Goguen, *L-fuzzy sets*, J. Math. Anal. Appl. **18** (1967), 145–174.
13. V. Gregori and A. Sapena, *On fixed point theorem in fuzzy metric spaces*, Fuzzy Sets Syst. **125** (2002), 245–252.
14. O. Kaleva and S. Seikkala, *On fuzzy metric spaces*, Fuzzy Sets Syst. **12** (1984), 215–229.
15. I. Kramosil and J. Michalek, *Fuzzy metric and statistical metric spaces*, Kybernetica **11** (1975), 326–334.
16. J. Rodríguez López and S. Ramaguera, *The Hausdorff fuzzy metric on compact sets*, Fuzzy Sets Syst. **147** (2004) 273–283.
17. D. Miheţ, *A Banach contraction theorem in fuzzy metric spaces*, Fuzzy Sets Syst. **144** (2004), 431–439.
18. E. Pap, O. Hadzic and R. Mesiar, *A fixed point theorem in probabilistic metric spaces and an application*, J. Math. Anal. Appl. **202** (1996), 433–449.
19. J. H. Park, *Intuitionistic fuzzy metric spaces*, Chaos, Solitons and Fractals **22** (2004), 1039–1046.
20. R. Saadati, A. Razani and H. Adibi, *A cmmon fixed point theorem in L-fuzzy metric spaces,* Chaos, Solitons and Fractals (doi:10.1016/j.chaos.2006.01.023)
21. R. Saadati and J. H. Park, *On the intuitionistic fuzzy topological spaces*, Chaos, Solitons and Fractals **27** (2006), 331–344.
22. R. Saadati and J. H. Park, *Intuitionistic fuzzy Euclidean normed spaces*, Commun. Math. Anal. **1(2)** (2006), 1–6.
23. R. Saadati, S. Sedghi and N. Shobe, *Modified intuitionistic fuzzy metric spaces and some fixed point theorems*, Chaos, Solitons and Fractals **38** (2008), 36–47.
24. S. Sedghi, D. Turkoglu and N. Shobe, *Generalization common fixed point theorem in complete fuzzy metric spaces*, J. Comput. Anal. Appl. **9(3)** (2007), 337–348.
25. S. Sedghi and N. Shobe, *Common Fixed Point Theorems for Multi-valued Contractions*, International Mathematical Forum, **2(31)** (2007), 1499–1506.
26. B. Singh and S. Jain, *A fixed point theorem in Menger space through weak compatibility*, J. Math. Anal. Appl. **301** (2005), 439-448.
27. L. A. Zadeh, *Fuzzy sets*, Inform. and control **8** (1965), 338–353.

Nonlinear Functional Analysis and Applications, Volume 1, 61–70

FIXED POINT THEOREMS IN
M-FUZZY METRIC SPACES WITH PROPERTY (E)

Nabi Shobe[1]

ABSTRACT. In this paper, a common fixed point theorem in M-fuzzy metric spaces for property (E) is proved.

1. Introduction and Preliminaries

The concept of fuzzy sets was introduced initially by Zadeh [19] in 1965. Since then, to use this concept in topology and analysis many authors have expansively developed the theory of fuzzy sets and application. George and Veeramani [7] and Kramosil and Michalek [9] have introduced the concept of fuzzy topological spaces induced by fuzzy metric which have very important applications in quantum particle physics particularly in connections with both string and E-infinity theory which were given and studied by El Naschie [3-5, 6, 18]. Many authors [1, 8, 11-15, 17] have proved fixed point theorem in fuzzy (probabilistic) metric spaces.One should there exists a space between spaces. And one such generalization is generalized metric space or D-metric space initiated by Dhage [2] in 1992. He proved some results on fixed points for a self-map satisfying a contraction for complete and bounded D-metric spaces. Rhoades [10] generalized Dhage's contractive condition by increasing the number of factors and proved the existence of unique fixed point of a self-map in D-metric space. Recently, motivated by the concept of compatibility for metric space, Singh and Sharma [16] introduced the concept of D-compatibility of maps in D-metric space and proved some fixed point theorems using a contractive condition.

In what follows (X, D) will denote a D-metric space, $\mathbf{N}$ the set of all natural numbers, and $\mathbf{R}^+$ the set of all positive real numbers.

Received August 19, 2007.

2000 *Mathematics Subject Classification.* 54E40, 54E35, 54H25.

Key words and phrases. M-Fuzzy contractive mapping, Complete M-fuzzy metric space, Common fixed point theorem.

[1] Department of Mathematics, Islamic Azad University-Babol Branch, Babol, Iran (*E-email:* nabi_shobe@yahoo.com)

Definition 1.1. Let X be a nonempty set. A *generalized metric* (or *D-metric*) on X is a function: $D : X^3 \to \mathbf{R}^+$ that satisfies the following conditions for each $x, y, z, a \in X$,

(1) $D(x, y, z) \geq 0$,

(2) $D(x, y, z) = 0$ if and only if $x = y = z$,

(3) $D(x, y, z) = D(p\{x, y, z\})$ (symmetry), where p is a permutation function,

(4) $D(x, y, z) \leq D(x, y, a) + D(a, z, z)$.

The pair (X, D) is called a *generalized metric* (or *D-metric*) *space*.

Immediate examples of such a function are

(a) $D(x, y, z) = \max\{d(x, y), d(y, z), d(z, x)\}$,

(b) $D(x, y, z) = d(x, y) + d(y, z) + d(z, x)$.

Here, d is the ordinary metric on X.

(c) If $X = \mathbf{R^n}$, then we define

$$D(x, y, z) = (||x - y||^p + ||y - z||^p + ||z - x||^p)^{\frac{1}{p}}$$

for every $p \in \mathbf{R}^+$.

(d) If $X = \mathbf{R}^+$, then we define

$$D(x, y, z) = \begin{cases} 0, & \text{if } x = y = z, \\ \max\{x, y, z\}, & \text{otherwise.} \end{cases}$$

Remark 1.2. In a D-metric space, we prove that $D(x, x, y) = D(x, y, y)$. For

(1) $D(x, x, y) \leq D(x, x, x) + D(x, y, y) = D(x, y, y)$

and, similarly,

(2) $D(y, y, x) \leq D(y, y, y) + D(y, x, x) = D(y, x, x)$.

Hence, by (1) and (2), we get $D(x, x, y) = D(x, y, y)$.

Let (X, D) be a D-metric space. For $r > 0$, define

$$B_D(x, r) = \{y \in X : D(x, y, y) < r\}$$

Example 1.3. Let $X = \mathbf{R}$. Denote $D(x, y, z) = |x - y| + |y - z| + |z - x|$ for all $x, y, z \in \mathbf{R}$. Thus

$$B_D(1, 2) = \{y \in \mathbf{R} : D(1, y, y) < 2\}$$
$$= \{y \in \mathbf{R} : |y - 1| + |y - 1| < 2\}$$
$$= \{y \in \mathbf{R} : |y - 1| < 1\} = (0, 2).$$

Definition 1.4. Let (X, D) be a D-metric space and $A \subset X$.

(1) If, for every $x \in A$, there exist $r > 0$ such that $B_D(x, r) \subset A$, then A is called *open subset* of X.

(2) A is said to be *D-bounded* if there exists $r > 0$ such that $D(x, y, y) < r$ for all $x, y \in A$.

(3) A sequence $\{x_n\}$ in X *converges* to x if and only if $D(x_n, x_n, x) = D(x, x, x_n) \to 0$ as $n \to \infty$. That is, for each $\epsilon > 0$, there exist $n_0 \in \mathbf{N}$ such that

(1.1) $$\forall n \geq n_0 \implies D(x, x, x_n) < \epsilon.$$

This is equivalent to the condition: for each $\epsilon > 0$ there exist $n_0 \in \mathbf{N}$ such that

$$(1.2) \qquad \forall n, m \geq n_0 \Longrightarrow D(x, x_n, x_m) < \epsilon$$

Indeed, if we have (1.1), then

$$D(x_n, x_m, x) = D(x_n, x, x_m) \leq D(x_n, x, x) + D(x, x_m, x_m) < \frac{\epsilon}{2} + \frac{\epsilon}{2} = \varepsilon.$$

Conversely, set $m = n$ in $(**)$ we have $D(x_n, x_n, x) < \epsilon$.

(4) A sequence $\{x_n\}$ in X is called a *Cauchy sequence* if, for each $\epsilon > 0$, there exits $n_0 \in \mathbf{N}$ such that $D(x_n, x_n, x_m) < \epsilon$ for each $n, m \geq n_0$. The D-metric space (X, D) is said to be *complete* if every Cauchy sequence is convergent.

Let τ be the set of all $A \subset X$ with $x \in A$ if and only if there exist $r > 0$ such that $B_D(x, r) \subset A$. Then τ is a topology on X (induced by the D-metric D).

Lemma 1.5. *Let (X, D) be a D-metric space. If $r > 0$, then ball $B_D(x, r)$ with center $x \in X$ and radius r is open ball.*

Proof. Let $z \in B_D(x, r)$, Then $D(x, z, z) < r$. If set $D(x, z, z) = \delta$ and $r' = r - \delta$ then we prove that $B_D(z, r') \subseteq B_D(x, r)$. Let $y \in B_D(z, r')$. Then, by triangular inequality, we have

$$D(x, y, y) = D(y, y, x) \leq D(y, y, z) + D(z, x, x) < r' + \delta = r.$$

Hence $B_D(z, r') \subseteq B_D(x, r)$. That is ball $B_D(x, r)$ is open ball. $\square$

Lemma 1.6. *Let (X, D) be a D-metric space. If a sequence $\{x_n\}$ in X converges to x, then x is unique.*

Proof. Let $x_n \to y$ and $y \neq x$. Since $\{x_n\}$ converges to x and y, for each $\epsilon > 0$, there exist $n_1 \in \mathbf{N}$ such that

$$n \geq n_1 \Longrightarrow D(x, x, x_n) < \frac{\epsilon}{2}$$

and $n_2 \in \mathbf{N}$ such that

$$n \geq n_2 \Longrightarrow D(y, y, x_n) < \frac{\epsilon}{2}.$$

If set $n_0 = \max\{n_1, n_2\}$, then, for every $n \geq n_0$, by the triangular inequality we have

$$D(x, x, y) \leq D(x, x, x_n) + D(x_n, y, y) < \frac{\epsilon}{2} + \frac{\epsilon}{2} = \varepsilon.$$

Hence $D(x, x, y) = 0$ is a contradiction. So, $x = y$. This completes the proof. $\square$

Lemma 1.7. *Let (X, D) be a D-metric space. If sequence $\{x_n\}$ in X is converges to x, then $\{x_n\}$ is a Cauchy sequence.*

Proof. Since $x_n \to x$ for each $\epsilon > 0$ there exists $n_1 \in \mathbf{N}$ such that

$$n \geq n_1 \Longrightarrow D(x_n, x_n, x) < \frac{\epsilon}{2}$$

and $n_2 \in \mathbf{N}$ such that

$$m \geq n_2 \Longrightarrow D(x, x_m, x_m) < \frac{\epsilon}{2}.$$

If set $n_0 = \max\{n_1, n_2\}$, then, for every $n, m \geq n_0$ by the triangular inequality, we have

$$D(x_n, x_n, x_m) \leq D(x_n, x_n, x) + D(x, x_m, x_m) < \frac{\epsilon}{2} + \frac{\epsilon}{2} = \epsilon.$$

Hence $\{x_n\}$ is a Cauchy sequence.   □

Definition 1.8. A binary operation $* : [0,1] \times [0,1] \to [0,1]$ is a *continuous t-norm* if it satisfies the following conditions:
  (1) $*$ is associative and commutative,
  (2) $*$ is continuous,
  (3) $a * 1 = a$ for all $a \in [0,1]$,
  (4) $a * b \leq c * d$ whenever $a \leq c$ and $b \leq d$ for each $a, b, c, d \in [0,1]$.

Two typical examples of the continuous t-norm are $a * b = ab$ and $a * b = \min(a, b)$.

Definition 1.9. A 3-tuple $(X, \mathcal{M}, *)$ is called a $\mathcal{M}$-*fuzzy metric space* if X is an arbitrary (non-empty) set, $*$ is a continuous t-norm and $\mathcal{M}$ is a fuzzy set on $X^3 \times (0, \infty)$, satisfying the following conditions: for each $x, y, z, a \in X$ and $t, s > 0$,
  (1) $\mathcal{M}(x, y, z, t) > 0$,
  (2) $\mathcal{M}(x, y, z, t) = 1$ if and only if $x = y = z$,
  (3) $\mathcal{M}(x, y, z, t) = \mathcal{M}(p\{x, y, z\}, t)$,(symmetry) where p is a permutation function,
  (4) $\mathcal{M}(x, y, a, t) * \mathcal{M}(a, z, z, s) \leq \mathcal{M}(x, y, z, t + s)$,
  (5) $\mathcal{M}(x, y, z, .) : (0, \infty) \to [0,1]$ is continuous.

Remark 1.10. Let $(X, \mathcal{M}, *)$ be a $\mathcal{M}$-fuzzy metric space. We prove that for every $t > 0$, $\mathcal{M}(x, x, y, t) = \mathcal{M}(x, y, y, t)$. Because, for each $\epsilon > 0$, by the triangular inequality, we have
  (1) $\mathcal{M}(x, x, y, \epsilon + t) \geq \mathcal{M}(x, x, x, \epsilon) * \mathcal{M}(x, y, y, t) = \mathcal{M}(x, y, y, t)$,
  (2) $\mathcal{M}(y, y, x, \epsilon + t) \geq \mathcal{M}(y, y, y, \epsilon) * \mathcal{M}(y, x, x, t) = \mathcal{M}(y, x, x, t)$.
By taking limits of (1) and (2) when $\epsilon \to 0$, we obtain $\mathcal{M}(x, x, y, t) = \mathcal{M}(x, y, y, t)$.

Let $(X, \mathcal{M}, *)$ be a $\mathcal{M}$-fuzzy metric space. For $t > 0$, the open ball $B_{\mathcal{M}}(x, r, t)$ with center $x \in X$ and radius $0 < r < 1$ is defined by

$$B_{\mathcal{M}}(x, r, t) = \{y \in X : \mathcal{M}(x, y, y, t) > 1 - r\}.$$

A subset A of X is called an *open set* if, for each $x \in A$, there exist $t > 0$ and $0 < r < 1$ such that $B_{\mathcal{M}}(x, r, t) \subseteq A$. A sequence $\{x_n\}$ in X *converges* to x if and only if $\mathcal{M}(x, x, x_n, t) \to 1$ as $n \to \infty$ for each $t > 0$. It is called a *Cauchy sequence* if, for each $0 < \epsilon < 1$ and $t > 0$, there exist $n_0 \in \mathbf{N}$ such that $\mathcal{M}(x_n, x_n, x_m, t) > 1 - \epsilon$ for each $n, m \geq n_0$. The $\mathcal{M}$-fuzzy metric $(X, \mathcal{M}, *)$ is said to be *complete* if every Cauchy sequence is convergent.

Example 1.11. Let X is a nonempty set and D be the D-metric on X. Denote $a * b = ab$ for all $a, b \in [0,1]$. For each $t \in]0, \infty[$, define

$$\mathcal{M}(x, y, z, t) = \frac{t}{t + D(x, y, z)}$$

for all $x, y, z \in X$. It is easy to see that $(X, \mathcal{M}, *)$ is a $\mathcal{M}$-fuzzy metric space.

Lemma 1.12. *Let $(X, M, *)$ is a fuzzy metric space. If we define $\mathcal{M} : X^3 \times (0, \infty) \to [0, 1]$ by*

$$\mathcal{M}(x, y, z, t) = M(x, y, t) * M(y, z, t) * M(z, x, t)$$

for every x, y, z in X, then $(X, \mathcal{M}, *)$ is a $\mathcal{M}$-fuzzy metric space.

Proof. (1) It is easy to see that, for every $x, y, z \in X$ and $t > 0$, $\mathcal{M}(x, y, z, t) > 0$.

(2) $\mathcal{M}(x, y, z, t) = 1$ if and only if $M(x, y, t) = M(y, z, t) = M(z, x, t) = 1$ if and only if $x = y = z$.

(3) $\mathcal{M}(x, y, z, t) = \mathcal{M}(p\{x, y, z\}, t)$, where p is a permutation function.

(4) We have

$$
\begin{aligned}
&\mathcal{M}(x, y, z, t + s) \\
&= M(x, y, t + s) * M(y, z, t + s) * M(z, x, t + s) \\
&\geq M(x, y, t) * M(y, a, t) * M(a, z, s) * M(z, a, s) * M(a, x, t) \\
&= M(x, y, a, t) * M(a, z, s) * M(z, a, s) * M(z, z, s) \\
&= \mathcal{M}(x, y, a, t) * \mathcal{M}(a, z, z, s)
\end{aligned}
$$

for every $s > 0$.   □

Lemma 1.13. *Let $(X, \mathcal{M}, *)$ be a $\mathcal{M}$-fuzzy metric space. Then $\mathcal{M}(x, y, z, t)$ is non-decreasing with respect to t for all x, y, z in X.*

Proof. By Definition 1.9(4), for each $x, y, z, a \in X$ and $t, s > 0$ we have

$$\mathcal{M}(x, y, a, t) * \mathcal{M}(a, z, z, s) \leq \mathcal{M}(x, y, z, t + s).$$

If set $a = z$ we get

$$\mathcal{M}(x, y, z, t) * \mathcal{M}(z, z, z, s) \leq \mathcal{M}(x, y, z, t + s),$$

that is, $\mathcal{M}(x, y, z, t + s) \geq \mathcal{M}(x, y, z, t)$.   □

Definition 1.14. *Let $(X, \mathcal{M}, *)$ be a $\mathcal{M}$-fuzzy metric space. $\mathcal{M}$ is said to be continuous function on $X^3 \times (0, \infty)$ if*

$$\lim_{n \to \infty} \mathcal{M}(x_n, y_n, z_n, t_n) = \mathcal{M}(x, y, z, t)$$

whenever a sequence $\{(x_n, y_n, z_n, t_n)\}$ in $X^3 \times (0, \infty)$ converges to a point $(x, y, z, t) \in X^3 \times (0, \infty)$, i.e.

$$\lim_{n \to \infty} x_n = x, \quad \lim_{n \to \infty} y_n = y, \quad \lim_{n \to \infty} z_n = z, \quad \lim_{n \to \infty} \mathcal{M}(x, y, z, t_n) = \mathcal{M}(x, y, z, t).$$

Lemma 1.15. *Let $(X, \mathcal{M}, *)$ be a $\mathcal{M}$-fuzzy metric space. Then $\mathcal{M}$ is continuous function on $X^3 \times (0, \infty)$.*

Proof. Let $x, y, z \in X$ and $t > 0$ and (x'_n, y'_n, z'_n, t'_n) be a sequence in $X^3 \times (0, \infty)$ that converges to (x, y, z, t). Since $(\mathcal{M}(x'_n, y'_n, z'_n, t'_n))$ is a sequence in $(0, 1]$, there is a subsequence (x_n, y_n, z_n, t_n) of the sequence (x'_n, y'_n, z'_n, t'_n) such that the sequence

$(\mathcal{M}(x_n, y_n, z_n, t_n))$ converges to some point of $[0,1]$. Fix $\delta > 0$ such that $\delta < \frac{t}{2}$. Then there exists $n_0 \in \mathbf{N}$ such that $|t - t_n| < \delta$ for every $n \geq n_0$. Hence we have

$$\mathcal{M}(x_n, y_n, z_n, t_n)$$
$$\geq \mathcal{M}(x_n, y_n, z_n, t - \delta)$$
$$\geq \mathcal{M}\left(x_n, y_n, z, t - \frac{4\delta}{3}\right) * \mathcal{M}\left(z, z_n, z_n, \frac{\delta}{3}\right)$$
$$\geq \mathcal{M}\left(x_n, z, y, t - \frac{5\delta}{3}\right) * \mathcal{M}\left(y, y_n, y_n, \frac{\delta}{3}\right) * \mathcal{M}\left(z, z_n, z_n, \frac{\delta}{3}\right)$$
$$\geq \mathcal{M}(z, y, x, t - 2\delta) * \mathcal{M}\left(x, x_n, x_n, \frac{\delta}{3}\right) * \mathcal{M}\left(y, y_n, y_n, \frac{\delta}{3}\right) * \mathcal{M}\left(z, z_n, z_n, \frac{\delta}{3}\right)$$

and

$$\mathcal{M}(x, y, z, t + 2\delta)$$
$$\geq \mathcal{M}(x, y, z, t_n + \delta)$$
$$\geq \mathcal{M}\left(x, y, z_n, t_n + \frac{2\delta}{3}\right) * \mathcal{M}\left(z_n, z, z, \frac{\delta}{3}\right)$$
$$\geq \mathcal{M}(x, z_n, y_n, t_n + \frac{\delta}{3}) * \mathcal{M}(y_n, y, y, \frac{\delta}{3}) * \mathcal{M}(z_n, z, z, \frac{\delta}{3})$$
$$\geq \mathcal{M}(z_n, y_n, x_n, t_n) * \mathcal{M}\left(x_n, x, x, \frac{\delta}{3}\right) * \mathcal{M}\left(y_n, y, y, \frac{\delta}{3}\right) * \mathcal{M}\left(z_n, z, z, \frac{\delta}{3}\right)$$

for all $n \geq n_0$. By taking limits $n \to \infty$, we obtain

$$\lim_{n \to \infty} \mathcal{M}(x_n, y_n, z_n, t_n) \geq \mathcal{M}(x, y, z, t - 2\delta) * 1 * 1 * 1 = \mathcal{M}(x, y, z, t - 2\delta)$$

and

$$\mathcal{M}(x, y, z, t + 2\delta) \geq \lim_{n \to \infty} \mathcal{M}(x_n, y_n, z_n, t_n) * 1 * 1 * 1 = \lim_{n \to \infty} \mathcal{M}(x_n, y_n, z_n, t_n),$$

respectively.

So, by continuity of the function $t \mapsto \mathcal{M}(x, y, z, t)$, we immediately deduce that

$$\lim_{n \to \infty} \mathcal{M}(x_n, y_n, z_n, t_n) = \mathcal{M}(x, y, z, t).$$

Therefore $\mathcal{M}$ is continuous on $X^3 \times (0, \infty)$. This completes the proof. $\qquad \square$

Definition 1.16. Let A and B be two self-mappings of a $\mathcal{M}$-fuzzy metric space $(X, \mathcal{M}, *)$. We say that A and B *satisfy the property* (E) if there exists a sequence $\{x_n\}$ such that

$$\lim_{n \to \infty} \mathcal{M}(Ax_n, u, u, t) = \lim_{n \to \infty} \mathcal{M}(Bx_n, u, u, t) = 1$$

for some $u \in X$ and for every $t > 0$.

Example 1.17. Let $X = \mathbf{R}$ and $\mathcal{M}(x, y, z, t) = \dfrac{t}{t + |x - y| + |y - z| + |x - z|}$ for every $x, y, z \in X$ and $t > 0$. Let A and B be mappings defined by

$$Ax = 2x + 1, \quad Bx = x + 2.$$

Consider the sequence $x_n = \frac{1}{n} + 1$ for $n = 1, 2, \cdots$. Thus we have

$$\lim_{n\to\infty} \mathcal{M}(Ax_n, 3, 3, t) = \lim_{n\to\infty} \mathcal{M}(Bx_n, 3, 3, t) = 1$$

for every $t > 0$. Then A and B satisfying in the property (E).

In the next example, we show that there are some mappings that have not property (E).

Example 1.18. Let $X = \mathbf{R}$ and $\mathcal{M}(x, y, z, t) = \dfrac{t}{t + |x - y| + |y - z| + |x - z|}$ for every $x, y, z \in X$ and $t > 0$. Let $Ax = x + 1$ and $Bx = x + 2$ and suppost that there exist a sequence $\{x_n\}$ such that

$$\lim_{n\to\infty} \mathcal{M}(Ax_n, u, u, t) = \lim_{n\to\infty} \mathcal{M}(Bx_n, u, u, t) = 1$$

for some $u \in X$ and for every $t > 0$. Therefore, we have

$$\lim_{n\to\infty} \mathcal{M}(Ax_n, u, u, t) = \lim_{n\to\infty} \mathcal{M}(x_n + 1, u, u, t) = \lim_{n\to\infty} \mathcal{M}(x_n, u - 1, u - 1, t) = 1$$

and

$$\lim_{n\to\infty} \mathcal{M}(Bx_n, u, u, t) = \lim_{n\to\infty} \mathcal{M}(x_n + 2, u, u, t) = \lim_{n\to\infty} \mathcal{M}(x_n, u - 2, u - 2, t) = 1.$$

Thus we conclude that $x_n \to u - 1$ and $x_n \to u - 2$, which is a contradiction. Hence A and B do not satisfy the property (E).

Definition 1.19. Let A and S be mappings from a $\mathcal{M}$-fuzzy metric space $(X, \mathcal{M}, *)$ into itself. Then the mappings are said to be *weak compatible* if they commute at their coincidence point, that is, $Ax = Sx$ implies that $ASx = SAx$.

Definition 1.20. Let A and S be mappings from a $\mathcal{M}$-fuzzy metric space $(X, \mathcal{M}, *)$ into itself. Then the mappings are said to be *compatible* if

$$\lim_{n\to\infty} \mathcal{M}(ASx_n, SAx_n, SAx_n, t) = 1$$

for every $t > 0$ whenever $\{x_n\}$ is a sequence in X such that

$$\lim_{n\to\infty} Ax_n = \lim_{n\to\infty} Sx_n = x \in X.$$

2. The Main Results

Let Φ denotes a family of mappings such that a function $\phi : [0, 1]^{12} \to [0, 1]$ is continuous and increasing in each co-ordinate variable and $\phi(s, s, \cdots, s) > s$ for every $s \in [0, 1)$.

Example 2.1. Let $\phi : [0, 1]^{12} \to [0, 1]$ be a function defined by

$$\phi(x_1, x_2, \cdots, x_{12}) = (\min\{x_i\})^h$$

for some $0 < h < 1$.

We begin by recalling some basic concepts of the main theory of this paper.

Theorem 2.2. *Let A, B, S and T be self-mappings of a $\mathcal{M}$-fuzzy metric space $(X, \mathcal{M}, *)$ satisfying:*

(a) $A(X) \subseteq T(X)$, $B(X) \subseteq S(X)$ and $T(X)$ or $S(X)$ is a complete fuzzy metric subspace of X,

(b) the pairs (A, S) and (B, T) are weakly compatible and (A, S) or (B, T) satisfy the property (E),

(c)

$$\mathcal{M}(Ax, By, Bz, t)$$

$$\geq \phi \left(\begin{array}{l} \mathcal{M}(Sx, Ty, Tz, kt), \mathcal{M}(Sx, By, Tz, kt), \mathcal{M}(Sx, Ty, Bz, kt), \\ \mathcal{M}(Sx, By, By, kt), \mathcal{M}(Ty, By, Bz, kt), \mathcal{M}(Ty, Ty, Bz, kt), \\ \mathcal{M}(Ty, By, By, kt), \mathcal{M}(Ty, Bz, Bz, kt), \mathcal{M}(By, Ty, Tz, kt), \\ \mathcal{M}(By, By, Tz, kt), \mathcal{M}(By, Tz, Tz, kt), \mathcal{M}(Tz, Bz, Bz, kt) \end{array} \right).$$

Then A, B, S and T have a unique common fixed point in X.

Proof. Let the pair (B, T) satisfy in property (E). Then there exist a sequence $\{x_n\}$ such that

$$\lim_{n \to \infty} \mathcal{M}(Bx_n, u, u, t) = \lim_{n \to \infty} \mathcal{M}(Tx_n, u, u, t) = 1$$

for some $u \in X$ and every $t > 0$. As $BX \subseteq SX$, there exist a sequence $\{y_n\}$ such that $Bx_n = Sy_n$ and $\lim_{n \to \infty} \mathcal{M}(Sy_n, u, u, t) = 1$.

Now, we prove that $\lim_{n \to \infty} \mathcal{M}(Ay_n, u, u, t) = 1$. Since

$$\mathcal{M}(Ay_n, Bx_n, Bx_{n+1}, t)$$

$$\geq \phi \left(\begin{array}{l} \mathcal{M}(Sy_n, Tx_n, Tx_{n+1}, kt), \mathcal{M}(Sy_n, Bx_n, Tx_{n+1}, kt), \\ \mathcal{M}(Sy_n, Tx_n, Bx_{n+1}, kt), \mathcal{M}(Sy_n, Bx_n, Bx_n, kt), \\ \mathcal{M}(Tx_n, Bx_n, Bx_{n+1}, kt), \mathcal{M}(Tx_n, Tx_n, Bx_{n+1}, kt), \\ \mathcal{M}(Tx_n, Bx_n, Bx_n, kt), \mathcal{M}(Tx_n, Bx_{n+1}, Bx_{n+1}, kt), \\ \mathcal{M}(Bx_n, Tx_n, Tx_{n+1}, kt), \mathcal{M}(Bx_n, Bx_n, Tx_{n+1}, kt), \\ \mathcal{M}(Bx_n, Tx_{n+1}, Tx_{n+1}, kt), \mathcal{M}(Tx_{n+1}z, Bx_{n+1}, Bx_{n+1}, kt) \end{array} \right),$$

Letting $n \to \infty$ in the above inequality, we get

$$\lim_{n \to \infty} \mathcal{M}(Ay_n, Bx_n, Bx_{n+1}, t)$$
$$\geq \phi(\mathcal{M}(u, u, u, kt), \mathcal{M}(u, u, u, kt), \cdots, \mathcal{M}(u, u, u, kt))$$
$$= 1.$$

Therefore, we have $\lim_{n \to \infty} \mathcal{M}(Ay_n, u, u, t) = 1$ and hence

$$\lim_{n \to \infty} Ay_n = \lim_{n \to \infty} Sy_n = \lim_{n \to \infty} Bx_n = \lim_{n \to \infty} Tx_n = u.$$

Since $S(X)$ is a complete $\mathcal{M}$-fuzzy metric space, there exist $x \in X$ such that $Sx = u$. If $Ax \neq u$, then we have

$$M(Ax, Bx_n, Bx_{n+1}, t)$$

$$\geq \phi \left(\begin{array}{l} \mathcal{M}(Sx, Tx_n, Tx_{n+1}, kt), \mathcal{M}(Sx, Bx_n, Tx_{n+1}z, kt), \\ \mathcal{M}(Sx, Tx_n, Bx_{n+1}, kt), \mathcal{M}(Sx, Bx_n, Bx_n, kt), \\ \mathcal{M}(Tx_n, Bx_n, Bx_{n+1}, kt), \mathcal{M}(Tx_n, Tx_n, Bx_{n+1}, kt), \\ \mathcal{M}(Tx_n, Bx_n, Bx_n, kt), \mathcal{M}(Tx_n, Bx_{n+1}, Bx_{n+1}, kt), \\ \mathcal{M}(Bx_n, Tx_n, Tx_{n+1}, kt), \mathcal{M}(Bx_n, Bx_n, Tx_{n+1}, kt), \\ \mathcal{M}(Bx_n, Tx_{n+1}, Tx_{n+1}, kt), \mathcal{M}(Tx_{n+1}z, Bx_{n+1}, Bx_{n+1}, kt) \end{array} \right),$$

Letting $n \to \infty$ we get $\mathcal{M}(Ax, u, u, t) = 1$, hence $Ax = u = Sx$. By (A, S) be weakly compatible , we have $ASx = SAx$, so

$$AAx = ASx = SAx = SSX.$$

Since $AX \subset TX$, there exist $v \in X$ such that $Ax = Tv$. We prove that $Tv = Bv$. If $Tv \neq Bv$, then

$$\mathcal{M}(Ax, Bv, Bv, t)$$
$$\geq \phi \left(\begin{array}{c} \mathcal{M}(Sx, Tv, Tv, kt), \mathcal{M}(Sx, Bv, Tv, kt), \mathcal{M}(Sx, Tv, Bv, kt), \\ \mathcal{M}(Sx, Bv, Bv, kt), \mathcal{M}(Tv, Bv, Bv, kt), \mathcal{M}(Tv, Tv, Bv, kt), \\ \mathcal{M}(Tv, Bv, Bv, kt), \mathcal{M}(Tv, Bv, Bv, kt), \mathcal{M}(Bv, Tv, Tv, kt), \\ \mathcal{M}(Bv, Bv, Tv, kt), \mathcal{M}(Bv, Tv, Tv, kt), \mathcal{M}(Tv, Bv, Bv, kt) \end{array} \right).$$

If $Bv \neq u$, then

$$\mathcal{M}(Ax, Bv, Bv, t) > \mathcal{M}(Ax, Bv, Bv, kt),$$

which is a contradiction. Thus $Tv = Bv = u$. By B and T be weakly compatible, we get $TTv = TBv = BTv = BBv$ and so $Tu = Bu$. We prove $Au = u$. In fact,

$$\mathcal{M}(Au, u, u, t)$$
$$= \mathcal{M}(Au, Bv, Bv, t)$$
$$\geq \phi \left(\begin{array}{c} \mathcal{M}(Su, Tv, Tv, kt), \mathcal{M}(Su, Bv, Tv, kt), \mathcal{M}(Su, Tv, Bv, kt), \\ \mathcal{M}(Su, Bv, Bv, kt), \mathcal{M}(Tv, Bv, Bv, kt), \mathcal{M}(Tv, Tv, Bv, kt), \\ \mathcal{M}(Tv, Bv, Bv, kt), \mathcal{M}(Tv, Bv, Bv, kt), \mathcal{M}(Bv, Tv, Tv, kt), \\ \mathcal{M}(Bv, Bv, Tv, kt), \mathcal{M}(Bv, Tv, Tv, kt), \mathcal{M}(Tv, Bv, Bv, kt) \end{array} \right).$$

If $Au \neq u$, then

$$\mathcal{M}(Au, u, u, t) > \mathcal{M}(Au, u, u, kt),$$

which is a contradiction. Thus $Au = Su = u$.

Now, we prove $Bu = u$. In fact,

$$\mathcal{M}(u, Bu, Bu, t)$$
$$= \mathcal{M}(Au, Bu, Bu, t)$$
$$\geq \phi \left(\begin{array}{c} \mathcal{M}(Su, Tu, Tu, kt), \mathcal{M}(Su, Bu, Tu, kt), \mathcal{M}(Su, Tu, Bu, kt), \\ \mathcal{M}(Su, Bu, Buv, kt), \mathcal{M}(Tu, Bu, Bu, kt), \mathcal{M}(Tu, Tu, Bu, kt), \\ \mathcal{M}(Tu, Bu, Bu, kt), \mathcal{M}(Tu, Bu, Bu, kt), \mathcal{M}(Bu, Tu, Tu, kt), \\ \mathcal{M}(Bu, Bu, Tu, kt), \mathcal{M}(Bu, Tu, Tu, kt), \mathcal{M}(Tu, Bu, Bu, kt) \end{array} \right).$$

If $Bu \neq u$, then

$$\mathcal{M}(u, Bu, Bu, t) > \mathcal{M}(u, Bu, Bu, kt),$$

which is a contradiction. Thus $Au = Bu = Su = Tu = u$. So, A, B, S and T have a fixed common point u.

Finally, to prove the uniqueness of the point u, let $v \neq u$ be another common fixed point of A, B, S and T. Then

$$\mathcal{M}(v, u, u, t)$$
$$= \mathcal{M}(Av, Bu, Bu, t)$$
$$\geq \phi \left(\begin{array}{l} \mathcal{M}(Sv, Tu, Tu, kt), \mathcal{M}(Sv, Bu, Tu, kt), \mathcal{M}(Sv, Tu, Bu, kt), \\ \mathcal{M}(Sv, Bu, Bu, kt), \mathcal{M}(Tu, Bu, Bu, kt), \mathcal{M}(Tu, Tu, Bu, kt), \\ \mathcal{M}(Tu, Bu, Bu, kt), \mathcal{M}(Tu, Bu, Bu, kt), \mathcal{M}(Bu, Tu, Tu, kt), \\ \mathcal{M}(Bu, Bu, Tu, kt), \mathcal{M}(Bu, Tu, Tu, kt), \mathcal{M}(Tu, Bu, Bu, kt) \end{array} \right)$$
$$> \mathcal{M}(v, u, u, kt),$$

which is a contradiction. This completes the proof. $\qquad\square$

References

1. Y. J. Cho, S. Sedghi and N. Shobe, *Generalized fixed point theorems for compatible mappings with some types in fuzzy metric spaces*, to appear in Chaos, Solitons and Fractals

2. B. C. Dhage, *Generalised metric spaces and mappings with fixed point*, Bull. Calcutta Math. Soc. **84(4)** (1992), 329–336.

3. El Naschie MS, *On the uncertainty of Cantorian geometry and two-slit experiment*, Chaos, Solitons and Fractals **9** (1998), 517–529.

4. El Naschie MS, *A review of E-infinity theory and the mass spectrum of high energy particle physics*, Chaos, Solitons and Fractals **19** (2004), 209–236.

5. El Naschie MS, *On a fuzzy Kahler-like Manifold which is consistent with two-slit experiment*, Int J of Nonlinear Science and Numerical Simulation **6** (2005), 95–98.

6. El Naschie MS, *The idealized quantum two-slit gedanken experiment revisited-?Criticism and reinterpretation*, Chaos, Solitons and Fractals **27** (2006), 9–13.

7. A. George and P. Veeramani, *On some result in fuzzy metric space*, Fuzzy Sets Systems **64** (1994), 395–399.

8. V. Gregori and A. Sapena, *On fixed-point theorem in fuzzy metric spaces*, Fuzzy Sets and Systems **125** (2002), 245–252.

9. I. Kramosil and J. Michalek, *Fuzzy metric and statistical metric spaces*, Kybernetica **11** (1975), 326–334.

10. B. E. Rhoades, *A fixed point theorem for generalized metric spaces*, Int. J. Math. Math. Sci. **19(1)** (1996), 145–153.

11. D. Miheţ, *A Banach contraction theorem in fuzzy metric spaces*, Fuzzy Sets Systems **144** (2004), 431–439.

12. R. Saadati and J. H. Park, *On the intuitionistic fuzzy topological spaces*, Chaos, Solitons and Fractals **27** (2006), 331–344.

13. R. Saadati, S. Sedghi and N. Shobe, *Modified intuitionistic fuzzy metric spaces and some fixed point theorems*, Chaos, Solitons and Fractals **38** (2008), 36–47.

14. S. Sedghi, D. Turkoglu and N. Shobe, *Generalization common fixed point theorem in complete fuzzy metric spaces*, J. Comput. Anal. Appl. **9(3)** (2007), 337–348.

15. S. Sedghi and N. Shobe, *Common fixed point theorems for multi-valued contractions*, Int. Math. Forum **2(31)** (2007), 1499–1506.

16. B. Singh and R. K. Sharma, *Common fixed points via compatible maps in D-metric spaces*, Rad. Mat. **11(1)** (2002), 145–153.

17. B. Schweizer, H. Sherwood and R. M. Tardiff, *Contractions on PM-space examples and counterexamples*, Stochastica **1** (1988), 5–17.

18. Y. Tanaka, Y. Mizno and T. Kado, *Chaotic dynamics in Friedmann equation*, Chaos, Solitons and Fractals **24** (2005), 407–422.

19. L. A. Zadeh, *Fuzzy sets*, Inform and Control **8** (1965), 338–353.

Nonlinear Functional Analysis and Applications, Volume 1, 71–83

ON FIXED POINT THEORY
IN WEAKLY CAUCHY NORMED SPACES

SAHAR MOHAMED ALI[1]

ABSTRACT. Let X be a weakly Cauchy normed space and C be a closed convex subset of X.It is proved that any contraction mapping T on C into itself has a unique fixed point and, moreover, if $y \in C$ is the unique fixed point of T, the sequence of iterates $\{T^n(x)\}_{n \in \mathcal{N}}$ is strongly convergent to the point y for every $x \in C$. If the parallelogram law holds in the given normed space, then any descending sequence of bounded closed convex subsets of X has a nonempty intersection, depending on this fact the asymptotic center of any bounded sequence with respect to any bounded closed convex subset C of X is singleton, accordingly, the existence of fixed points of the nonexpansive mapping and another version of Browder's strong convergence theorem for mappings on C into itself. We also study some facts concerning the $\{a, b, c\}$-generalized contraction and $\{a, b, c\}$-generalized nonexpansive mappings. We commented that the given weakly Cauchy normed space is not necessarily Hilbert or uniformly convex Banach space in general.

1. Introduction

The following is the Banach contraction principle which is the basis theorem of fixed point theory long time ago.

Theorem 1.1. *Let X be a complete metric space and T be a contraction of X into itself. Then T has a unique fixed point $y \in X$. Moreover, the sequence of iterates $\{T^n(x)\}_{n \in \mathcal{N}}$ is strongly convergent to the point y for every $x \in X$.*

Received August 21, 2007.

2000 *Mathematics Subject Classification.* 54H26, 47H10.

Key words and phrases. Fixed Point, contraction mapping, non-expansive mapping, asymptotic center, asymptotic Radius, weakly Cauchy normed space.

[1] Department of Mathematics, Faculty of Science, Ain Shams University, Cairo, Egypt (*E-mail:* saharm_ali@yahoo.com)

Mathematicians in the field of fixed point theory try to improve the results of this theorem in which changing the assumptions imposed on the contraction mapping or on the given topological space, as the case of

(1) contractive mappings while compact metric space is necessarily,

(2) nonexpansive mappings while bounded closed convex subsets of uniformly convex Banach space is necessarily [5], or

(3) as in the case of Caristi's fixed point theorem using a proper convex lower semi-continuous mappings, (for other trials, one can see [4]).

The nonexpansive mappings may not have fixed point even on uniformly convex (not bounded) Banach spaces in general.

Depending only on the parallelogram law, we have the following lemma:

Lemma 1.1. ([9]) *Let X be a normed space in which parallelogram law is hold and C be a closed convex subset of X. If T is a nonexpansive mapping on C into C and the set of fixed points $F(T)$ is nonempty, then $F(T)$ is closed convex subset of X.*

We also will be in need of the following interesting theorems:

Theorem 1.2. ([1]) *Let X be a weakly Cauchy normed space in which the parallelogram law holds. Then every nonempty closed convex subset of X is Chebyshev. Equivalently, the metric projection P_C exists on every nonempty closed convex subset C of X. Equivalently, for every $x \in X$, there exists a unique element $P_C(x) = y \in C$, called the best approximation element of x in C such that, $\|x - y\| = dist(x, C) =: \inf\{\|x - z\| : z \in C\}$.*

Theorem 1.3. ([7]) *Let X be weakly Cauchy normed space in which parallelogram law holds. If $\{C_n\}_{n \in \mathcal{N}}$ is descending sequence of closed bounded convex subsets of X, then the intersection $\cap_{n \in \mathcal{N}} C_n$ is not empty.*

2. Preliminaries

Let X be a linear space. Then a function $f : X \to (-\infty, \infty]$ is said to be proper lower semicontinuous and convex function if it satisfies the following conditions ([9], [5], and [8]):

(1) There is $x \in X$ such that $f(x) < \infty$ (proper).

(2) For any real number α, the set $\{x \in X : f(x) \le \alpha\}$ is closed convex subset of X (lower semicontinuous).

(3) For any $x, y \in X$ and $t \in [0, 1]$, $f(tx + (1 - t)y) \le tf(x) + (1 - t)f(y)$ (convex function).

Let X be a normed space and A be a mapping from X into itself. Then

(1) A is said to be *r-contraction*, where $0 \le r < 1$, if and only if

$$\|A(x) - A(y)\| \le r\|x - y\|, \quad \forall x, y \in X.$$

(2) A is said to be *contractive* if and only if

$$\|A(x) - A(y)\| < \|x - y\|, \quad \forall x, y \in X \ (x \ne y).$$

(3) A is said to be *nonexpansive* if and only if

$$\|A(x) - A(y)\| \leq \|x - y\|, \quad \forall x, y \in X.$$

(4) A is said to be $\{a, b, c\}$-*generalized contraction*, where $a, b,$ and c are real numbers, $0 \leq a, b, c < 1$ and $a + b + c < 1$, if and only if

$$\|A(x) - A(y)\| \leq a\|x - y\| + b\|A(x) - x\| + c\|A(y) - y\|, \quad \forall x, y \in X.$$

(5) A is said to be $\{a, b, c\}$-*generalized nonexpansive*, where $a, b,$ and c are real numbers, $0 \leq a, b, c \leq 1$ and $a + b + c = 1$, if and only if

$$\|A(x) - A(y)\| \leq a\|x - y\| + b\|A(x) - x\| + c\|A(y) - y\| \quad \forall x, y \in X.$$

A point $y \in X$ is said to be *fixed point* with respect to the operator T if and only if $T(y) = y$.

Remark 2.1. (1) Every r-contraction mapping is $\{r, 0, 0\}$-contraction.
(2) Every nonexpansive mapping is $\{1, 0, 0\}$-nonexpansive.

As T is contraction, it preserves strong convergence, but it does not preserve weak convergence in general.

A normed space X is said to be *weakly Cauchy normed space* if and only if every Cauchy sequence in X is weakly convergent to an element in X.

Definition 2.1. Let C be a nonempty bounded closed subset of X, $\{x_n\}_{n \in \mathcal{N}}$ be a bounded sequence in X, and define

$$f(x) := \limsup_{n \to \infty} \|x_n - x\|.$$

(1) The asymptotic radius of $\{x_n\}_{n \in \mathcal{N}}$ is denoted by $r(C, \{x_n\}_{n \in \mathcal{N}})$ and is defined as follows:

$$r(C, \{x_n\}_{n \in \mathcal{N}}) := \inf\{f(x) : x \in C\}.$$

(2) The asymptotic center of $\{x_n\}_{n \in \mathcal{N}}$ is denoted by $A(C, \{x_n\}_{n \in \mathcal{N}})$ and is defined as follows:

$$A(C, \{x_n\}_{n \in \mathcal{N}}) := \{y \in C : f(y) = r(C, \{x_n\}_{n \in \mathcal{N}})\}.$$

Some of the properties of the asymptotic radius and the asymptotic center of a bounded sequence in a uniformly convex Banach space is given in [5]. It is shown that

Theorem 2.1. ([5]) *Every bounded sequence in a uniformly convex Banach space has a unique asymptotic center with respect to any closed convex subset of X.*

3. Fixed Point Theory of Contraction and Nonexpansive Mappings in Weakly Cauchy Normed Spaces

In [6], we proved the existence of a unique fixed point of contraction mappings defined on a closed convex subset of a weakly Cauchy normed space. In this section, we display the proof of this theorem that will be needed in the next section.

In [7], we proved an interesting results for the nonexpansive mappings which was proved in Banach spaces, but, in weakly Cauchy normed space, proved the existence of fixed point of a nonexpansive mapping defined on a bounded closed convex subset C of a weakly Cauchy normed space X in which parallelogram law holds.

In this section, we also show that the asymptotic center of any bounded sequence with respect to a closed convex subset is singleton but in the case of weakly Cauchy normed space in which parallelogram law holds.

We have the following:

Lemma 3.1. *Let X be a weakly Cauchy normed space in which the parallelogram law holds, C be a closed convex subset of X. If T is a nonexpansive mapping on C into C and the set of fixed point $F(T)$ is nonempty, then $F(T)$ is Chebyshev subset of X.*

Proof. Using Theorem 1.2 and Lemma 1.1, we get the proof.   □

Theorem 3.1. *Every bounded sequence in a weakly Cauchy normed space X in which parallelogram law holds has a unique asymptotic center with respect to any closed convex bounded subset of X.*

Proof. Let C be a bounded closed convex subset of X, first we will show that the function f is attaining its minimum at some $y \in C$. Let $\{t_n\}_{n \in \mathcal{N}}$ be a converging to zero sequence of positive real numbers converging to 0, according to Theorem 1.3, the following sequence of descending bounded closed convex subsets of X has the nonempty intersection:

$$C_n := \{x : f(x) \leq \inf_{x \in C} f(x) + t_n\}.$$

Thus there is $y \in X$ such that $f(y) \leq \inf_{x \in C} f(x) + t_n$ for each $n \in \mathcal{N}$ and so f attains its minimum at y.

Using the parallelogram law, for every x and $y \in X$, we have the following strict inequality:

$$\left\| x_n - \frac{x+y}{2} \right\| = \left\| \frac{x_n - x}{2} + \frac{x_n - y}{2} \right\| < \max\{\|x_n - x\|, \|x_n - y\|\}.$$

This in turn implies that, for every y and $z \in X$ with $y \neq z$,

(3.1) $f\left(\frac{y+z}{2}\right) < \max\{f(y), f(z)\}.$

To show that such y is unique. Consider $z \in X$ such that $y \neq z$ and $f(z) = \inf\{f(x) : x \in C\}$. Using the inequality (3.1) gives the following obvious contradiction:

$$f\left(\frac{y+z}{2}\right) < \max\{f(y), f(z)\} = f(y) = f(z) = \inf_{x \in C} f(x).$$

                                                                                □

Remark 3.1. Let X be a metric space with metric d, T be a nonexpansive mapping on X into itself and $\{x_n\}_{n \in \mathcal{N}}$ be a sequence in X such that the sequence $\{d(x_n, T(x_n))\}_{n \in \mathcal{N}}$ converges to zero. If $\{x_n\}_{n \in \mathcal{N}}$ is strongly convergent to a point y in X, then y is a fixed point of T.

In fact, if y is the limit point of $\{x_n\}_{n \in \mathcal{N}}$, we have

$$d(y, T(y)) \leq d(x_n, y) + d(x_n, T(x_n)) + d(T(x_n), T(y))$$
$$\leq 2d(x_n, y) + d(x_n, T(x_n)) \to 0.$$

This proved that $T(y) = y$.

The following is an interesting result:

Theorem 3.2. ([6]) *Let X be a weakly Cauchy space, C be a closed convex subset of X and T a contraction mapping from C into C. Then T has a unique fixed point $y \in C$. Moreover, the sequence $\{T^n(x)\}_{n \in \mathcal{N}}$ of iterates is strongly convergent to the point y for every $x \in C$.*

$$\lim_{n \to \infty} T^n(x) = y, \quad \forall x \in C.$$

Proof. Let $x \in C$. Since T is contraction on C, the inequalities

$$\|T^{n+1}(x) - T^n(x)\| \leq r^n \|T(x) - x\|,$$

insures that, if for all $l, m \in \mathcal{N}$ with $l \leq m$,

$$\|T^m(x) - T^l(x)\| \leq \frac{r^l}{1-r} \|T(x) - x\|$$

and in turns shows that the sequence $\{T^n(x)\}_{n \in \mathcal{N}}$ is a Cauchy sequence in C and so, since X is a weakly Cauchy space, the Cauchy sequence $\{T^n(x)\}_{n \in \mathcal{N}}$ is weakly convergent to some element $y \in X$. Since C is convex closed, it contains all its weak limits as well as all its strong limits and hence $y \in C$.

Now, we claim that y is the unique fixed point of T. To prove that, let l be arbitrarily natural number and define the proper lower-semicontinuous convex real valued function ϕ_l on C by the following formula:

$$\phi_l(x) = \|T^l(y) - x\|$$

for any real number α. Then the set $G_{l\alpha} := \{x \in C : \phi_l(x) \leq \alpha\}$ is a closed convex subset of C and, in particular, for any $\epsilon > 0$, the set

$$G_{l(\|T^l(y)-y\|-\epsilon)} := \{x \in C : \|T^l(y) - x\| \leq \|T^l(y) - y\| - \epsilon\}$$

is closed convex and hence it is closed in the weak topology. Therefore, its complement $G^c_{l(\|T^l(y)-y\|-\epsilon)} := \{x \in C : \|T^l(y)-x\| > \|T^l(y)-y\|-\epsilon\}$ is weakly open set which containing y and then there is a neighborhood $N(y)$ of y such that $N(y) \subset G^c_{l(\|T^l(y)-y\|-\epsilon)}$. On the other side, there is $n_0 \in \mathcal{N}$ such that $T^n(x) \in N(y)$ for every $n \geq n_0$ and so it follows that

$$\|T^l(y) - y\| - \epsilon < \|T^l(y) - T^n(x)\|, \quad \forall n \geq n_0.$$

We have two cases: For $l \geq n_0$, it follows that

$$\|T^l(y) - y\| - \epsilon < \|T^l(y) - T^l(x)\|,$$

which in turns shows that

$$\|T^l(y) - y\| - \epsilon \leq r^l \|y - x\|.$$

For $I < n_0$, it follows that

$$\|T^l(y) - T^n(x)\| \le r^l \left[\frac{r}{1-r} \|T(x) - x\| + \|y - x\| \right].$$

In both cases, taking the limit as $l \to \infty$, we see that

$$0 \le \lim_{l \to \infty} \|T^l(y) - y\| < \epsilon.$$

Since ϵ is arbitrarily, one get

$$\lim_{l \to \infty} \|T^l(y) - y\| = 0$$

and thus the sequence $\{T^n(y)\}_{n \in \mathcal{N}}$ converges strongly to y. Since T preserves strong convergence, we have

$$T(y) = T\left(\lim_{n \to \infty} T^n(y) \right) = \lim_{n \to \infty} T(T^n(y)) = \lim_{n \to \infty} T^{n+1}(y) = y.$$

This shows that y is a fixed point of T.

To show that y is unique, let z be another fixed point. Then we have the following contradiction,

$$\|y - z\| = \|T(y) - T(z)\| \le r\|y - x\| < \|y - z\|.$$

To show that the weak convergent is a strong convergent, it is enough to notice that

$$\|T^n(x) - y\| = \|T^n(x) - T^n(y))\| \le r^n \|y - x\|.$$

This completes the proof.   □

In the case of weakly Cauchy normed spaces, it is still true that:

Lemma 3.2. *Let X be a weakly Cauchy normed space and C be a bounded closed convex subset of X. If T is a nonexpansive mapping C into itself, then $\inf\{\|x - T(x)\| : x \in C\} = 0$.*

Proof. Fix a point $y_0 \in C$, let $\{r_n\}_{n \in \mathcal{N}}$ be a sequence of nonnegative real numbers converging to one and $0 \le r_n < 1$. Define the corresponding sequence of contraction operators on C as follows:

(3.2) $$T_n(x) := (1 - r_n)y_0 + r_n T(x).$$

In fact,

$$\|T_n(x) - T_n(z)\| = r_n \|T(x) - T(z)\| \le r_n \|x - z\|, \quad \forall x, y \in C.$$

Using Theorem 3.2, for each $n \in \mathcal{N}$, that is, T_n has a unique fixed point $x_n \in C$, $T_n(x_n) = x_n$. Hence a sequence $\{x_n\}_{n \in \mathcal{N}}$ in C i defined by

(3.3) $$x_n = (1 - r_n)y_0 + r_n T(x_n).$$

We have

$$\|x_n - T(x_n)\| = (1 - r_n)\|y_0 - T(x_n)\| \le (1 - r_n)Diam(C).$$

Taking the limit as n tends to ∞, $\|x_n - T(x_n)\| \to 0$. This limit insures that

$$\inf\{\|x - T(x)\| : x \in C\} = 0.$$

This completes the proof.   □

Remark 3.1. If X is a weakly Cauchy normed space, C is a bounded closed convex subset of X, T is given in Lemma 3.2 is a contraction from C into itself, then the sequence given in equation (3.3) is not only a Cauchy sequence, but also strongly convergent to the unique fixed point of T.

In fact, we show that $\{x_n\}_{n \in \mathcal{N}}$ is a Cauchy sequence. It follows that

$$\|x_n - x_m\| = \|r_n[T(x_n) - T(x_m)] + (r_n - r_m)T(x_m) + (r_m - r_n)y_0\|$$
$$\leq r_n\|T(x_n) - T(x_m)\| + |r_n - r_m|\|T(x_m) - y_0\|$$
$$\leq r_n r\|x_n - x_m\| + |r_n - r_m|Diam(C),$$

which implies $(1 - r_n r)\|x_n - x_m\| \leq |r_n - r_m|Diam(C)$, that is,

$$\|x_n - x_m\| \leq \frac{|r_n - r_m|Diam(C)}{(1 - r_n r)}.$$

Since $\{r_n\}_{n \in \mathcal{N}}$ is a Cauchy sequence, $\{x_n\}_{n \in \mathcal{N}}$ is a Cauchy sequence.

Finally, if y is the unique fixed point of T, then

$$d(x_n, y) \leq d(x_n, T(x_n)) + d(T(x_n), y)$$
$$= d(x_n, T(x_n)) + d(T(x_n), T(y))$$
$$\leq d(x_n, T(x_n)) + rd(x_n, y),$$

which implies, as $n \to \infty$,

$$d(x_n, y) \leq \frac{d(x_n, T(x_n))}{1 - r} \to 0.$$

This completes the proof. □

Remark 3.2. If C is a bounded closed convex subset of a normed space X, $\{r_n\}_{n \in \mathcal{N}}$ a sequence given in the proof of Lemma 3.2 is such that $0 \leq r_n < 1$ and $\{r_n\}_{n \in \mathcal{N}}$ converges to a real number t, where $t < 1$, then the sequence given in (3.3) is a Cauchy sequences. In fact, we have $\|x_n - x_m\| \leq \frac{|r_n - r_m|Diam(C)}{(1 - r_n)}$, which tends to zero as $n, m \to \infty$.

For the nonexpansive mappings, we have the following:

Theorem 3.3. *Let X be a weakly Cauchy normed space in which parallelogram law holds and C be a bounded closed convex subset of X. If T is a nonexpansive mapping from C into itself, then T has fixed points. Moreover, if $C_1[0,1]$ is the set of all nonnegative sequences $\{t_n\}_{n \in \mathcal{N}}$ of real numbers, $0 \leq t_n < 1$, then $C \times C_n[0,1] \subset F(T)$.*

Proof. Using Theorem 3.1, the asymptotic center of the sequence $\{x_n\}_{n \in \mathcal{N}}$ given in the equation (3.3) in Lemma 3.2 is a fixed point of T. This completes the proof. □

We also have the following version of Browder's strong convergence theorem:

Theorem 3.4. *Let X be a normed space in which every bounded sequence has a weakly convergent subsequence and the parallelogram law hold in X. If C is a bounded closed convex subset of X and T is a nonexpansive operator from C into itself, then, for every $y_0 \in C$, the sequence given in (3.3) converges strongly to the best approximation element of y_0 in $F(T)$.*

Proof. Using Lemma 3.1, Theorem 3.3 and the same steps of [9, Theorem 3.2.1 p. 61] with the given assumptions, we have a complete proof. □

4. On $\{a, b, c\}$-Generalized-Contraction and Generalized-Nonexpansive Mappings in Weakly Cauchy Normed Spaces

A generalization of Banach's contraction principle in two directions was given in [2], more explicitly, it is proved that

Theorem 4.1. *Let X be a Cauchy normed space, C be a closed convex subset of X and T be an $\{a, b, c\}$-generalized contraction mapping from C into C. Then T has a unique fixed point $y \in C$. Moreover, for every $x \in C$, the sequence $\{T^n(x)\}_{n \in N}$ of iterates is convergent strongly to y, that is,*

$$\lim_{n \to \infty} T^n(x) = y.$$

The problem weather the $\{a, b, c\}$-generalized-nonexpansive mapping defined on a bounded closed convex subset C of a weakly Cauchy normed space X in which parallelogram law holds or even in uniformly convex Banach space is open and has no affirmative solution. Therefore, we extend our results in this field, we show that if T is a $\{a, b, c\}$-generalized-nonexpansive mapping defined on a bounded closed convex subset of a weakly Cauchy normed space, then $\inf\{\|x - T(x)\| : x \in C\} = 0$ provided contracting point in C.

We have the following definition:

Definition 4.1. Let X be a normed space, C be a subset of X and $0 \le t$, $a, b, c < 1$ with $b, c \neq 1$ and $a + b + c = 1$. Then a point $x_0 \in C$ is said to be the *abct-contracting point* with respect to C if there is $\lambda < \frac{1 - \{ta + b + c\}}{1 - t}$ such that

$$b\|x_0 - x\| + c\|x_0 - y\| \le \lambda\|x - y\|, \quad \forall x, y, \in C \ (x \neq y).$$

We have the following lemma:

Lemma 4.1. *Let X be a weakly Cauchy normed space in which the parallelogram law holds, C be a bounded closed convex subset of X, T be a mapping from C into itself for which there are real numbers a, b, $a + b = 1$, such that*

$$\|T(x) - T(y)\| \le a\|x - y\| + b\|T(x) - x\|$$

and, in addition, $\{x_n\}_{n \in N}$ be a bounded sequence in X such that $\{\|T(x_n) - x_n\|\}_{n \in N}$ is converging to zero. Then T has fixed points.

Proof. Using Theorem 3.1, the sequence $\{x_n\}_{n \in N}$ has a unique asymptotic center. Let y be the unique asymptotic center of $\{x_n\}_{n \in N}$, then we have the following inequality:

$$\|x_n - T(y)\| \le \|x_n - T(x_n)\| + \|T(x_n) - T(y)\|$$
$$\le \|x_n - T(x_n)\| + a\|x_n - y\| + b\|x_n - T(x_n)\|$$
$$= (1 + b)\|x_n - T(x_n)\| + a\|x_n - y\|$$

and so Thus $T(T(y)) = T(y)$. Since such y is unique, we see that $T(y) = y$. This completes the proof. □

Theorem 4.2. *Let X be a weakly Cauchy normed space, C be a closed convex subset of X, T be an $\{a, b, c\}$-nonexpansive mapping from C into C and $b, c \neq 1$. Assume, in addition, that x_0 is abct-contracting point with respect to C and A is the mapping from C into itself defined by*

$$A(x) = (1 - t)x_0 + tT(x).$$

Then there is a real number r, $0 < r < 1$, for which A fulfils the following inequalities

(a) *For every natural number n, we have*

$$\|A^n(x) - A^{n-1}(x)\| \leq r^{n-1}\|A(x) - x\|.$$

(b) *For given natural numbers n and m, $n < m$, we have*

$$\|A^m(x) - A^n(x)\| \leq \frac{r^n}{1 - r}\|A(x) - x\|.$$

(c) *For any natural number l and x, $y \in C$, we have*

$$\|A^l(x) - A^l(y)\|$$
$$\leq [1 - (b + c)]^l\|x - y\|$$
$$+ \frac{r}{r - [1 - (b + c)]}[r^l - \{1 - (b + c)\}^l][b\|A(x) - x\| + c\|A(y) - y\|],$$

(d) *For natural numbers l, m, $l \leq m$ and x, $y \in C$, we have*

$$\|A^m(x) - A^l(y)\| \leq \sum_{k=l-1}^{m-1} r^k\|A(x) - x\| + \|A^l(x) - A^l(y)\|$$

Proof. (a) Let $x \in C$ If $A(x) = x$ then the proof is completed. So, let $A(x) \neq x$. Since T is $\{a, b, c\}$-nonexpansive on C, we have

$$\|A^2(x) - A^1(x)\| = t\|T(A(x)) - T(x)\|$$
$$\leq t\{a\|A(x) - x\| + b\|T(A(x)) - A(x)\| + c\|T(x) - x\|\}$$
$$\leq (ta)\|A(x) - x\| + b\|tT(A(x)) - tA(x)\| + c\|tT(x) - tx\|$$
$$\leq (ta)\|A(x) - x\| + b\|(A^2(x) - A(x)) + (1 - t)(x_0 - A(x))\|$$
$$+ c\|(A(x) - x) + (1 - t)(x_0 - x)\|$$
$$\leq (ta + c)\|A(x) - x\| + b\|A^2(x) - A(x)\|$$
$$+ (1 - t)\{b\|x_0 - A(x)\| + c\|(x_0 - x)\|\},$$

which gives

$$\|A^2(x) - A^1(x)\| \leq \left(\frac{ta + c}{1 - b}\right)\|A(x) - x\|$$
$$+ \left(\frac{1 - t}{1 - b}\right)\{b\|x_0 - A(x)\| + c\|(x_0 - x)\|\}.$$

Since x_0 is the *abct*-contracting point with respect to C, there is λ, $\lambda < \frac{1-\{ta+b+c\}}{1-t}$, such that

$$b\|x_0 - A(x)\| + c\|x_0 - x\| \leq \lambda\|A(x) - x\|.$$

Thus we obtain

$$\|A^2(x) - A^1(x)\| \le \left(\frac{ta+c}{1-b}\right)\|A(x) - x\| + \left(\frac{\lambda(1-t)}{1-b}\right)\|A(x) - x\|$$

$$= \left(\frac{ta+c}{1-b} + \frac{\lambda(1-t)}{1-b}\right)\|A(x) - x\|$$

$$= \left(\frac{ta+c+(1-t)\lambda}{1-b}\right)\|A(x) - x\|.$$

Let $r = \frac{ta+c+(1-t)\lambda}{1-b}$, then, clearly, $r < 1$ and

$$\|A^2(x) - A^1(x)\| \le r\|A(x) - x\|.$$

Repeating the last steps, we see that

$$\|A^n(x) - A^{n-1}(x)\| \le r^{n-1}\|A(x) - x\|.$$

(b) The above inequalities yield that, for any $n, m \in \mathcal{N}$ with $n \le m$

$$\|A^m(x) - A^n(x)\|$$
$$\le \|A^m(x) - A^{m-1}(x)\| + \|A^{m-1}(x) - A^{m-2}(x)\| + \cdots + \|A^{n+1}(x) - A^n(x)\|$$
$$\le \{r^n + r^{n+1} + \cdots + r^{m-2} + r^{m-1}\}\|A(x) - x\|$$
$$= \frac{r^n(1 - r^{m-(n+1)})}{1-r}\|A(x) - x\|$$
$$\le \frac{r^n}{1-r}\|A(x) - x\|.$$

(c) We have

$$\|A^l(x) - A^l(y)\|$$
$$\le t[a\|A^{l-1}(x) - A^{l-1}(y)\| + b\|T(A^{l-1}(x)) - A^{l-1}(x)\|$$
$$\quad + c\|T(A^{l-1}(y)) - A^{l-1}(y)\|]$$
$$\le (ta)\|A^{l-1}(x) - A^{l-1}(y)\| + b\|A^l(x)) - A^{l-1}(x)\|$$
$$\quad + c\|A^l(y)) - A^{l-1}(y)\| + b(1-t)\|x_0 - A^{l-1}(x)\|$$
$$\quad + c(1-t)\|x_0 - A^{l-1}(y)\|$$
$$\le (ta)\|A^{l-1}(x) - A^{l-1}(y)\| + r^l[b\|A(x) - (x)\|$$
$$\quad + c\|A(y) - y\|] + (1-t)\lambda\|A^{l-1}(x) - A^{l-1}(y)\|$$
$$\le (1 - (b+c))\|A^{l-1}(x) - A^{l-1}(y)\| + r^l[b\|A(x) - (x)\| + c\|A(y) - y\|].$$

Repeating the last steps with the term $\|A^{l-1}(x) - A^{l-1}(y)\|$ gives

$$\|A^l(x) - A^l(y)\| \le (1 - (b+c))\big[(1-(b+c))\|A^{l-2}(x) - A^{l-2}(y)\|$$
$$+ r^{l-1}\big[b\|A(x) - (x)\| + c\|A(y) - y\|\big]\big]$$
$$+ r^l\big[b\|A(x) - (x)\| + c\|A(y) - y\|\big]$$
$$= (1 - (b+c))^2\|A^{l-2}(x) - A^{l-2}(y)\| +$$
$$+ ((1 - (b+c))r^{l-1} + r^l)\big[b\|A(x) - (x)\| + c\|A(y) - y\|\big].$$

Repeating the last step again $(l-2)$-times gives

$$\|A^l(x) - A^l(y)\|$$
$$\le [1 - (b+c)]^l\|x - y\| + \big[[1 - (b+c)]^l + [1 - (b+c)]^{l-1}r$$
$$+ \cdots + [1 - (b+c)]r^{l-1} + r^l\big]\big[b\|A(x) - (x)\| + c\|A(y) - y\|\big]$$
$$= [1 - (b+c)]^l\|x - y\|$$
$$+ \frac{r^l[1 - (\frac{1-(b+c)}{r})^l]}{1 - (\frac{1-(b+c)}{r})}\big[b\|A(x) - (x)\| + c\|A(y) - y\|\big]$$
$$= [1 - (b+c)]^l\|x - y\|$$
$$+ \frac{r}{r - [1 - (b+c)]}\big[r^l - [1 - (b+c)]^l\big]\big[b\|A(x) - (x)\| + c\|A(y) - y\|\big].$$

(d) We have

$$\|A^m(x) - A^l(y)\| \le \|A^m(x) - A^{m-1}(x)\| + \|A^{m-1}(x) - A^{m-2}(x)\|$$
$$+ \cdots + \|A^{l-2}(x) - A^{l-1}(x)\| + \|A^l(x) - A^l(y)\|$$

and so, using (c), we see that

$$\|A^m(x) - A^l(y)\| \le \big[r^{m-1} + r^{m-2} + \cdots + r^{l-1}\big]\|A(x) - x\| + \|A^l(x) - A^l(y)\|.$$

This completes the proof.   □

Remark. Using Theorem 4.2, we have the following limits and equations:

(4.1) $$\lim_{n \to \infty} \|A^n(x) - A^{n-1}(x)\| = 0, \quad \forall x \in C.$$

(4.2) $$\lim_{n,m \to \infty} \|A^m(x) - A^n(x)\| = 0, \quad \forall x \in C.$$

(4.3) $$\lim_{l \to \infty} \|A^l(y) - A^l(x)\| = 0, \quad \forall x, y \in C.$$

(4.4) $$\lim_{m,l \to \infty} \|A^m(y) - A^l(x)\| = 0, \quad \forall x, y \in C.$$

Theorem 4.3. *Let X be a weakly Cauchy normed space, C be a closed convex subset of X and T be an $\{a, b, c\}$-generalized-nonexpansive mapping from C into C. Assume, in addition, that x_0 is the abct-contracting point with respect to C. Then the mapping A*

from C into itself defined by

$$A(x) = (1 - t)x_0 + tT(x)$$

has a unique fixed point $y \in C$. Moreover the sequence $\{A^n(x)\}_{n \in \mathcal{N}}$ of iterates is strongly convergent to the point y for every $x \in C$, that is,

$$\lim_{n \to \infty} A^n(x) = y, \quad \forall x \in C.$$

Proof. Using the equation (4.2), the sequence $\{A^n(x)\}_{n \in \mathcal{N}}$ is a Cauchy sequence in X. From the same steps of Theorem 3.2 with the help of Theorem 4.2, the equations (4.3) and (4.4), the sequence $\{A^n(y)\}_{n \in \mathcal{N}}$ converges strongly to y. Since

$$\|A(x) - A(z)\| = t\|T(x) - T(z)\| \leq [ta + (1 - t)\lambda]\|x - z\| + b\|A(x) - x\|$$
$$+ c\|A(z) - z\|, \quad \forall x, y \in C,$$

the inequality holds:

$$\|A(y) - y\| = \lim_{n \to \infty} \|A(y) - A^n(y)\| \leq [ta + (1 - t)\lambda] \lim_{n \to \infty} \|y - A^{n-1}(y)\|$$
$$+ b\|A(y) - y\| + c \lim_{n \to \infty} \|A^n(y) - A^{n-1}(y)\|$$
$$= b\|A(y) - y\|,$$

which proves that $A(y) = y$ and so y is fixed point of A.

Using the equation (4.3) shows that

$$\lim_{n \to \infty} \|A^n(x) - y\| = \lim_{n \to \infty} \|A^n(x) - A^n(y)\| = 0$$

and in turns shows that the weak limit of a sequence $\{A^n(x)\}_{n \in \mathcal{N}}$ is actually a strong limit.

Finally, to show that y is unique, let z be another fixed point of A. Then we have the following contradiction,

$$\|y - z\| = \|A(y) - A(z)\| = t\|T(y) - T(z)\| \leq [ta + (1 - t)\lambda]\|y - z\| < \|y - z\|.$$

This completes the proof.   □

We also have the following:

Theorem 4.4. *Let X be a weakly Cauchy normed space, C be a bounded closed convex subset of X, T be an $\{a, b, c\}$-generalized-nonexpansive mapping from C into itself and $b, c \neq 1$. Assume, in addition, that C has the contracting point. Then*

$$\inf\{\|x - T(x)\| : x \in C\} = 0.$$

Proof. Let $x_0 \in C$ be the contacting point of C, $\{t_n\}_{n \in \mathcal{N}}$ be a sequence of non negative real numbers such that $\lim_{n \to \infty} t_n = 1$ and $0 \leq t_n < 1$. Define the corresponding sequence of operators $\{A_n\}_{n \in \mathcal{N}}$ from C into C as follows:

(4.5) $A_n(x) := (1 - t_n)x_0 + t_n T(x).$

Using Theorem 4.2, for each $n \in \mathcal{N}$, A_n has a unique fixed point $x_n \in C$, that is, $A_n(x_n) = x_n$. Hence a sequence $\{x_n\}_{n \in \mathcal{N}}$ in C is defined by

(4.6) $x_n = (1 - t_n)x_0 + t_n T(x_n).$

We have
$$\|x_n - T(x_n)\| = (1 - t_n)\|x_0 - T(x_n)\| \le (1 - t_n)Diam(C).$$
Taking the limit as n tends to ∞, $\|x_n - T(x_n)\| \to 0$. This limit insures that
$$\inf\{\|x - T(x)\| : x \in C\} = 0.$$
This completes the proof. $\qquad\square$

Using Lemma 4.1 and Theorem 4.4, we have the following:

Theorem 4.5. *Let X be a weakly Cauchy normed space in which parallelogram law holds, C be a bounded closed convex subset of X which containing ab0r-contracting point and T be an $\{a, b, 0\}$-generalized-nonexpansive mapping from C into itself. Then T has fixed points.*

Finally, we have the following another version of Browder's strong convergence theorem.

Theorem 4.6. *Let X be a normed space in which every bounded sequence has a weakly convergent subsequence and the parallelogram law hold in X. If C contains the ab0r-contracting point and T is an $\{a, b, 0\}$-generalized-nonexpansive mapping from C into itself, then, for every ab0r-contracting point x_0 of C, the sequence given in (4.6) converges strongly to the best approximation element of x_0 in $F(T)$.*

Proof. Using Lemma 3.1, Theorem 4.5 and the same steps of [9, Theorem 3.2.1 p. 61] with the given assumptions, we have a complete proof. $\qquad\square$

REFERENCES

1. E. M. El-Shobaky, S. M. Ali and W. Takahashi, *On the projection constant problems and the existence of the metric projections in normed spaces,* Abstr. Appl. Anal. **6(7)** (2001), 401–410.

2. E. M. El-Shobaky, S. M. Ali and M. S. Ali, *Generalization of Banach contraction principle in two directions* , J. Math. Stat. **3(3)** (2007), 112–115.

3. E. M. El-Shobaky, S. M. Ali and M. S. Ali, *Abstract fixed point theory of set-valued mappings*, J. Math. Stat. **3(2)** (2007), 49–53.

4. W. A. Kirk and B. Sims, *Handbook of Metric Fixed Point Theory*, Kluwer Academic Puplishers, Dordrecht, Boston, London, 2001.

5. K. Goebel and S. Riech, *Uniform Convexity, Hyperbolic Geometry, and Nonexpansive Mapping*, Marcel Dekker, Inc. New York and Basel. 1984.

6. S. M. Ali, *Reduced assumption in the Banach contraction principle,* J. Math. Stat. **2(1)**, (2006), 343–345.

7. S. M. Ali, *Fixed points of nonexpansive operators on weakly Cauchy normed sapces,* J. Math. Stat. **3(2)**, (2007), 54–57.

8. S. M. Ali, *On $\{a, b, c\}$-nonexpansive mappings in normed spaces,* to appear in Internat. J. Math. Stat.

9. W. Takahashi. *Nonlinear Functional Analysis, Fixed Point Theory and its Applications,* Yokohama Publishers, Yokohama, 2000.

Nonlinear Functional Analysis and Applications, Volume 1, 85–94

A STRONG CONVERGENCE THEOREM FOR NONEXPANSIVE MAPPINGS

YISHENG SONG[1,*] AND SUMEI XU[2]

ABSTRACT. Let E be a real reflexive Banach space which admits a weakly continuous duality mapping J_φ with a gauge function φ, and K be a nonempty closed convex subset of E. Suppose that $T : K \to K$ is a nonexpansive mapping with a fixed point. For arbitrary initial value $x_0 \in K$ and a fixed weak contraction $f : K \to K$, define iteratively a sequence $\{x_n\}$ as follows:

$$x_{n+1} = \alpha_n f(x_n) + \beta_n x_n + \gamma_n T x_n, \quad \forall n \geq 0,$$

where $\{\alpha_n\}, \{\beta_n\}, \{\gamma_n\} \subset (0,1)$ satisfy proper conditions. The strong convergence of $\{x_n\}$ is proved. Our results unify and develop the corresponding ones of T. H. Kim and H. K. Xu [Nonlinear Anal. 61(2005) 51-60] and C.E. Chidume and C. O. Chidume [J. Math. Anal Appl. 318(2006) 288-295] and many other existing literatures.

1. Introduction and Preliminaries

Let K be a nonempty closed convex subset of a Banach space E and let T be nonexpansive mappings from K into itself (recall that a mapping $T : K \to K$ is nonexpansive if $\|Tx - Ty\| \leq \|x - y\| \quad \forall x, y \in K$). In 1967, Halpern [10] firstly introduced the following iteration scheme in Hilbert spaces: $y, x_0 \in K$, $\alpha_n \in [0,1]$,

(1.1) $$x_{n+1} = \alpha_{n+1} y + (1 - \alpha_{n+1}) T x_n, \quad \forall n \geq 0.$$

He pointed out that the control conditions $(C1)$ $\lim_{n \to \infty} \alpha_n = 0$, $(C2)$ $\sum_{n=1}^{\infty} \alpha_n = \infty$ are necessary for the convergence of the iteration scheme (1.1) to a fixed point of T. Subsequently, many mathematicians studied the strong convergence for the iteration scheme (1.1) (see [13, 14, 16, 18-22, 24-28).

Received August 21, 2007. * Corresponding author.

2000 *Mathematics Subject Classification.* 47H06, 47H09, 7H10, 47H14.

Key words and phrases. Reflexive Banach space, weakly continuous duality mapping, strong convergence.

[1] College of Mathematics and Information Science, Henan Normal University, Henan 453007, People's Republic of China (*E-mails:* songyisheng123@yahoo.com.cn and songyisheng123@163.com)

[2] School of Mathematical Science, AnYang Normal University, People's Republic of China (*E-mail:* xusumei123@yahoo.com.cn)

It is our purpose to unify the iteration schemes dealt by Kim-Xu [11] and Chidume-Chidume [4] and many other existing literatures.

Recently, Kim and Xu [11] dealt with the following iterative scheme for one nonexpansive mapping T in a uniformly smooth Banach space: for $x_0,\ u \in K$,

$$(1.2) \qquad \begin{cases} x_{n+1} = \alpha_n u + (1 - \alpha_n)y_n \\ y_n = \beta_n x_n + (1 - \beta_n)T x_n, \quad \forall n \geq 0. \end{cases}$$

where $\alpha_n \in (0,1), \beta_n \in [0,1)$, and α_n satisfies $(C1)$, $(C2)$ and $(C4)$ $\sum_{n=1}^{\infty} |\alpha_{n+1} - \alpha_n| < +\infty$, and β_n satisfies $(B2)$ $\sum_{n=1}^{\infty} |\beta_n - \beta_{n+1}| < \infty$. And proved several strong convergence theorems.

Very recently, Chidume-Chidume [4] studied the following: for $x_0,\ u \in K$,

$$(1.3) \qquad \begin{cases} x_{n+1} = \alpha_n u + (1 - \alpha_n)S x_n, \\ S x_n = (1 - \delta)x_n + \delta T x_n, \quad \forall n \geq 0, \end{cases}$$

where $\delta \in (0,1)$ and α_n satisfies $(C1)$ and $(C2)$ and proved several strong convergence theorems in a real Banach space E which has uniformly Gâteaux differentiable norm.

In this paper, we will deal with the iterative sequence $\{x_n\}$ given by (1.4) and obtain the strong convergence results. For a nonexpansive mapping $T : K \to K$ and a fixed weak contraction $f : K \to K$, $\{\alpha_n\}, \{\beta_n\}, \{\gamma_n\} \subset [0,1]$ with $\alpha_n + \beta_n + \gamma_n = 1$, define iteratively the sequence $\{x_n\}$ as follows:

$$(1.4) \qquad x_{n+1} = \alpha_n f(x_n) + \beta_n x_n + \gamma_n T x_n. \quad \forall n \geq 0,$$

Throughout this paper, let E be a real Banach space and E^* be its dual space. Let K be a nonempty subset of E and $T : K \to K$ be a mapping. The fixed point set of T is denoted by $F(T) := \{x \in K : Tx = x\}$. We write $x_n \rightharpoonup x$ (respectively, $x_n \overset{*}{\rightharpoonup} x$) to indicate that the sequence x_n weakly (respectively, weak*) converges to x; as usual $x_n \to x$ will symbolize strong convergence.

By a gauge function φ, we mean a continuous strictly increasing function $\varphi : [0, \infty) \to [0, \infty)$ such that $\varphi(0) = 0$ and $\lim_{r \to \infty} \varphi(r) = \infty$. The mapping $J_\varphi : E \to 2^{E^*}$ defined by

$$J_\varphi(x) = \{f \in E^* : \langle x, f \rangle = \|x\|\|f\|, \|f\| = \varphi(\|x\|)\}, \quad \forall x \in E,$$

is called the duality mapping with the gauge function φ. In particular, the duality mapping with the gauge function $\varphi(t) = t$, denoted by J, is referred to as the normalized duality mapping. Clearly, there holds the relation $J_\varphi(x) = \frac{\varphi(\|x\|)}{\|x\|} J(x), \forall x \neq 0$ (see [3, 28]). Browder [3] initiated the study of certain classes of nonlinear operators by means of the duality mapping J_φ.

Following Browder [3], we say that a Banach space E has *a weakly continuous duality mapping* if there exists a gauge function φ for which the duality mapping J_φ is single-valued and weak-weak* sequentially continuous (that is, if $x_n \rightharpoonup x$, then $J_\varphi(x_n) \overset{*}{\rightharpoonup} x$). It is well known that $l^p(1 < p < \infty)$ has a weakly continuous duality map with a gauge function $\varphi(t) = t^{p-1}$. Set

$$\Phi(t) = \int_0^t \varphi(\tau)d\tau, t \geq 0 \ \text{ and } \ \phi(x) = \Phi(\|x\|), \quad \forall x \in E.$$

Then

$$J_\varphi(x) = \partial\phi(x), \quad \forall x \in E,$$

where ∂ denotes the subdifferential in the sense of convex analysis (recall that the subdifferential of the convex function $\phi : E \to R$ at $x \in E$ is the set $\partial\phi(x) = \{x^* \in E^* : \phi(y) \geq \phi(x) + \langle x^*, y - x \rangle, \forall y \in E\}$ [5]).

The first part of the following lemma is an immediate consequence of the subdifferential inequality and the proof of the second part can be found in [8, Theorem 5]; see also [6].

Lemma 1.1. *Assume that E has a weakly continuous duality mapping J_φ with a gauge function φ. Then we havbe the following:*

(a) *The inequality holds*

$$\Phi(\|x + y\|) \leq \Phi(\|x\|) + \langle y, J_\varphi(x + y) \rangle, \forall x, y \in E.$$

(b) *Assume a sequence $\{x_n\}$ in E is weakly convergent to a point x. Then there holds the identity*

$$\limsup_{n \to \infty} \Phi(\|x_n - y\|) = \limsup_{n \to \infty} \Phi(\|x_n - x\|) + \Phi(\|y - x\|), \quad \forall x, y \in E.$$

In particular, E satisfies Opial's condition; that is, if $\{x_n\}$ is a sequence weakly convergent to x, then the inequality holds

$$\limsup_{n \to \infty} \|x_n - x\| < \limsup_{n \to \infty} \|x_n - y\|, \quad y \in E, quad\forall y \neq x.$$

We also need the demiclosedness principle for non-expansive mappings.

Lemma 1.2. *([7, Lemma 4]) Let E be a Banach space satisfying Opial's condition, and K be a nonempty closed convex subset of E. If $T : K \to K$ is a non-expansive mapping, then $(I - T)$ is demiclosed at zero. That is, if $\{x_n\} \subset K$ with $x_n \rightharpoonup x$ and $x_n - Tx_n \to 0$, then $x - Tx = 0$.*

If C is a nonempty convex subset of a Banach space E and D is a nonempty subset of C, then a mapping $P : C \to D$ is called to be a *retraction* if P is continuous with $F(P) = D$. A mapping $P : C \to D$ is called to be *sunny* if $P(Px + t(x - Px)) = Px$, $\forall x \in C$ whenever $Px + t(x - Px) \in C$ and $t > 0$. A subset D of C is said to be a *sunny non-expansive retract* of C if there exists a sunny non-expansive retraction of C onto D. For more details, see [9, 12, 14, 15, 23]. The following Lemma is well known [9, 23].

Lemma 1.3. *Let C be a nonempty convex subset of a smooth Banach space E, $\emptyset \neq D \subset C$, $J : E \to E^*$ the normalized duality mapping of E, and $P : C \to D$ be a retraction. Then P is both sunny and non-expansive if and only if there holds the inequality:*

(1.5) $\langle x - Px, J(y - Px) \rangle \leq 0, \quad \forall x \in C, y \in D.$

Hence there is at most one sunny non-expansive retraction from C onto D.

Note that the inequality (2.1) is equivalent to the inequality (2.2)

(1.6) $\langle x - Px, J_\varphi(y - Px) \rangle \leq 0, \quad \forall x \in C, y \in D,$

where φ is an arbitrary gauge function. This is because there holds the relation $J_\varphi(x) = \frac{\varphi(\|x\|)}{\|x\|} J(x), \forall x \neq 0$.

The following result can be obtained easily using the same proof technique of [28, Proposition 3.2] together with Song and Chen [20, Theorem 2.2]; also see [14, Theorem 3.2] and [15, Corollary].

Lemma 1.4. *Let E be a reflexive Banach space having a weakly continuous duality mapping J_φ with a gauge function φ. Suppose K is a nonempty closed convex subset of E, and $T : K \to K$ is a non-expansive mapping with $F(T) \neq \emptyset$. Then the fixed point set $F(T)$ of T is a sunny non-expansive retract of K.*

Lemma 1.5. *(Xu [27, Lemma 2.5]) Let $\{a_n\}$ be a sequence of nonnegative real numbers satisfying the property*

$$a_{n+1} \leq (1 - \alpha_n)a_n + \alpha_n \theta_n + \delta_n, \quad \forall n \geq 0,$$

where $\{\alpha_n\} \subset (0,1)$, $\{\delta_n\} \subset \mathbb{R}$ and $\{\theta_n\} \subset \mathbb{R}$ such that
(a) $\sum_{n=0}^\infty \alpha_n = \infty$;
(b) $\limsup_{n\to\infty} \theta_n \leq 0$;
(c) $\sum_{n=0}^\infty |\delta_n| < +\infty$.
Then $\{a_n\}$ converges to zero, as $n \to \infty$.

2. Main Results

In this section, for a nonexpansive mapping $T : K \to K$, we firstly show the strong convergence of $\{x_n\}$ defined by (2.1). For an arbitrary initial value $x_0 \in K$ and fixed anchor $u \in K$, define iteratively the sequence $\{x_n\}$ as follows:

(2.1) $$x_{n+1} = \alpha_n u + \beta_n x_n + \gamma_n T x_n, \quad \forall n \geq 0,$$

where $\{\alpha_n\}, \{\beta_n\}, \{\gamma_n\} \subset [0,1]$ with $\alpha_n + \beta_n + \gamma_n = 1$. We shall study the iterative scheme (2.1) under the following conditions.

Condition 2.1.
(C1) $\lim_{n\to\infty} \alpha_n = 0$;
(C2) $\sum_{n=1}^\infty \alpha_n = \infty$;
(C3) $\lim_{n\to\infty} \dfrac{\alpha_n}{\alpha_{n+1}} = 1$ *or* (C4) $\sum_{n=1}^\infty |\alpha_{n+1} - \alpha_n| < +\infty$;
(B1) $\lim_{n\to\infty} \beta_n = \beta \in [0,1)$;
(B2) $\sum_{n=1}^\infty |\beta_{n+1} - \beta_n| < +\infty$ *or*
(B3) $\lim_{n\to\infty} \dfrac{|\beta_n - \beta_{n-1}|}{\alpha_n} = 0$.

Clearly, $\alpha_n = \beta_n = \frac{1}{n+1}$ fulfils the above requirement. If $\beta_n = (1 - \alpha_n)(1 - \delta)$ and $\gamma_n = (1 - \alpha_n)\delta$ ($\delta \in (0,1)$ is a constant), then (2.1) turns into the iteration studied by Chidume-Chidume [4]. When $\beta_n = (1 - \alpha_n)\delta_n$ and $\gamma_n = (1 - \alpha_n)(1 - \delta_n)$, (2.1) is the iteration researched by Kim-Xu [11]. Thus Condition 2.1 contains theirs as the special cases.

Theorem 2.1. *Let E be a reflexive Banach space having a weakly continuous duality mapping J_φ with a gauge function φ. Suppose that K is a nonempty closed convex subset of E, and T is a non-expansive mapping from K into itself with $F(T) \neq \emptyset$. Let $\{x_n\}$ be the sequence generated by (2.1). Suppose that $\{\alpha_n\}$ and $\{\beta_n\}$ satisfy the Condition 2.1. Then $\{x_n\}$ converges strongly to Pu, where P is the unique sunny non-expansive retraction from K onto $F(T)$.*

Proof. Firstly, we prove that $\{x_n\}$ is bounded. In fact, take $p \in F(T)$, by the nonexpansivity of T and $\alpha_n + \beta_n + \gamma_n = 1$, we have

$$
\begin{aligned}
\|x_{n+1} - p\| &= \|\alpha_n(u - p) + \beta_n(x_n - p) + \gamma_n(Tx_n - p)\| \\
&\leq \alpha_n \|u - p\| + (\beta_n + \gamma_n)\|x_n - p\| \\
&\leq \alpha_n \|u - p\| + (1 - \alpha_n)\|x_n - p\| \\
&\leq \max\{\|x_n - p\|, \|u - p\|\}.
\end{aligned}
$$

By induction,

$$
\|x_n - p\| \leq \max\{\|x_0 - p\|, \|u - p\|\}, \quad \forall n \geq 0.
$$

So the set $\{x_n\}$ is bounded, and that so is $\{Tx_n\}$ since

$$
\|Tx_n\| \leq \|Tx_n - p\| + \|p\| \leq \|x_n - p\| + \|p\|.
$$

Set $y_n = \dfrac{\beta_n}{1 - \alpha_n}x_n + \dfrac{\gamma_n}{1 - \alpha_n}Tx_n$. By (2.1) and $\alpha_n + \beta_n + \gamma_n = 1$, we have

$$
(2.2) \qquad\qquad x_{n+1} = \alpha_n u + (1 - \alpha_n)y_n,
$$

$$
(2.3) \qquad\qquad y_n = \frac{\beta_n}{1 - \alpha_n}x_n + \left(1 - \frac{\beta_n}{1 - \alpha_n}\right)Tx_n.
$$

Putting $C = \sup_{n \in \mathbb{N}}\{\|u\|, \|x_n\|, \|Tx_n\|\}$ and $\lambda_n = \frac{\beta_n}{1 - \alpha_n}$, then, by the condition $(C1)$ and $(B1)$, we have $\lambda_n \in (0, 1)$ and $\lambda_n \to \beta$.

Let $A_n = \lambda_n I + (1 - \lambda_n)T$ and $A = \beta I + (1 - \beta)T$. Then it is obvious that $A_n, A : K \to K$ are two nonexpansive mappings and $F(A) = F(T) = F(A_n)$ and $y_n = A_n x_n$. We also have $\|Ax_n\| \leq \beta\|x_n\| + (1 - \beta)\|Tx_n\| \leq C$. Using (2.2), we obtain

$$
\begin{aligned}
\|x_{n+1} - Ax_n\| &\leq \alpha_n\|u - Ax_n\| + (1 - \alpha_n)\|A_n x_n - Ax_n\| \\
&\leq \alpha_n(\|u - Ax_n\| + (1 - \alpha_n)|\lambda_n - \beta|\|x_n - Tx_n\| \\
&\leq 2C\alpha_n + 2C|\lambda_n - \beta|.
\end{aligned}
$$

By the assumption $(C1)$ and $\lambda_n \to \beta$, we obtain

$$
(2.4) \qquad\qquad \lim_{n \to \infty}\|x_{n+1} - Ax_n\| \to 0.
$$

Now we show that $\|x_{n+1} - x_n\| \to 0$ as $n \to \infty$.

From (2.1) and $\alpha_n + \beta_n + \gamma_n = 1$, we can we get the following inequality:

$$\|x_{n+1} - x_n\|$$
$$= \|\alpha_n u + \beta_n x_n + \gamma_n T x_n - \alpha_{n-1} u + \beta_{n-1} x_{n-1} + \gamma_{n-1} T x_{n-1}\|$$
$$\leq |\alpha_n - \alpha_{n-1}|\|u\| + \|\beta_n x_n - \beta_n x_{n-1} + \beta_n x_{n-1} - \beta_{n-1} x_{n-1}\|$$
$$\quad + \|\gamma_n T x_n - \gamma_n T x_{n-1} + \gamma_n T x_{n-1} - \gamma_{n-1} T x_{n-1}\|$$
$$\leq |\alpha_n - \alpha_{n-1}|\|u\| + \beta_n\|x_n - x_{n-1}\| + \gamma_n\|T x_n - T x_{n-1}\|$$
$$\quad + |\gamma_n - \gamma_{n-1}|\|T x_{n-1}\| + |\beta_n - \beta_{n-1}|\|x_{n-1}\|$$
$$\leq (\beta_n + \gamma_n)\|x_n - x_{n-1}\| + |\alpha_n - \alpha_{n-1}|\|u\|$$
$$\quad + |(\alpha_{n-1} - \alpha_n) + (\beta_{n-1} - \beta_n)|\|T x_{n-1}\| + |\beta_n - \beta_{n-1}|\|x_{n-1}\|$$
$$\leq (1 - \alpha_n)\|x_n - x_{n-1}\| + |\alpha_n - \alpha_{n-1}|(\|u\| + \|T x_{n-1}\|)$$
$$\quad + |\beta_n - \beta_{n-1}|(\|x_{n-1}\| + \|T x_{n-1}\|)$$
$$\leq (1 - \alpha_n)\|x_n - x_{n-1}\| + 2C(|\alpha_n - \alpha_{n-1}| + |\beta_n - \beta_{n-1}|).$$

We easily know that

(2.5)
$$\|x_{n+1} - x_n\|$$
$$\leq (1 - \alpha_n)\|x_n - x_{n-1}\| + 2C(|\alpha_n - \alpha_{n-1}| + |\beta_n - \beta_{n-1}|)$$
$$= (1 - \alpha_n)\|x_n - x_{n-1}\| + 2C\alpha_n|1 - \frac{\alpha_{n-1}}{\alpha_n}| + 2C|\beta_n - \beta_{n-1}|$$
$$= (1 - \alpha_n)\|x_n - x_{n-1}\| + 2C\alpha_n\left(\left|1 - \frac{\alpha_{n-1}}{\alpha_n}\right| + \frac{|\beta_n - \beta_{n-1}|}{\alpha_n}\right)$$
$$= (1 - \alpha_n)\|x_n - x_{n-1}\| + 2C|\alpha_n - \alpha_{n-1}| + \alpha_n\left|\frac{2C|\beta_n - \beta_{n-1}|}{\alpha_n}\right|.$$

Case(1) By the conditions $(C4)$ and $(B2)$, we have

$$\sum_{n=1}^{\infty} 2C(|\alpha_n - \alpha_{n-1}| + |\beta_n - \beta_{n-1}|) < +\infty.$$

So, adding to condition $(C2)$, inequality (2.5) satisfies Lemma 1.5 ($\theta_n \equiv 0$).

Case(2) By the condition $(C3)$, we have

$$\theta_n = 2C|1 - \frac{\alpha_{n-1}}{\alpha_n}| \to 0, \quad \delta_n = 2C|\beta_n - \beta_{n-1}|.$$

So, adding to the the conditions $(C2)$ and $(B2)$, the inequality (2.5) also satisfies Lemma 1.5.

Case(3) By the conditions $(C3)$ and $(B3)$, we get

$$\theta_n = 2C\left(\left|1 - \frac{\alpha_{n-1}}{\alpha_n}\right| + \frac{|\beta_n - \beta_{n-1}|}{\alpha_n}\right) \to 0.$$

So, adding to the condition $(C2)$, the inequality (3.5) also satisfies Lemma 1.5 ($\delta_n \equiv 0$).

Case(4) By the condition $(C4)$, we get that

$$\theta_n = \frac{2C|\beta_n - \beta_{n-1}|}{\alpha_n} \to 0 \quad and \quad \delta_n = 2C|\alpha_n - \alpha_{n-1}|.$$

So, adding to the conditions $(C2)$ and $(B3)$, the inequality (2.5) also satisfies Lemma 1.5.

Hence we have $\lim_{n \to \infty} \|x_{n+1} - x_n\| = 0$. Combining (2.4), we get

$$(2.6) \qquad \lim_{n \to \infty} \|x_n - Ax_n\| = 0.$$

It follows from Lemma 1.4 that $F(T)$ is the sunny nonexpansive retract of K. Denote by P the unique sunny non-expansive retraction of K onto $F(T)$. We next show that

$$(2.7) \qquad \limsup_{n \to \infty} \langle u - Pu, J_\varphi(x_n - Pu) \rangle \leq 0.$$

Indeed, we can take a subsequence $\{x_{n_k}\}$ of $\{x_n\}$ such that

$$\limsup_{n \to \infty} \langle u - Pu, J_\varphi(x_n - Pu) \rangle = \lim_{k \to \infty} \langle u - Pu, J_\varphi(x_{n_k} - Pu) \rangle.$$

We may assume that $x_{n_k} \rightharpoonup x^*$ by the reflexivity of E and the boundedness of $\{x_n\}$. It follows from Lemma 1.2 and (2.6) that $x^* \in F(T)$. From the weak continuity of the duality mapping J_φ and (1.6), it follows that

$$\limsup_{n \to \infty} \langle u - Pu, J_\varphi(x_n - Pu) \rangle = \langle u - Pu, J_\varphi(x^* - Pu) \rangle \leq 0.$$

Finally, we show that $x_n \to Pu$. As a matter of fact, apply Lemma 1.1(i) to get

$$\begin{aligned}
\Phi(\|x_{n+1} - Pu\|) \\
&= \Phi(\|\beta_n(x_n - Pu) + \gamma_n(Tx_n - Pu) + \alpha_n(u - Pu)\|) \\
&\leq \Phi(\beta_n\|x_n - Pu\| + \gamma_n\|Tx_n - Pu\|) + \alpha_n\langle u - Pu, J_\varphi(x_{n+1} - Pu) \rangle \\
&\leq (\beta_n + \gamma_n)\Phi(\|x_n - Pu\|) + \alpha_n\langle u - Pu, J_\varphi(x_{n+1} - Pu) \rangle.
\end{aligned}$$

By $\alpha_n + \beta_n + \gamma_n = 1$, we have

$$(2.8) \qquad \Phi(\|x_{n+1} - Pu\|) \leq (1 - \alpha_n)\Phi(\|x_n - Pu\|) + \alpha_n\langle u - Pu, J_\varphi(x_{n+1} - Pu) \rangle.$$

Using Lemma 1.5, $(C2)$, (2.7) and (2.8), we conclude that $x_n \to Pu$. This completes the proof. $\qquad \square$

Let $\beta_n \equiv 0$, then we get the following result which is a development and complementarity of the corresponding ones given [16, 24, 27].

Corollary 2.2. *Let E be a reflexive Banach space having a weakly continuous duality mapping J_φ with a gauge function φ. Suppose that K is a nonempty closed convex subset of E and T is a non-expansive mapping from K into itself with $F(T) \neq \emptyset$. Let $\{x_n\}$ be the sequence generated by*

$$x_{n+1} = \alpha_n u + (1 - \alpha_n)Tx_n, \quad \forall n \geq 0,$$

where $\{\alpha_n\} \subset [0, 1]$. Suppose that $\{\alpha_n\} \subset [0, 1]$ satisfies the conditions $(C1)$, $(C2)$ and $(C3)$. Then $\{x_n\}$ converges strongly to Pu, where P is the unique sunny non-expansive retraction from K onto $F(T)$.

Recall a mapping $f : K \to K$ is called to be *weakly contractive* if

$$\|f(x) - f(y)\| \leq \|x - y\| - \psi(\|x - y\|), \quad \forall x, y \in K,$$

where $\psi : [0, +\infty) \to [0, +\infty)$ is a continuous and strictly increasing function such that ψ is positive on $(0, +\infty)$ and $\psi(0) = 0$. Clearly, the mapping contains contractive mapping as a special case ($\psi(t) = (1 - \beta)t$). Rhoades [17] obtained the result-like Banach's Contraction Mapping Principle for the weakly contractive mapping.

Theorem R. ([17, Theorem 2]) *Let (X, d) be a complete metric space and A a weakly contractive mapping on X. Then A has a unique fixed point p in X. Moreover, for $x \in X$, $\{A^n x\}$ strongly converges to p.*

We will use the following facts concerning numerical recursive inequalities (see [1, 2]).

Lemma 2.3. *Let $\{\lambda_n\}$ and $\{\beta_n\}$ be two sequences of nonnegative real numbers and $\{\alpha_n\}$ a sequence of positive numbers and $\{\alpha_n\}$ a sequence of positive numbers satisfying the conditions $\sum_{n=0}^{\infty} \gamma_n = \infty$ and $\lim_{n \to \infty} \dfrac{\beta_n}{\alpha_n} = 0$. Let the recursive inequality*

$$\lambda_{n+1} \leq \lambda_n - \alpha_n \psi(\lambda_n) + \beta_n, \quad \forall n \geq 0,$$

be given, where $\psi(\lambda)$ is a continuous and strict increasing function for all $\lambda \geq 0$ with $\psi(0) = 0$. Then $\{\lambda_n\}$ converges to zero as $n \to \infty$.

Theorem 2.4. *Let E be a reflexive Banach space having a weakly continuous duality mapping J_φ with a gauge function φ. Suppose that K is a nonempty closed convex subset of E and T is a non-expansive mapping from K into itself with $F(T) \neq \emptyset$. For arbitrary initial value $x_0 \in K$ and a fixed weak contractive mapping $f : K \to K$ with a function ψ, let $\{x_n\}$ be the sequence generated by*

$$(2.9) \qquad\qquad x_{n+1} = \alpha_n f(x_n) + \beta_n x_n + \gamma_n T x_n, \quad \forall n \geq 0,$$

where $\{\alpha_n\}, \{\beta_n\}, \{\gamma_n\} \subset [0, 1]$ with $\alpha_n + \beta_n + \gamma_n = 1$. Suppose that $\{\alpha_n\}$ and $\{\beta_n\}$ satisfy the Condition 2.1. Then $\{x_n\}$ converges strongly to $z = Pf(z)$, where P is the unique sunny non-expansive retraction from K onto $F(T)$.

Proof. It follows from Lemma 1.4 that $F(T)$ is the sunny nonexpansive retract of K. Denote by P the unique sunny non-expansive retraction of K onto $F(T)$. For any $x, y \in K$, we have

$$\|P(f(x)) - P(f(y))\| \leq \|f(x) - f(y)\| \leq \|x - y\| - \psi(\|x - y\|).$$

So, Pf is a weakly contractive mapping with a function ψ. Then, by Theorem R, there exists a unique element $z \in K$ such that $z = P(f(z)) \in F(T)$. Thus we may define a sequence $\{y_n\}$ in K by

$$y_{n+1} = \alpha_n f(z) + \beta_n y_n + \gamma_n T y_n, \quad \forall n \geq 0.$$

Then Theorem 2.1 assures $y_n \rightarrow P(f(z)) = z$ as $n \rightarrow \infty$. For all $n \geq 0$, noting $\alpha_n + \beta_n + \gamma_n = 1$, we have

$$\|x_{n+1} - y_{n+1}\|$$
$$\leq \alpha_n \|f(x_n) - f(z)\| + \beta_n \|x_n - y_n\| + \gamma_n \|Tx_n - Ty_n\|$$
$$\leq \alpha_n (\|f(x_n) - f(y_n)\| + \|f(y_n) - f(z)\|) + (\beta_n + \gamma_n) \|x_n - y_n\|$$
$$\leq \|x_n - y_n\| - \alpha_n \psi(\|x_n - y_n\|) + \alpha_n (\|y_n - z\| - \psi(\|y_n - z\|))$$
$$\leq \|x_n - y_n\| - \alpha_n \psi(\|x_n - y_n\|) + \alpha_n \|y_n - z\|.$$

Thus, putting $\lambda_n = \|x_n - y_n\|$, the following recursive inequality:

$$\lambda_{n+1} \leq \lambda_n - \alpha_n \psi(\lambda_n) + \alpha_n \|y_n - z\|.$$

Since $\|y_n - z\| \rightarrow 0$, then it follows from Lemma 2.3 that $\lim_{n \rightarrow \infty} \|x_n - y_n\| = 0$. Hence

$$\lim_{n \rightarrow \infty} \|x_n - z\| \leq \lim_{n \rightarrow \infty} (\|x_n - y_n\| + \|y_n - z\|) = 0.$$

Consequently, we obtain the strong convergence of $\{x_n\}$ to $z = Pf(z)$. This completes the proof. $\qquad\square$

Corollary 2.5. *Let E be a reflexive Banach space having a weakly continuous duality mapping J_φ with a gauge function φ. Suppose that K is a nonempty closed convex subset of E and T is a non-expansive mapping from K into itself with $F(T) \neq \emptyset$. For arbitrary initial value $x_0 \in K$ and a fixed weak contractive mapping $f : K \rightarrow K$ with a function ψ, let $\{x_n\}$ be the sequence defined by*

$$x_{n+1} = \alpha_n f(x_n) + (1 - \alpha_n)Tx_n, \quad \forall n \geq 0,$$

where $\{\alpha_n\} \subset [0,1]$. Suppose that $\{\alpha_n\} \subset [0,1]$ satisfies the condition $(C1)$, $(C2)$ and $(C3)$. Then $\{x_n\}$ converges strongly to $z = Pf(z)$, where P is the unique sunny non-expansive retraction from K onto $F(T)$.

Remark. If $\psi(t) = (1 - k)t$, then f is a contractive mapping with the contractive coefficient k. Hence our iteration contains the viscosity approximation methods ([13, 26]) as a special case.

REFERENCES

1. Ya. I. Alber and A. N. Iusem, *Extension of subgradient techniques for nonsmooth optimization in Banach spaces,* Set-valued Anal. **9(4)** (2001), 315–335.
2. Y. Alber, S. Reich and J. C. Yao, *Iterative methods for solving fixed-point problems with nonself-mappings in Banach spaces,* Abstract Appl. Anal. **4** (2003), 193–216.
3. F. E. Browder, *Convergence theorems for sequences of nonlinear operators in Banach spaces,* Math. Z. **100** (1967), 201–225.
4. C. E. Chidume and C. O. Chidume, *Iterative approximation of fixed points of nonexpansive mappings,* J. Math. Anal Appl. **318** (2006), 288–295.
5. K. Deimling, *Nonlinear Functional Analysis,* 1988 Springer-Verlag New Tork, Inc.
6. N. M. Gulevich, *Fixed points of nonexpansive mappings,* Journal of Mathematical Sciences, **79(1)** (1986), 755–815.
7. J. Gornicki, *Weak convergence theorems for asymptotically nonexpansive mappings in uniformly convex Banach spaces,* Comment. Math. Univ. Carolin. **30** (1989), 249–252.
8. J. P. Gossez and E. L. Dozo, *Some geometric properties related to the fixed point theory for nonexpansive mappings,* Pacfic J. Math. **40** (1972), 565–573.

9. K. Goebel and S. Reich, *Uniformly Convexity, Hyperbolic Geometry, and Nonexpansive mappings,* Marcel Dekker, New York and Basel, 1984.

10. B. Halpen, *Fixed points of nonexpansive maps.* Bull. Amer. Math. Soc. **73**(1967), 957–961.

11. T. H. Kim and H. K. Xu, *Strong convergence of modified Mann iterations,* Nonlinear Anal. **61** (2005), 51–60.

12. E. Kopeckà and S. Reich, *Nonexpansive retractions in Banach spaces,* Erwin Schroedinger Institute Preprint No. 1787, 2006.

13. A. Moudafi, *Viscosity Approximation Methods for Fixed-Points Problems,* J. Math. Anal. Appl. **241** (2000), 46–55.

14. S. Reich, *Asymptotic behavior of contractions in Banach spaces,* J. Math. Anal Appl. **44** (1973), 57–70.

15. S. Reich, *Approximating zeros of accretive operators,* Proc. Amer. Math. Soc. **51** (1975), 381–384.

16. S. Reich, *Strong convergence theorems for resolvents of accretive operators in Banach spaces* J. Math. Anal Appl. **75** (1980), 287–292.

17. B. E. Rhoades, *Some theorems on weakly contractive maps,* Nonlinear Anal. **47** (2001), 2683–2693.

18. Y. Song, *Viscosity approximation for nonexpansive nonself-mappings in reflexive Banach spaces,* J. of Systems Science and Mathematical Science, **27(4)** (2007) 1–7.

19. Y. Song and R. Chen, *Strong convergence theorems on an iterative method for a family of finite nonexpansive mappings,* Applied Mathematics and Computation, **180** (2006), 275–287.

20. Y. Song and R. Chen, *Viscosity approximation methods for nonexpansive nonself-mappings,* J. Math. Anal. Appl. **321** (2006), 316–326.

21. Y. Song and R. Chen, *Viscosity approximative methods to Cesàro means for non-expansive mappings,* Applied Mathematics and Computation **186** (2007), 1120–1128.

22. N. Shioji and W. Takahashi, *Strong convergence of approximated sequences for nonexpansive mappings in Banach spaces,* Proc. Amer. Math. Soc. **125** (1997), 3641–3645.

23. W. Takahashi, *Nonlinear Functional Analysis– Fixed Point Theory and its Applications,* Yokohama Publishers inc, Yokohama, 2000 (Japanese).

24. W. Takahashi and Y. Ueda, *On Reich's strong convergence theorems for resolvents of accretive operators,* J. Math. Anal Appl. **104**(1984), 546–553.

25. R. Wittmann, *Approximation of fixed points of nonexpansive mappings,* Arch. Math. **59**(1992), 486–491.

26. H. K. Xu, *Viscosity approximation methods for nonexpansive mappings,* J. Math. Anal. Appl. **298** (2004), 279–291.

27. H. K. Xu, *Iterative algorithms for nonlinear operators,* J. London Math. Soc. **66** (2002), 240–256.

28. H. K. Xu, *A strong convergence theorem for contraction semigroups in Banach spaces,* Bull. Austral. Math. Soc. **72** (2005), 371–379.

Nonlinear Functional Analysis and Applications, Volume 1, 95–101

THE CONVERGENCE OF ITERATIVE SEQUENCE FOR ASYMPTOTICALLY DEMICONTRACTIVE MAPS IN BANACH SPACES

Xin-feng He[1], Yin-ying Zhou[2] and Zhen He[1,*]

Abstract. Let E be a real Banach space and $T : E \to E$ an asymptotically demicontractive and uniformly L-Lipschitzian map with $F(T) := \{x \in E : x = Tx\} \neq \emptyset$. We prove necessary and sufficient conditions for the strong convergence of the Ishikawa iterative sequence to a fixed point of T.

1. Introduction

Let E be a real normed linear space, E^* its dual and let $\langle \cdot, \cdot \rangle$ denote the generalized duality pairing between E and E^*. Let $J : E \to 2^{E^*}$ be the normalized duality mapping defined for each $x \in E$ by

$$J(x) = \{f^* \in E : \langle x, f \rangle = \|x\|^2 = \|f^*\|^2\}.$$

It is well known that if E is smooth then J is single-valued. In the sequel we shall denote the single-valued normalized duality map by j.

The map T is called *asymptotically nonexpansive* with sequence $\{k_n\} \subset [1, \infty)$ if $\lim_{n \to \infty} k_n = 1$ and $\forall x, y \in E$,

(1.1) $$\|T^n x - T^n y\| \leq k_n \|x - y\|, \quad \forall n \in N,$$

Received August 22, 2007.   * Corresponding author.

2000 *Mathematics Subject Classification.* 47H07, 47H09, 47H10.

Key words and phrases. Asymptotically demicontractive maps, strong convergence, necessary and sufficient condition, Banach spaces.

This work was supported by the Key Project of Chinese Ministry of Education (No.207104) and The National Natural Science Foundation of China (Grant No. 60873203).

[1] College of Mathematics and Computer, Hebei University, Baoding 071002, People's Republic of China (*E-mails:* hxf@mail.hbu.cn (X.F. He) and zhen_he@163.com (Z. He))

[2] School of Mathematics and Information Sciences, Langfang Normal College, Langfang 065000, People's Republic of China (*E-mail:* zhouyinying_hbu@163.com)

and is called *asymptotically pseudocontractive* with sequence $\{k_n\}$ if $\lim_{n\to\infty} k_n = 1$ and for all $x, y \in E$ there exists $j(x - y) \in J(x - y)$ such that

$$(1.2) \qquad \langle T^n x - T^n y, j(x - y) \rangle \leq k_n \|x - y\|^2, \quad \forall n \in N.$$

The map T is said to be *uniformly L-Lipschitzian* if $\exists L > 0$, a constant, such that $\forall x, y \in E$ and $\forall n \in N$,

$$(1.3) \qquad \|T^n x - T^n y\| \leq L\|x - y\|.$$

Let $F(T) := \{x \in E : x = Tx\} \neq \emptyset$ denote the set of fixed points of T . If Eqs. (1.1) and (1.2) hold $\forall x \in E$ and $\forall y = x^* \in F(T)$, then the map T is said to be, respectively, *asymptotically quasi-nonexpansive*, and *asymptotically hemicontractive*.

Let $E = H$ (the Hilbert space). A map $T : E \to E$ is said to be *k-strictly asymptotically pseudocontractive* (see, e.g., [3,5]) if there exists a sequence $\{a_n\}$ with $\lim_{n\to\infty} a_n = 1$ such that

$$(1.4) \qquad \|T^n x - T^n y\|^2 \leq a_n^2 \|x - y\|^2 + k\|(I - T^n)x - (I - T^n)y\|^2$$

for some $k \in [0, 1)$ and for all $x, y \in E$ and $n \in N$. If $a_n \equiv 1$, $T : E \to E$ is said to be *k-strictly pseudocontractive* (see, e.g., [8,9]). $T : E \to E$ is called *asymptotically demicontractive* (see, e.g., [3,5]) if there exists a sequence $\{a_n\} \subset [1, \infty)$ with $\lim_{n\to\infty} a_n = 1$ such that

$$(1.5) \qquad \|T^n x - x^*\|^2 \leq a_n^2 \|x - x^*\|^2 + k\|x - T^n x\|^2$$

for some $k \in [0, 1)$ and for all $x \in E$, $x^* \in F(T)$ and $n \in N$. The class of *k*-strictly asymptotically pseudocontractive and asymptotically demicontractive maps were introduced by Liu [3]. If $k = 0$ in (1.5), then $T : E \to E$ is *asymptotically quasi-nonexpansive*. Let E be an arbitrary Banach space. $T : E \to E$ is *k-strictly asymptotically pseudocontractive* (see, e.g., [4,5]) if $\forall x, y \in E$ there exists $j(x - y) \in J(x - y)$ and a constant $k \in [0, 1)$ such that

$$
(1.6) \qquad
\begin{aligned}
&Re\langle (I - T^n)x - (I - T^n)y, j(x - y) \rangle \\
&\geq \frac{1}{2}(1 - k)\|(I - T^n)x - (I - T^n)y\|^2 - \frac{1}{2}(a_n^2 - 1)\|x - y\|^2.
\end{aligned}
$$

Furthermore, it follows from (1.6) that $T : E \to E$ is asymptotically demicontractive if $F(T) \neq \emptyset$ and for all $x \in E$ and $x^* \in F(T)$ there exists $j(x - x^*) \in J(x - x^*)$ and a constant $k \in [0, 1)$ such that

$$(1.7) \qquad Re\langle (I - T^n)x, j(x - x^*) \rangle \geq \frac{1}{2}(1 - k)\|(I - T^n)x\|^2 - \frac{1}{2}(a_n^2 - 1)\|x - x^*\|^2.$$

In Hilbert space using (1.6) we can get

$$
\begin{aligned}
&\|T^n x - T^n y\|^2 \\
&= \|x - T^n x - y + T^n y - (x - y)\|^2 \\
&= \|x - T^n x - (y - T^n y)\|^2 + \|x - y\|^2 \\
&\quad - 2\langle x - T^n x - (y - T^n y), x - y \rangle \\
&= \|x - T^n x - (y - T^n y)\|^2 + \|x - y\|^2 \\
&\quad - (1 - k)\|x - T^n x - (y - T^n y)\|^2 + (a_n^2 - 1)\|x - y\|^2 \\
&\leq a_n^2 \|x - y\|^2 + k\|(I - T^n)x - (I - T^n)y\|^2.
\end{aligned}
$$

(1.8)

This is (1.4). Specially, put $y = x^*$, (1.8) becomes

$$
\begin{aligned}
\|T^n x - x^*\|^2 &= \|x - T^n x\|^2 + \|x - x^*\|^2 - 2\langle x - T^n x, x - x^* \rangle \\
&= \|x - T^n x\|^2 + \|x - x^*\|^2 - (1 - k)\|x - T^n x\|^2 \\
&\quad + (a_n^2 - 1)\|x - x^*\|^2 \\
&\leq a_n^2 \|x - x^*\|^2 + k\|x - T^n x\|^2.
\end{aligned}
$$

In 1973, Petryshyn and Williamson [6] proved a necessary and sufficient condition for the Picard and the Mann iterative schemes to converge strongly to fixed points of quasi-nonexpansive mappings in Hilbert spaces. Liu [1,2] extended the above results and obtained some necessary and sufficient conditions for an Ishikawa-type iterative scheme with errors to converge to fixed points of asymptotically quasi-nonexpansive maps.

It is our purpose in this paper to prove necessary and sufficient conditions for the strong convergence of the Ishikawa-type iteration process to a fixed point of an asymptotically demicontractive map in (real) Banach spaces. Our theorems thus improve and extend the results of Liu [1], [2], Osilike [5] , Moore [7] and several others.

In the sequel we shall make use of the following lemma.

Lemma 1.1. *Let $\{a_n\}$ and $\{u_n\}$ be sequences of nonnegative real numbers satisfying the inequality*

$$a_{n+1} \leq (1 + u_n)a_n + b_n, \quad n \geq 0.$$

If $\sum_{n \geq 0} b_n < \infty$, $\sum_{n \geq 0} u_n < \infty$, then,
(a) $\lim_{n \to \infty} a_n$ exists.
(b) If there exists subsequence of $\{a_n\}$ converges to 0, then $\lim_{n \to \infty} a_n = 0$.

Lemma 1.2. *Let E be a real normed linear space. Then $\forall x, y \in E$ and for $j(x - y) \in J(x - y)$ the following inequality holds:*

(1.9) $$\|x + y\|^2 \leq \|x\|^2 + 2\langle y, j(x + y) \rangle.$$

2. Main Results

Lemma 2.1. *Let E be a real normed linear space and $T : E \to E$ a uniformly L-Lipschitzian asymptotically demicontractive map with sequence $\{a_n^2\} \subset [1, \infty)$, $\lim_{n \to \infty} a_n^2 = 1$ and $F(T) \neq \emptyset$. Let $\{\alpha_n\}, \{\beta_n\} \subset [0, 1]$ be two real sequence such that $\sum_{n=1}^{\infty} \alpha_n^2 < \infty$,*

$\sum_{n=1}^{\infty} \alpha_n \beta_n < \infty$, $\sum_{n=1}^{\infty} (a_n^2 - 1) < \infty$. *Let* $\{x_n\}$ *be the sequence generated from an arbitrary* $x_0 \in E$ *by*

(2.1)
$$\begin{cases} y_n = (1 - \beta_n)x_n + \beta_n T^n x_n, \\ x_{n+1} = (1 - \alpha_n)x_n + \alpha_n T^n y_n, \end{cases}$$

Then $\forall x^* \in F(T)$ *and* $\forall n, m \in N$,
(a) there exists $M > 0$ *such that* $\|x_n - x^*\| \leq M$.
(b) $\lim_{n \to \infty} \|x_n - x^*\|$ *exists.*
(c) $\|x_{n+1} - x^*\| \leq (1 + \alpha_n^2)\|x_n - x^*\| + \mu_n$ *for some* $\{\mu_n\}$ *with* $\sum_{n \geq 0} \mu_n < \infty$.
(d) $\|x_{n+m} - x^*\| \leq D\|x_n - x^*\| + D\sum_{i=n}^{n+m-1} \mu_i$, *where* $D = e^{\sum_{i=n}^{n+m-1} \alpha_i^2}$.

Proof. (a) and (b). Using (1.8), (1.9) and (2.1) we get that

(2.2)
$$\begin{aligned} &\|x_{n+1} - x^*\|^2 \\ &\leq (1 - \alpha_n)^2\|x_n - x^*\|^2 + 2\alpha_n\langle T^n y_n - x^*, j(x_{n+1} - x^*)\rangle \\ &= (1 - \alpha_n)^2\|x_n - x^*\|^2 - 2\alpha_n\langle x_{n+1} - T^n x_{n+1}, j(x_{n+1} - x^*)\rangle \\ &\quad + 2\alpha_n\langle x_{n+1} - x^*, j(x_{n+1} - x^*)\rangle + 2\alpha_n\langle T^n y_n - T^n x_{n+1}, j(x_{n+1} - x^*)\rangle \\ &\leq (1 - \alpha_n)^2\|x_n - x^*\|^2 + \alpha_n(a_n^2 - 1)\|x_{n+1} - x^*\|^2 \\ &\quad + 2\alpha_n\|x_{n+1} - x^*\|^2 + 2\alpha_n\|x_{n+1} - x^*\|\|T^n y_n - T^n x_{n+1}\|. \end{aligned}$$

Moreover,

(2.3)
$$\begin{aligned} \|y_n - x^*\| &\leq (1 - \beta_n)\|x_n - x^*\| + \beta_n\|T^n x - x^*\| \\ &\leq [1 - \beta_n + L\beta_n]\|x_n - x^*\| \leq (1 + L\beta_n)\|x_n - x^*\| \\ &\leq (1 + L)\|x_n - x^*\|. \end{aligned}$$

and

(2.4)
$$\begin{aligned} \|x_{n+1} - x^*\| &\leq (1 - \alpha_n)\|x_n - x^*\| + \alpha_n\|T^n y - x^*\| \\ &\leq [1 - \alpha_n + L(1 + L)\alpha_n]\|x_n - x^*\| \\ &= (1 + 2\alpha_n L + \alpha_n L^2)\|x_n - x^*\| \\ &\leq (1 + L)^2\|x_n - x^*\|, \end{aligned}$$

and

(2.5)
$$\begin{aligned} \|T^n y_n - T x_{n+1}\| &\leq L\|x_{n+1} - y_n\| \leq L(1 - \alpha_n)\|x_n - y_n\| + L\alpha_n\|T^n y - y_n\| \\ &\leq L(1 - \alpha_n)\beta_n\|T^n x_n - x_n\| + L(1 + L)\alpha_n\|y_n - x^*\| \\ &\leq (1 + L)(1 - \alpha_n)\beta_n\|x_n - x^*\| + L(1 + L)^2\alpha_n\|x_n - x^*\| \\ &\leq L(1 + L)^2(\alpha_n + \beta_n)\|x_n - x^*\|, \end{aligned}$$

and hence,

$$\begin{aligned}
\|x_{n+1} - x^*\|^2 &\leq (1-\alpha_n)^2\|x_n - x^*\|^2 + \alpha_n(a_n^2-1)(1+L)^4\|x_n - x^*\|^2 \\
&\quad + 2\alpha_n(1 + 2\alpha_n L + \alpha_n L^2)^2\|x_n - x^*\|^2 \\
&\quad + 2\alpha_n(\alpha_n + \beta_n)L(1+L)^4\|x_n - x^*\| \\
&\leq (1+\alpha_n^2)\|x_n - x^*\|^2 + \alpha_n(a_n^2-1)(1+L)^4\|x_n - x^*\|^2 \\
&\quad + 8\alpha_n^2(2L + L^2)^2\|x_n - x^*\|^2 + 2\alpha_n^2 L(1+L)^4\|x_n - x^*\|^2 \\
&\quad + 2\alpha_n\beta_n L(1+L)^4\|x_n - x^*\|^2 \\
&= (1+u_n)\|x_n - x^*\|^2,
\end{aligned}$$

(2.6)

where $u_n = \alpha_n^2 + \alpha_n(a_n^2-1)(1+L)^4 + 8\alpha_n^2(2L+L^2)^2 + 2(\alpha_n^2 + \alpha_n\beta_n)L(1+L)^4$. Observe that $\sum_{n\geq 0} u_n < \infty$. From (2.6) we get

$$(2.7) \qquad \|x_{n+1} - x^*\|^2 \leq \prod_{i=0}^{n}(1+u_i)\|x_1 - x^*\|^2 \leq e^{\sum_{n\geq 0} u_n}\|x_0 - x^*\|^2.$$

So that $\|x_n - x^*\| \leq M$ for some $M > 0$. If we set $a_n = \|x_n - x^*\|^2$, and $b_n = 0$ then, by Lemma 1.1, $\lim_{n\to\infty}\|x_n - x^*\|$ exists. From (2.3) we have $\|y_n - x^*\| \leq (1+L)\|x_n - x^*\|$, thus $\|y_n - x^*\|$ also is bounded.

(c). From (2.6) we get

$$\|x_{n+1} - x^*\|^2 \leq (1 + \alpha_n^2 + \lambda_n)\|x_n - x^*\|,$$

where $\lambda_n = u_n - \alpha_n^2$. Moreover,

$$(2.8) \qquad \begin{aligned}
\|x_{n+1} - x^*\| &\leq (1 + \alpha_n^2 + \lambda_n)\|x_n - x^*\| \\
&\leq (1 + \alpha_n^2)\|x_n - x^*\| + \lambda_n M = (1 + \alpha_n^2)\|x_n - x^*\| + \mu_n,
\end{aligned}$$

where $\mu_n = \lambda_n M = (u_n - \alpha_n^2)$. Observe that $\sum_{n\geq 0}\mu_n < \infty$.

(d). From (c) and $\forall n, m \in N$ we get

$$\begin{aligned}
\|x_{n+m} - x^*\| &\leq (1 + \alpha_{n+m-1}^2)\|x_{n+m-1} - x^*\| + \mu_{n+m-1} \\
&\leq (1 + \alpha_{n+m-1}^2)(1 + \alpha_{n+m-2}^2)\|x_{n+m-2} - x^*\| \\
&\quad + (1 + \alpha_{n+m-1}^2)\mu_{n+m-2} + \mu_{n+m-1}
\end{aligned}$$

$$\vdots \qquad\qquad \vdots$$

$$(2.9) \qquad \begin{aligned}
&\leq \prod_{i=n}^{n+m-1}(1+\alpha_i^2)\|x_n - x^*\| + \prod_{i=n}^{n+m-1}(1+\alpha_i^2)\sum_{i=n}^{n+m-1}\mu_n \\
&\leq e^{\sum_{i=n}^{n+m-1}\alpha_i^2}\|x_n - x^*\| + e^{\sum_{i=n}^{n+m-1}\alpha_i^2}\sum_{i=n}^{n+m-1}\mu_n \\
&\leq D\|x_n - x^*\| + D\sum_{i=n}^{n+m-1}\mu_i.
\end{aligned}$$

where $D = e^{\sum_{i=n}^{n+m-1}\alpha_i^2}$. This completes the proof. $\qquad\square$

Remark 2.2. If the condition $\sum_{n=1}^{\infty} \alpha_n \beta_n < \infty$ changes $\sum_{n=1}^{\infty} \beta_n < \infty$, then, we have

$$
\begin{aligned}
\|y_n - x^*\| &\geq (1 - \beta_n)\|x_n - x^*\| - \beta_n\|T^n x_n - x^*\| \\
&\geq (1 - \beta_n - \beta_n L)\|x_n - x^*\| \\
&= [1 - (1 + L)\beta_n]\|x_n - x^*\|.
\end{aligned}
$$

(2.10)

Combining (2.3) and $\lim_{n\to\infty} \beta_n = 0$, we can get $\lim_{n\to\infty} \|y_n - x^*\|$ exists, and $\lim_{n\to\infty} \|x_n - x^*\| = \lim_{n\to\infty} \|y_n - x^*\|$.

Theorem 2.3. *Let E be a real Banach space and $T : E \to E$ a uniformly L-Lipschitzian asymptotically demicontractive map with sequence $\{a_n^2\} \subset [1, \infty)$, $\lim_{n\to\infty} a_n^2 = 1$ and $F(T) \neq \emptyset$. Let $\{\alpha_n\}, \{\beta_n\} \subset [0, 1]$ be a real sequence such that $\sum_{n=1}^{\infty} \alpha_n^2 < \infty$, $\sum_{n=1}^{\infty} \alpha_n \beta_n < \infty$, $\sum_{n=1}^{\infty} (a_n^2 - 1) < \infty$. Let $\{x_n\}$ be the sequence generated from an arbitrary $x_0 \in E$ by (2.1). Then $\{x_n\}$ converges strongly to a fixed point of T if and only if $\liminf_{n\to\infty} d(x_n, F(T)) = 0$.*

Proof. The necessity of Theorem is clear. From (c) of Lemma 2.1 we obtain

$$
d(x_{n+1}, F(T)) \leq (1 + \alpha_n^2) d(x_n, F(T)) + \delta_n
$$

Since $\liminf_{n\to\infty} d(x_n, F(T)) = 0$, we have from (b) of Lemma 2.1 that

$$
\lim_{n\to\infty} d(x_n, F(T)) = 0.
$$

It now suffices to show that $\{x_n\}$ is Cauchy. For this, let $\varepsilon > 0$ be given. Since $\lim_{n\to\infty} d(x_n, F(T)) = 0$ and $\sum_{i=n}^{\infty} \delta_n < \infty$, there exists a positive integer N_1 such that $\forall n \geq N_1$,

$$
d(x_n, F(T)) \leq \frac{\varepsilon}{3D}, \quad and \quad \sum_{i=n}^{\infty} \delta_n < \frac{\varepsilon}{6D}.
$$

In particular there exists $x^* \in F(T)$ such that $d(x_{N_1}, x^*) \leq \frac{\varepsilon}{3D}$. Now from Lemma 2.1(d) we have that, $\forall n \geq N_1$, that

$$
\begin{aligned}
\|x_{n+m} - x_n\| &\leq \|x_{n+m} - x^*\| + \|x_n - x^*\| \\
&\leq D\|x_{N_1} - x^*\| + D\sum_{i=N_1}^{N_1+m-1} \delta_i + D\|x_{N_1} - x^*\| + D\sum_{i=N_1}^{N_1+m-1} \delta_i \\
&\leq \varepsilon.
\end{aligned}
$$

Hence $\lim_{n\to\infty} x_n$ exists (since E is complete). Suppose that $\lim_{n\to\infty} x_n = x^*$. We now show that $x^* \in F(T)$. But given any $\overline{\varepsilon} > 0$ there exists a positive integer $N_2 \geq N_1$ such that $\forall n \geq N_2$,

$$
\|x_n - x^*\| \leq \frac{\overline{\varepsilon}}{2(1 + L)} \quad and \quad d(x_n, F(T)) \leq \frac{\overline{\varepsilon}}{2(1 + L)}.
$$

Thus, there exists $y^* \in F(T)$ such that $\|x_{N_2} - y^*\| \leq \frac{\bar{\varepsilon}}{2(1+L)}$. We then have the following estimates:

$$\|Tx^* - x^*\| \leq \|Tx^* - y^*\| + \|x_{N_2} - y^*\| + \|x_{N_2} - x^*\|$$
$$\leq L\|x^* - y^*\| + \|x_{N_2} - y^*\| + \|x_{N_2} - x^*\|$$
$$\leq (L+1)\|x_{N_2} - y^*\| + (L+1)\|x_{N_2} - x^*\|$$
$$\leq \bar{\varepsilon}.$$

Since $\bar{\varepsilon} > 0$ is arbitrary we have that $Tx^* = x^*$. This completes the proof. □

Theorem 2.4. *Let E be a real Banach space and $T : E \to E$ a uniformly L-Lipschitzian asymptotically demicontractive map with sequence $\{a_n^2\} \subset [1, \infty)$, $\lim_{n \to \infty} a_n^2 = 1$ and $F(T) \neq \emptyset$. Let $\{\alpha_n\}, \{\beta_n\} \subset [0, 1]$ be a real sequence such that $\sum_{n=1}^{\infty} \alpha_n^2 < \infty$, $\sum_{n=1}^{\infty} \alpha_n \beta_n < \infty$, $\sum_{n=1}^{\infty}(a_n^2 - 1) < \infty$. Let $\{x_n\}$ be the sequence generated from an arbitrary $x_0 \in E$ by (2.1). Then $\{x_n\}$ converges strongly to $x^* \in F(T)$ if and only if there exists an infinite subsequence of $\{x_n\}$ which converges strongly to $x^* \in F(T)$.*

Proof. Let $x^* \in F(T)$ and $\{x_{n_j}\}$ a subsequence of $\{x_n\}$ such that

$$\lim_{n \to \infty} \|x_{n_j} - x^*\| = 0.$$

Since, by Lemma 2.1(b), $\lim_{n \to \infty} \|x_n - x^*\|$ exists then $\lim_{n \to \infty} \|x_n - x^*\| = 0$. □

Remark 2.5. Under the assumption of Theorem 2.3, 2.4 , our results also can discuss for k-strictly pseudocontractive, appearing in [8],[9].

Remark 2.6. If take $\beta_n = 0$, we can get all results in [7]. Appling the assumption of Remark 2.1, the convergence of $\{y_n\}$ also can obtain in Theorem 2.3 and 2.4.

References

1. Q. Liu, *Iterative sequences for asymptotically quasi-nonexpansive mappings*, J.Math. Anal. Appl. **259** (2001), 1–7.
2. Q. Liu, *Iterative sequences for asymptotically quasi-nonexpansive mappings with error members*, J. Math. Anal. Appl. **259** (2001), 18–24.
3. Q. Liu, *On Naimpally and Singhs open question*, J. Math. Anal. Appl. **124** (1987), 157–164.
4. C. Moore, *Iterative approximation of fixed points of asymptotically pseudocontractive maps*, submitted to Bull. Austral. Math. Soc.
5. M. O. Osilike, *Iterative approximation of fixed points of asymptotically demicontractive mappings*, Indian J. Pure Appl. Math. **29** (1998), 1–9.
6. W. V. Petryshyn and T. E. Williamson, *Strong and weak convergence of successive approximations for quasi-nonexpansive mappings*, J. Math. Anal. Appl. **43** (1973), 459–497.
7. Chika Moore and B. V. C. Nnoli, *Note iterative sequence for asymptotically demicontractive maps in Banach spaces*, J. Math. Anal. Appl. **302** (2005), 557–562.
8. Haiyun Zhou, *Strong convergence theorems for a family of Lipschitz quasi-pseudo-contractions in Hilbert spaces*, Nonlinear Analysis **71** (2009), 120–125.
9. Yeol Je Cho, Shin Min Kangb and Xiaolong Qin, *Some results on k-strictly pseudo-contractive mappings in Hilbert spaces*, Nonlinear Analysis **70** (2009), 1956–1964.

Nonlinear Functional Analysis and Applications, Volume 1, 103–110

A CHARACTERIZATION OF UPPER SEMI-CONTINUITY OF THE SOLUTION MAP TO THE VERTICAL IMPLICIT HOMOGENEOUS F-COMPLEMENTARITY PROBLEM WITH R_0-CONDITION

RONG HU[1] AND YA-PING FANG[2,*]

ABSTRACT. In this paper we introduce a class of vertical implicit F-complementarity problem and derive a necessary and sufficient condition for the upper semicontinuity of solution map to the vertical implicit homogeneous F-complementarity problem with R_0-condition.

1. Introduction

In the theory of complementarity problem, continuity properties of solution maps are important subjects and have been investigated by many authors. For the classical linear complementarity problem, the continuity of its solution map was considered in the papers [2, 8, 13, 17]. In [4], Fang and Huang investigated the upper semicontinuity of the solution map to the horizontal linear complementarity problem with R_0-condition. In [6], Fang and Huang studied the upper semicontinuity of the solution map to the vertical implicit linear complementarity problem with R_0-condition. In [7], Fang and Huang introduced the concept of R_0-condition for the mixed linear complementarity problem and studied the upper semicontinuity of the solution map in the mixed linear complementarity problem with R_0-condition. For other related results, one refers to [9, 14, 16, 18, 19].

Recently, several classes of F-complementarity problems were introduced and studied. Yin et al [20] first introduced the class of F-complementarity problems. Fang and Huang

Received August 24, 2007.   * Corresponding author.

2000 *Mathematics Subject Classification.* 90C33.

Key words and phrases. Vertical implicit F-complementarity problem, solution map, upper semicontinuity, homogeneity, type R_0.

This paper was supported by the Scientific Research Foundation of CUIT (CRF200704 and CSRF200601).

[1] Department of Computational Science, Chengdu University of Information Technology, Chengdu, Sichuan, People's Republic of China

[2] Department of Mathematics, Sichuan University, Chengdu, Sichuan, People's Republic of China (*E-mail*: fabhcn@yahoo.com.cn)

[3, 5] introduced several classes of vector F-complementarity problems. Li and Huang [15] further introduced and studied the class of vector F-implicit complementarity problems. For other results on F-complementarity problems, one refers to [10, 11]. Motivated and inspired by the above works, in this paper we introduce a new class of F-complementarity problems, named vertical implicit F-complementarity problems, which include as special cases vertical implicit complementarity problems [4], the F-complementarity problems [20], and implicit complementarity problems (see [12]). We derive a sufficient and necessary condition for the upper semicontinuity of the solution map to the vertical implicit homogeneous F-complementarity problem with R_0-condition.

2. Vertical Implicit F-Complementarity Problems

In this section we introduce a class of vertical implicit F-complementarity problems and give some definitions and notations.

Let R^m, R^n, R^l be three Euclidean spaces with $m \geq n \geq l$. We say that $G : R^l \to R^m$ is a vertical block function [6] of type $(m_1, m_2, \cdots, m_l)$ if G has the form $G = (G_1, G_2, \cdots, G_l)$, where $G_j : R^l \to R^{m_j}$ and $m = \sum_{j=1}^{l} m_j$, that is, G is partitioned into l blocks such that the jth block, G_j, of G, is a nonlinear function from R^l into R^{m_j}. Denote by G_j^i the ith of the jth block, G_j, of G. Let $T : R^l \to R^m$ and $S : R^l \to R^n$ be two vertical block functions of types $(m_1, \cdots, m_l)$ and $(n_1, \cdots, n_l)$, respectively, and $F : R^l \to R$ be a nonlinear function with the form

$$F(x) = F_1(x_1) + F_2(x_2) + \cdots + F_l(x_l), \quad \forall x = (x_1, \cdots, x_l),$$

where $F_j : R \to R$ is a nonlinear function.

Now we introduce the following vertical implicit F-complementarity problem: find $x = (x_1, \cdots, x_l) \in R^l$ such that

$$(VIFCP) \quad \begin{cases} S(x) \in R_+^n, \\ \left(\prod_{i=1}^{m_j} T_j^i(x)\right) \times \left(\prod_{i=1}^{n_j} S_j^i(y)\right) + F_j(y_j) \geq 0, \\ \quad \forall y = (y_1, \cdots, y_l) \in R_+^l, \quad j = 1, \cdots, l, \\ \left(\prod_{i=1}^{m_j} T_j^i(x)\right) \times \left(\prod_{i=1}^{n_j} S_j^i(x)\right) + F_j(x_j) = 0, \quad j = 1, \cdots, l, \end{cases}$$

where T_j^i and S_j^i denote the ith of T_j and S_j, respectively.

If $F \equiv 0$, then $(VIFCP)$ reduces to the vertical implicit complementarity problem [6]:

find $x = (x_1, \cdots, x_l) \in R^l$ such that

$$(VICP) \quad \begin{cases} S(x) \in R_+^n, \\ T(x) \in R_+^m, \\ \left(\prod_{i=1}^{m_j} T_j^i(x)\right) \times \left(\prod_{i=1}^{n_j} S_j^i(x)\right) = 0, \quad j = 1, \cdots, l, \end{cases}$$

If $m = n = l$, then $(VIFCP)$ reduces to the following implicit F-complementarity problem (named F-implicit complementarity problem in [11]):

find $x \in R^n$ such that

$$(IFCP) \quad \begin{cases} S(x) \in R_+^n, \\ \langle T(x), S(y) \rangle + F(y) \geq 0, \quad \forall y \in R_+^n, \\ \langle T(x), S(x) \rangle + F(x) = 0. \end{cases}$$

If $m = n = l$ and $S(x) \equiv x$, then problem $(VIFCP)$ collapses to the F-complementarity problem [20]:

find $x \in R_+^n$ such that

$$(FCP) \quad \langle T(x), x \rangle + F(x) = 0 \text{ and } \langle T(x), y \rangle + F(y) \geq 0, \quad \forall y \geq 0.$$

If $F \equiv 0$, $S(x) = Nx + c$, and $T(x) = Mx + d$, where $M \in R^{m \times l}$, $N \in R^{n \times l}$ are two vertical block matrices of type $(m_1, \cdots, m_l)$ and $(n_1, \cdots, n_l)$, respectively, $d \in R^m$ and $c \in R^n$ are two vector partitioned to conform to the entries in the blocks, M_j, of M, and N_j, of N, respectively, then problem $(VIFCP)$ reduces to the vertical implicit linear complementarity problem [4]:

find $x \in R^n$ such that

$$Nx + c \geq 0, \quad Mx + d \geq 0,$$

$$(VILCP) \quad \left(\prod_{i=1}^{m_j} (M_j x + d_j)^i \right) \times \left(\prod_{i=1}^{n_j} (N_j x + c_j)^i \right) = 0, \quad j = 1, \cdots, l,$$

which includes as a special case the vertical linear complementarity problem [1].

3. Main Results

In this section we give a characterization of the upper semicontinuity of the solution map to the vertical implicit homogeneous F-complementarity problem of type R_0.

Define $\mathcal{A}$, $\mathcal{B}$, and $\mathcal{F}$ as follows:

$\mathcal{A} = \{A|\ A : R^l \to R^n$ is a continuous vertical block function of type $(n_1, \cdots, n_l)$ and for any fixed i, j, $A_j^i(\lambda x) = \lambda^{\rho_j^i} A_j^i(x)$ for all $x \in R^l$ and $\lambda > 0\}$,

$\mathcal{B} = \{B|\ B : R^l \to R^m$ is a continuous vertical block function of type $(m_1, \cdots, m_l)$ and for any fixed i, j, $B_j^i(\lambda x) = \lambda^{\theta_j^i} B_j^i(x)$ for all $x \in R^l$ and $\lambda > 0\}$,

and

$\mathcal{F} = \{F|\ F : R^l \to R$ is a continuous function with the form $F(x) = F_1(x_1) + \cdots + F_l(x_l)$ and for any fixed j, $F_j(\lambda x_j) = \lambda^{\tau_j} F_j(x_j)$ for all $x = (x_1, \cdots, x_l) \in R^l$ and $\lambda > 0\}$,

where $\rho_j^i, \theta_j^i, \tau_j$ are nonnegative constants.

$\mathcal{A}, \mathcal{B}, \mathcal{F}$ can be normed by

$$\|A\| = \max_{\|x\|=1} \|A(x)\|, \quad \|B\| = \max_{\|x\|=1} \|B(x)\|, \quad \|F\| = \max_{\|x\|=1} |F(x)|.$$

Let

$$\mathcal{H} = \mathcal{A} \times \mathcal{B} \times \mathcal{F} \quad \text{and} \quad \mathcal{D} = R^n \times R^m.$$

For $(A, B, F) \in \mathcal{H}$ and $(c, d) \in \mathcal{D}$, consider the vertical implicit F-complementarity problem with

$$S(x) = A(x) + c, \quad T(x) = B(x) + d,$$

i.e., we consider the following vertical implicit homogeneous F-complementarity problem: find $x = (x_1, \cdots, x_l) \in R^l$ such that

$(VIHFCP)$
$$\begin{cases} A(x) + c \in R_+^n, \\ \left(\prod_{i=1}^{m_j}(B_j^i(x) + d_j^i)\right) \times \left(\prod_{i=1}^{n_j}(A_j^i(y) + c_j^i)\right) + F_j(y_j) \geq 0, \\ \quad \forall y \in R_+^l, \ j = 1, \cdots, l, \\ \left(\prod_{i=1}^{m_j}(B_j^i(x) + d_j^i)\right) \times \left(\prod_{i=1}^{n_j}(A_j^i(x) + c_j^i)\right) + F_j(x_j) = 0, \\ \quad j = 1, \cdots, l, \end{cases}$$

where $c = (c_1, \cdots, c_l)$ and $d = (d_1, \cdots, d_l)$ are partitioned to conform to the block functions A and B respectively.

Definition 3.1. Let $H = (A, B, F)$ and $q = (c, d)$. Denote by $\varphi(H, q)$ the solution set of $(VIHFCP)$. Let

$$\mathcal{H}_0 = \{H \in \mathcal{H} : x \in R^l \text{ and } x \neq 0 \text{ imply that } x \notin \varphi(H, 0)\}.$$

We say that $(VIHFCP)$ satisfies R_0-*condition* if and only if $H \in \mathcal{H}_0$. In the sequel we always consider $\varphi(\cdot, \cdot)$ as a multivalued mapping from $\mathcal{H} \times \mathcal{D}$ into R^l.

Remark 3.1. Definition 3.1 covers the definitions of R_0-conditions for the classical linear complementarity, the horizontal linear complementarity problem, the vertical linear complementarity problem, and the mixed linear complementarity problem. See, e.g., [2, 4, 6, 7, 17].

Definition 3.2. Let X, Y be two topological vector spaces. A multivalued map $G : X \to 2^Y$ is said to have a *closed graph* if the set $\{(x, y) \in X \times Y : y \in G(x)\}$ is closed in $X \times Y$. G is said to be upper semicontinuous at $x \in X$ if for any open set $\Omega \subset Y$ with $G(x) \subset \Omega$, there exists a neighborhood V of x such that $G(x') \subset \Omega$ for all $x' \in V$.

Proposition 3.1. φ *as a multivalued map from* $\mathcal{H} \times \mathcal{D}$ *into* R^l *has a closed graph.*

Proof. Let $H^k = (A^k, B^k, F^k) \to H = (A, B, F) \in \mathcal{H}$, $q^k = (c^k, d^k) \to q = (c, d) \in \mathcal{D}$, $x^k \to x$, and $x^k \in \varphi(H^k, q^k)$. It follows that

$$\begin{cases} A^k(x^k) + c^k \in R_+^n, \\ \left(\prod_{i=1}^{m_j}((B^k)_j^i(x^k) + (d^k)_j^i)\right) \times \left(\prod_{i=1}^{n_j}((A^k)_j^i(y) + (c^k)_j^i)\right) + F_j^k(y_j) \geq 0, \\ \quad \forall y \in R_+^l, \quad j = 1, \cdots, l, \\ \left(\prod_{i=1}^{m_j}((B^k)_j^i(x^k) + (d^k)_j^i)\right) \times \left(\prod_{i=1}^{n_j}((A^k)_j^i(x^k) + (c^k)_j^i)\right) + F_j^k(x_j^k) = 0, \\ \quad j = 1, \cdots, l. \end{cases}$$

Since $H^k \to H$ and $q^k \to q$, it is easy to see that

$$
\begin{cases}
A(x) + c \in R_+^n, \\
\left(\prod_{i=1}^{m_j}(B_j^i(x) + d_j^i)\right) \times \left(\prod_{i=1}^{n_j}(A_j^i(y) + c_j^i)\right) + F_j(y_j) \geq 0, \\
\quad \forall y \in R_+^l, \; j = 1, \cdots, l, \\
\left(\prod_{i=1}^{m_j}(B_j^i(x) + d_j^i)\right) \times \left(\prod_{i=1}^{n_j}(A_j^i(x) + c_j^i)\right) + F_j(x_j) = 0, \quad j = 1, \cdots, l.
\end{cases}
$$

This implies that $x \in \varphi(H, q)$ and so φ has a closed graph. $\qquad\square$

Theorem 3.1. *Let $H = (A, B, F) \in \mathcal{H}$. Assume that the following condition*

$$
(3.1) \qquad\qquad \sum_{i=1}^{n_j} \rho_j^i + \sum_{i=1}^{m_j} \theta_j^i = \tau_j
$$

holds for each j and that there exists $\bar{q} = (\bar{c}, \bar{d}) \in \mathcal{D}$ such that $\varphi(H, \bar{q})$ is bounded. Then $\varphi(\cdot, \cdot) : \mathcal{H} \times \mathcal{D} \to 2^{R^l}$ is upper semicontinuous at (H, q) for all $q \in \mathcal{D}$ if and only if $H \in \mathcal{H}_0$.

Proof. Let $H = (A, B, F) \in \mathcal{H}_0$. Suppose by contradiction that $\varphi(\cdot, \cdot)$ is not upper semicontinuous at (H, q) for some $q = (c, d) \in \mathcal{D}$. Then there exists an open set $\Omega \subset R^l$ with $\varphi(H, q) \subset \Omega$, and there exist sequences $\{H^k\} \subset \mathcal{H}$ and $\{q^k\} \subset \mathcal{D}$ such that $H^k \to H$, $q^k \to q$, $x^k \in \varphi(H^k, q^k)$, but $x^k \notin \Omega$ for all k. It follows that

$$
(3.2) \quad
\begin{cases}
A^k(x^k) + c^k \in R_+^n, \\
\left(\prod_{i=1}^{m_j}((B^k)_j^i(x^k) + (d^k)_j^i)\right) \times \left(\prod_{i=1}^{n_j}((A^k)_j^i(y) + (c^k)_j^i)\right) \\
\quad + F_j^k(y_j) \geq 0, \quad \forall y \in R_+^l, \; j = 1, \cdots, l, \\
\left(\prod_{i=1}^{m_j}((B^k)_j^i(x^k) + (d^k)_j^i)\right) \times \left(\prod_{i=1}^{n_j}((A^k)_j^i(x^k) + (c^k)_j^i)\right) \\
\quad + F_j^k(x_j^k) = 0, \quad j = 1, \cdots, l.
\end{cases}
$$

We claim that $\{x^k\}$ has no convergent subsequence. Indeed, if there exists $\{x^{k_j}\} \subset \{x_k\}$ such that $x_{k_j} \to \bar{x}$ as $k \to \infty$. It follows from Proposition 3.1 that $\bar{x} \in \varphi(H, q)$ and so $\bar{x} \in \Omega$. But this is impossible since $x^k \notin \Omega$ and Ω is open. Hence $\{x^k\}$ has no convergent subsequence. Without loss of generality, we may assume that $\|x^k\| \to \infty$ and $x^k/\|x^k\| \to \bar{v} \neq 0$. By (3.1), (3.2) and the definition of $\mathcal{H}$, it is easy to see that

$$
\frac{x^k}{\|x^k\|} \in \varphi(H^k, (e^k, f^k)),
$$

where

$$
e^k = \begin{pmatrix} (e^k)_1 \\ \vdots \\ (e^k)_l \end{pmatrix}, \quad
f^k = \begin{pmatrix} (f^k)_1 \\ \vdots \\ (f^k)_l \end{pmatrix},
$$

$$
(e^k)_j^i = \left(\frac{1}{\|x^k\|}\right)^{\rho_j^i} (c^k)_j^i, \quad (f^k)_j^i = \left(\frac{1}{\|x^k\|}\right)^{\theta_j^i} (d^k)_j^i.
$$

Since φ has a closed graph, letting $k \to \infty$, one has $\bar{v} \in \varphi(H, 0)$. This contradicts $H \in \mathcal{H}_0$ and so $\varphi(\cdot, \cdot)$ is upper semicontinuous at (H, q) for all $q \in \mathcal{D}$.

Conversely, let φ be upper semicontinuous at (H, q) for all $q \in \mathcal{D}$. Assume, by absurd, that $H \notin \mathcal{H}_0$. Then there exists $\bar{x} \neq 0$ such that

$$
(3.3) \quad
\begin{cases}
A(\bar{x}) \in R_+^n, \\
\left(\prod_{i=1}^{m_j} B_j^i(\bar{x})\right) \times \left(\prod_{i=1}^{n_j} A_j^i(y)\right) + F_j(y_j) \geq 0, \\
\quad \forall y \in R_+^l, \ j = 1, \cdots, l, \\
\left(\prod_{i=1}^{m_j} B_j^i(\bar{x})\right) \times \left(\prod_{i=1}^{n_j} A_j^i(\bar{x})\right) + F_j(\bar{x}_j) \geq 0, \quad j = 1, \cdots, l.
\end{cases}
$$

Choose $l \in R^l$ such that $\langle l, \bar{x} \rangle \neq 0$. For any given $t > 0$, define $x_t = x/t$, and A_t, B_t as follows:

$$
(3.4) \quad
\begin{cases}
A_t = ((A_t)_1, \cdots, (A_t)_l), \quad B_t = ((B_t)_1, \cdots, (B_t)_l), \\
(A_t)_j^i(x) = A_j^i(x) - \frac{\|t\langle l, x\rangle\|^{\rho_j^i}}{\|t\langle l, \bar{x}\rangle\|^{\rho_j^i}} \bar{c}_j^i, \\
(B_t)_j^i(x) = B_j^i(x) - \frac{\|t\langle l, x\rangle\|^{\theta_j^i}}{\|t\langle l, \bar{x}\rangle\|^{\theta_j^i}} \bar{d}_j^i.
\end{cases}
$$

By (3.4) and the homogeneity of A_j^i and B_j^i,

$$
(3.5) \quad
\begin{cases}
(A_t)_j^i(x_t) + \bar{c}_j^i = (\tfrac{1}{t})^{\rho_j^i} A_j^i(\bar{x}), \\
(B_t)_j^i(x_t) + \bar{d}_j^i = (\tfrac{1}{t})^{\theta_j^i} B_j^i(\bar{x})
\end{cases}
$$

for all i, j. It follows from (3.1), (3.3), and (3.5) that

$$
\begin{cases}
A_t(x_t) + \bar{c} \in R_+^n, \\
\left(\prod_{i=1}^{m_j}((B_t)_j^i(x_t) + \bar{d}_j^i)\right) \times \left(\prod_{i=1}^{n_j}((A_t)_j^i(y) + \bar{c}_j^i)\right) + F_j(y_j) \geq 0, \\
\quad \forall y \in R_+^l, \quad j = 1, \cdots, l, \\
\left(\prod_{i=1}^{m_j}((B_t)_j^i(x_t) + \bar{d}_j^i)\right) \times \left(\prod_{i=1}^{n_j}((A_t)_j^i(x_t) + \bar{c}_j^i)\right) + F_j((x_t)_j) = 0, \\
\quad j = 1, \cdots, l.
\end{cases}
$$

This means that $x_t \in \varphi((A_t, B_t, F), \bar{q})$. Moreover, from (3.4), $(A_t, B_t, F) \to (A, B, F)$ as $t \to 0$. Since $\varphi(H, \bar{q})$ is bounded, there exists a bounded open set $\Omega \subset R^l$ such that $\varphi(H, \bar{q}) \subset \Omega$. Since φ is upper semi-continuous, $x_t \in \Omega$ for all sufficiently small t. But it is impossible since $\|x_t\| \to \infty$ as $t \to 0$. Thus $H \in \mathcal{H}_0$. This completes the proof. $\qquad \square$

Remark 3.2. Theorem 3.1 generalizes Theorem 2.1 of Oettli and Yen [17] and Theorem 3.1 of Fang and Huang [6].

Theorem 3.2. *Let $H \in \mathcal{H}$ and relation (3.1) hold. Then $H \in \mathcal{H}_0$ if and only if $\varphi(H, q)$ is bounded for all $q \in \mathcal{D}$.*

Proof. Let $\varphi(H, q)$ be bounded for all $q \in \mathcal{D}$. Then $\varphi(H, 0)$ is bounded. Suppose, for contradiction, that $H \notin \mathcal{H}_0$. Then there exists $x \neq 0$ such that $x \in \varphi(H, 0)$. Since relation (1) holds, by the definition of $\mathcal{H}$ and simple arguments, one has $\lambda x \in \varphi(H, 0)$ for all $\lambda > 0$. This contradicts the fact that $\varphi(H, 0)$ is bounded. Thus $H \in \mathcal{H}_0$.

Conversely, let $H \in \mathcal{H}_0$. If there exists $\bar{q} = (\bar{c}, \bar{d}) \in \mathcal{D}$ such that $\varphi(H, \bar{q})$ is unbounded. Without loss of generality, choose $x^k \in \varphi(H, \bar{q})$ such that $\|x^k\| \to \infty$ and $x^k/\|x^k\| \to \bar{x} \neq 0$. As proved in Theorem 3.1,

$$\frac{x^k}{\|x^k\|} \in \varphi(H, (e, f)),$$

where

$$e = \begin{pmatrix} e_1 \\ \vdots \\ e_l \end{pmatrix}, \quad f = \begin{pmatrix} f_1 \\ \vdots \\ f_l \end{pmatrix},$$

$$e_j^i = \left(\frac{1}{\|x\|}\right)^{\rho_j^i} \bar{c}_j^i, \quad f_j^i = \left(\frac{1}{\|x\|}\right)^{\theta_j^i} \bar{d}_j^i.$$

Since φ has a closed graph, $\bar{x} \in \varphi(H, 0)$, which contradicts $H \in \mathcal{H}_0$. This implies that $\varphi(H, q)$ is bounded for all $q \in \mathcal{D}$. This completes the proof. $\square$

References

1. R. W. Cottle and G. B. Dantzig, *A generalization of the linear complementarity problem*, J. Combinatorial Theory **8**(1970), 79–90.
2. R. W. Cottle, J. S. Pang, and R. E. Stone, *The Linear Complementarity Problems*, Academic Press, New York, 1992.
3. Y. P. Fang and N. J. Huang, *The vector F-complementary problems with demipseudomonotone mappings in Banach spaces*, Appl. Math. Lett. **16 (7)**(2003), 1019–1024.
4. Y. P. Fang and N. J. Huang, *A characterization of upper semi-continuity of the solution map to the horizontal linear complementarity problem of type R_0*, Z. Angew. Math. Mech. **85(12)** (2005), 904–907.
5. Y. P. Fang and N. J. Huang, *Least element problems of feasible sets for vector F-complementarity problems with pseudomonotonicity*, Acta Mathematica Sinica, Chinese Series **48(3)** (2005), 499–508.
6. Y.P. Fang and N.J. Huang, *On the upper semi-continuity of the solution map to the vertical implicit homogeneous complementarity problem of type R_0*, Positivity **10** (2006), 95–104.
7. Y. P. Fang and N. J. Huang, *The equivalence of upper semi-continuity of the solution map and R_0-condition in the mixed linear complementarity Problem*, Appl. Math. Lett. **19** (2006), 667–672.
8. M. S. Gowda, *On the continuity of the solution map in linear complementarity problems*, SIAM J. Optim. **2** (1992), 619–634.
9. N. J. Huang and Y. P. Fang, *The upper semicontinuity of the solution maps in vector implicit quasicomplementarity problems of type R_0*, Appl. Math. Lett. **16** (2003), 1151–1156.
10. N. J. Huang and Y. P. Fang, *Strong vector F-complementary problem and least element problem of feasible set*, Nonlinear Anal. TMA **61(6)** (2005), 901–918.
11. N. J. Huang and J. Li, *F-implicit complementarity problems in Banach spaces*, Z. Anal. Anwen. **23(2)** (2004), 293–302.
12. G. Isac, *Topological Methods in Complementarity Theory*, Kluwer Academic Publishers, Dordrecht, 2000.
13. M. J. M. Jansen and S. H. Tijs, *Robustness and nondegenerateness for linear complementarity problems*, Math. Program. **37(3)** (1987), 293–308.
14. C. Jones and M. Gowda, *On the connectedness of solution sets in linear complementarity problems*, Linear Algebra Appl. **272** (1998), 33–44.
15. J. Li and N. J. Huang, *Vector F-implicit complementarity problems in Banach spaces*, Appl. Math. Lett. **19(5)** (2006), 464–471.

Nonlinear Functional Analysis and Applications, Volume 1, 111–117

ISOMORPHISMS IN BANACH ALGEBRAS

CHOONKIL PARK[1]

ABSTRACT. Using the Hyers-Ulam-Rassias stability method, we investigate isomorphisms in Banach algebras and derivations on Banach algebras associated with the following functional equation

$$\frac{1}{q}f(qx + qy + qz) = f(x) + f(y) + f(z)$$

for a fixed nonzero rational number q. The concept of Hyers-Ulam-Rassias stability originated from the Th.M. Rassias' stability theorem that appeared in his paper: On the stability of the linear mapping in Banach spaces, Proc. Amer. Math. Soc. **72** (1978), 297–300.

1. Introduction and Preliminaries

Ulam [32] gave a talk before the Mathematics Club of the University of Wisconsin in which he discussed a number of unsolved problems. Among these was the following question concerning the stability of homomorphisms.

We are given a group G and a metric group G' with metric $\rho(\cdot, \cdot)$. Given $\epsilon > 0$, does there exist a $\delta > 0$ such that if $f : G \to G'$ satisfies $\rho(f(xy), f(x)f(y)) < \delta$ for all $x, y \in G$, then a homomorphism $h : G \to G'$ exists with $\rho(f(x), h(x)) < \epsilon$ for all $x \in G$?

By now an affirmative answer has been given in several cases, and some interesting variations of the problem have also been investigated.

Hyers [8] considered the case of approximately additive mappings $f : E \to E'$, where E and E' are Banach spaces and f satisfies *Hyers inequality*

$$\|f(x + y) - f(x) - f(y)\| \le \epsilon$$

Received August 30, 2007.

2000 *Mathematics Subject Classification.* 39B72.

Key words and phrases. additive functional equation, isomorphism, derivation, Banach algebra.
[1] Department of Mathematics, Hanyang University, Seoul 133-791, Korea (*E-mail:* baak@hanyang.ac.kr)

for all $x, y \in E$. It was shown that the limit

$$L(x) = \lim_{n \to \infty} \frac{f(2^n x)}{2^n}$$

exists for all $x \in E$ and that $L : E \to E'$ is the unique additive mapping satisfying

$$\|f(x) - L(x)\| \leq \epsilon.$$

Th. M. Rassias [23] provided a generalization of Hyers' Theorem which allows the *Cauchy difference to be unbounded.*

Theorem 1.1. (Th.M. Rassias) *Let $f : E \to E'$ be a mapping from a normed vector space E into a Banach space E' subject to the inequality*

(1.1) $$\|f(x + y) - f(x) - f(y)\| \leq \theta(\|x\|^p + \|y\|^p)$$

for all $x, y \in E$, where θ and p are positive real numbers with $p < 1$. Then the limit

$$L(x) = \lim_{n \to \infty} \frac{f(2^n x)}{2^n}$$

exists for all $x \in E$ and $L : E \to E'$ is the unique additive mapping which satisfies

$$\|f(x) - L(x)\| \leq \frac{2\theta}{2 - 2^p} \|x\|^p$$

for all $x \in E$. Also, if for each $x \in E$ the function $f(tx)$ is continuous in $t \in \mathbb{R}$, then L is $\mathbb{R}$-linear.

Th. M. Rassias [24] during the 27th International Symposium on Functional Equations asked the question whether such a theorem can also be proved for $p \geq 1$. Gajda [4] following the same approach as in Th. M. Rassias [23], gave an affirmative solution to this question for $p > 1$. It was shown by Gajda [4], as well as by Th. M. Rassias and Šemrl [29], that one cannot prove a Th. M. Rassias' type theorem when $p = 1$. The counterexamples of Gajda [4], as well as of Th. M. Rassias and Šemrl [29], have stimulated several mathematicians to invent new definitions of *approximately additive* or *approximately linear* mappings (cf. P. Găvruta [5], who among others studied the Hyers-Ulam stability of functional equations). The inequality (1.1) that was introduced for the first time by Th. M. Rassias [23] provided a lot of influence in the development of a generalization of the Hyers-Ulam stability concept. This new concept is known as *Hyers-Ulam-Rassias stability* of functional equations (cf. the books of P. Czerwik [1, 2], D. H. Hyers, G. Isac and Th. M. Rassias [9]).

Beginning around the year 1980, the topic of approximate homomorphisms and their stability theory in the field of functional equations and inequalities was taken up by several mathematicians (cf. D. H. Hyers and Th. M. Rassias [10], Th. M. Rassias [27] and the references therein).

J. M. Rassias [21] following the spirit of the innovative approach of Th. M. Rassias [23] for the unbounded Cauchy difference proved a similar stability theorem in which he replaced the factor $\|x\|^p + \|y\|^p$ by $\|x\|^p \cdot \|y\|^q$ for $p, q \in \mathbb{R}$ with $p + q \neq 1$ (see also [22] for a number of other new results).

Theorem 1.2. ([20, 21, 22]) *Let X be a real normed linear space and Y a real complete normed linear space. Assume that $f : X \to Y$ is an approximately additive mapping for which there exist constants $\theta \geq 0$ and $p \in \mathbb{R} - \{1\}$ such that f satisfies inequality*

$$\|f(x + y) - f(x) - f(y)\| \leq \theta \cdot \|x\|^{\frac{p}{2}} \cdot \|y\|^{\frac{p}{2}}$$

for all $x, y \in X$. Then there exists a unique additive mapping $L : X \to Y$ satisfying

$$\|f(x) - L(x)\| \leq \frac{\theta}{|2^p - 2|} \|x\|^p$$

for all $x \in X$. If, in addition, $f : X \to Y$ is a mapping such that the transformation $t \to f(tx)$ is continuous in $t \in \mathbb{R}$ for each fixed $x \in X$, then L is an $\mathbb{R}$-linear mapping.

Several mathematicians have contributed works on these subjects (see [11]-[16], [25]-[28], [31]). Gilányi [6] showed that if f satisfies the functional inequality

(1.2) $$\|2f(x) + 2f(y) - f(x - y)\| \leq \|f(x + y)\|$$

then f satisfies the Jordan-von Neumann functional inequality

$$2f(x) + 2f(y) = f(x + y) + f(x - y).$$

See also [30]. Fechner [3] and Gilányi [7] proved the Hyers-Ulam-Rassias stability of the functional inequality (1.2). Park, Cho and Han [18] proved the Hyers-Ulam-Rassias stability of functional inequalities associated with Jordan-von Neumann type additive functional equations.

Throughout this paper, assume that A is a Banach algebra with norm $\| \cdot \|_A$ and that B is a Banach algebra with -norm $\| \cdot \|_B$.

This paper is organized as follows: In Section 2, we investigate isomorphisms in Banach algebras associated with the functional equation

(1.3) $$\frac{1}{q} f(qx + qy + qz) = f(x) + f(y) + f(z)$$

for a fixed nonzero rational number q.

In Section 3, we investigate derivations on Banach algebras associated with the functional equation (1.3).

2. Isomorphisms in Banach Algebras

Definition 2.1. A $\mathbb{C}$-linear mapping $H : A \to B$ is called a *homomorphism in Banach algebras* if $H(xy) = H(x)H(y)$ for all $x, y \in A$. If, in addition, the mapping $H : A \to B$ is bijective, then the mapping $H : A \to B$ is called an *isomorphism in Banach algebras*.

In this section, we investigate isomorphisms in Banach algebras associated with the functional equation (1.3).

Proposition 2.2. *Let X and Y be normed spaces with norms $\| \cdot \|_X$ and $\| \cdot \|_Y$, respectively. Let $f : X \to Y$ be a mapping with $f(0) = 0$ such that*

(2.1) $$\|f(x) + f(y) + f(z)\|_Y \leq \|\frac{1}{q} f(qx + qy + qz)\|_Y$$

for all $x, y, z \in X$. Then f is Cauchy additive, i.e., $f(x + y) = f(x) + f(y)$.

Proof. Letting $z = 0$ and $y = -x$ in (2.1), we get

$$\|f(x) + f(-x)\|_Y \leq \left\|\frac{1}{q}f(0)\right\|_Y = 0$$

for all $x \in X$. Hence $f(-x) = -f(x)$ for all $x \in X$.

Letting $z = -x - y$ in (2.1), we get

$$\|f(x) + f(y) - f(x+y)\|_Y = \|f(x) + f(y) + f(-x-y)\|_Y$$
$$\leq \left\|\frac{1}{q}f(0)\right\|_Y = 0$$

for all $x, y \in X$. Thus

$$f(x+y) = f(x) + f(y)$$

for all $x, y \in X$, as desired.   □

Theorem 2.3. *Let* $r \neq 1$ *and* θ *be nonnegative real numbers, and* $f : A \rightarrow B$ *a bijective mapping with* $f(0) = 0$ *such that*

(2.2) $$\|\mu f(x) + f(y) + f(z)\|_B \leq \left\|\frac{1}{q}f(q\mu x + qy + qz)\right\|_B,$$

(2.3) $$\|f(xy) - f(x)f(y)\|_B \leq \theta(\|x\|_A^{2r} + \|y\|_A^{2r})$$

for all $\mu \in \mathbb{T}^1 := \{\lambda \in \mathbb{C}: |\lambda| = 1\}$ *and all* $x, y, z \in A$. *Then the bijective mapping* $f : A \rightarrow B$ *is an isomorphism in Banach algebras.*

Proof. Let $\mu = 1$ in (2.2). By Proposition 2.2 , the mapping $f : A \rightarrow B$ is Cauchy additive.

Letting $z = 0$ and $y = -\mu x$ in (2.2), we get

$$\mu f(x) - f(\mu x) = \mu f(x) + f(-\mu x) = 0$$

for all $x \in A$. So $f(\mu x) = \mu f(x)$ for all $x \in A$. By the same reasoning as in the proof of Theorem 2.1 of [14], the mapping $H : A \rightarrow B$ is $\mathbb{C}$-linear.

(a) Assume that $r < 1$. By (2.3),

$$\|f(xy) - f(x)f(y)\|_B = \lim_{n\to\infty} \frac{1}{4^n}\|f(4^n xy) - f(2^n x)f(2^n y)\|_B$$
$$\leq \lim_{n\to\infty} \frac{4^{nr}}{4^n}\theta(\|x\|_A^{2r} + \|y\|_A^{2r}) = 0$$

for all $x, y \in A$. So

$$f(xy) = f(x)f(y)$$

for all $x, y \in A$.

(b) Assume that $r > 1$. By a similar method to the proof of the case (i), one can prove that the mapping $f : A \rightarrow B$ satisfies

$$f(xy) = f(x)f(y)$$

for all $x, y \in A$. Therefore, the bijective mapping $f : A \rightarrow B$ is an isomorphism in Banach algebras, as desired.   □

Theorem 2.4. *Let $r \neq 1$ and θ be nonnegative real numbers, and $f : A \to B$ a bijective mapping satisfying $f(0) = 0$ and (2.2) such that*

$$(2.4) \qquad \|f(xy) - f(x)f(y)\|_B \leq \theta \cdot \|w\|_A^r \cdot \|x\|_A^r$$

for all $x, y \in A$. Then the bijective mapping $f : A \to B$ is an isomorphism in Banach algebras.

Proof. By the same reasoning as in the proof of Theorem 2.3, the mapping $f : A \to B$ is $\mathbb{C}$-linear.

(a) Assume that $r < 1$. By (2.4),

$$\|f(xy) - f(x)f(y)\|_B = \lim_{n \to \infty} \frac{1}{4^n} \|f(4^n xy) - f(2^n x)f(2^n y)\|_B$$
$$\leq \lim_{n \to \infty} \frac{4^{nr}}{4^n} \theta \cdot \|w\|_A^r \cdot \|x\|_A^r = 0$$

for all $x, y \in A$. So

$$f(xy) = f(x)f(y)$$

for all $x, y \in A$.

(b) Assume that $r > 1$. By a similar method to the proof of the case (i), one can prove that the mapping $f : A \to B$ satisfies

$$f(xy) = f(x)f(y)$$

for all $x, y \in A$. Therefore, the bijective mapping $f : A \to B$ is an isomorphism in Banach algebras, as desired. $\qquad \square$

3. Derivations on Banach Algebras

Definition 3.1. A $\mathbb{C}$-linear mapping $\delta : A \to A$ is called a *derivation* if $\delta(xy) = \delta(x)y + x\delta(y)$ for all $x, y \in A$.

We investigate derivations on Banach algebras associated with the functional equation (1.3).

Theorem 3.2. *Let $r \neq 1$ and θ be nonnegative real numbers, and $f : A \to A$ a mapping with $f(0) = 0$ such that*

$$(3.1) \qquad \|\mu f(x) + f(y) + f(z)\|_A \leq \left\| \frac{1}{q} f(q\mu x + qy + qz) \right\|_A,$$

$$(3.2) \qquad \|f(xy) - f(x)y - xf(y)\|_A \leq \theta(\|x\|_A^{2r} + \|y\|_A^{2r})$$

for all $\mu \in \mathbb{T}^1$ and all $x, y, z \in A$. Then the mapping $f : A \to A$ is a derivation on A.

Proof. By the same reasoning as in the proof of Theorem 2.3, the mapping $f : A \to A$ is $\mathbb{C}$-linear.

(a) Assume that $r < 1$. By (3.2),

$$\|f(xy) - f(x)y - xf(y)\|_A = \lim_{n \to \infty} \frac{1}{4^n} \|f(4^n xy) - f(2^n x) \cdot 2^n y - 2^n x f(2^n y)\|_A$$
$$\leq \lim_{n \to \infty} \frac{4^{nr}}{4^n} \theta(\|x\|_A^{2r} + \|y\|_A^{2r}) = 0$$

for all $x, y \in A$. So

$$f(xy) = f(x)y + xf(y)$$

for all $x, y \in A$.

(b) Assume that $r > 1$. By a similar method to the proof of the case (i), one can prove that the mapping $f : A \to A$ satisfies

$$f(xy) = f(x)y + xf(y)$$

for all $x, y \in A$. Therefore, the mapping $f : A \to A$ is a derivation on A, as desired.   □

Theorem 3.3. *Let $r \neq 1$ and θ be nonnegative real numbers, and $f : A \to A$ a mapping satisfying $f(0) = 0$ and (3.1) such that*

(3.3) $$\|f(xy) - f(x)y - xf(y)\|_A \leq \theta \cdot \|x\|_A^r \cdot \|y\|_A^r$$

for all $x, y \in A$. Then the mapping $f : A \to A$ is a derivation on A.

Proof. The proof is similar to the proofs of Theorems 2.3 and 3.2.   □

REFERENCES

1. S. Czerwik, *Functional Equations and Inequalities in Several Variables*, World Scientific Publishing Company, New Jersey, London, Singapore and Hong Kong, 2002.
2. S. Czerwik, *Stability of Functional Equations of Ulam-Hyers-Rassias Type*, Hadronic Press, Palm Harbor, Florida, 2003.
3. W. Fechner, *Stability of a functional inequalities associated with the Jordan-von Neumann functional equation*, Aequationes Math. **71** (2006), 149–161.
4. Z. Gajda, *On stability of additive mappings*, Int. J. Math. Math. Sci. **14** (1991), 431–434.
5. P. Găvruta, *A generalization of the Hyers-Ulam-Rassias stability of approximately additive mappings*, J. Math. Anal. Appl. **184** (1994), 431–436.
6. A. Gilányi, *Eine zur Parallelogrammgleichung äquivalente Ungleichung*, Aequationes Math. **62** (2001), 303–309.
7. A. Gilányi, *On a problem by K. Nikodem*, Math. Inequal. Appl. **5** (2002), 707–710.
8. D. H. Hyers, *On the stability of the linear functional equation*, Proc. Nat. Acad. Sci. U.S.A. **27** (1941), 222–224.
9. D. H. Hyers, G. Isac and Th. M. Rassias, *Stability of Functional Equations in Several Variables*, Birkhäuser, Basel, 1998.
10. D. H. Hyers and Th. M. Rassias, *Approximate homomorphisms*, Aequationes Math. **44** (1992), 125–153.
11. C. Park, *Lie *-homomorphisms between Lie C^*-algebras and Lie *-derivations on Lie C^*-algebras*, J. Math. Anal. Appl. **293** (2004), 419–434.
12. C. Park, *Homomorphisms between Poisson JC^*-algebras*, Bull. Braz. Math. Soc. **36** (2005), 79–97.
13. C. Park, *Homomorphisms between Lie JC^*-algebras and Cauchy-Rassias stability of Lie JC^*-algebra derivations*, J. Lie Theory **15** (2005), 393–414.
14. C. Park, *Isomorphisms between unital C^*-algebras*, J. Math. Anal. Appl. **307** (2005), 753–762.
15. C. Park, *Approximate homomorphisms on JB^*-triples*, J. Math. Anal. Appl. **306** (2005), 375–381.
16. C. Park, *Isomorphisms between C^*-ternary algebras*, J. Math. Phys. **47**, no. 10, 103512 (2006).
17. C. Park, *Hyers-Ulam-Rassias stability of a generalized Apollonius-Jensen type additive mapping and isomorphisms between C^*-algebras*, to appear in Math. Nachr..
18. C. Park, Y. Cho and M. Han, *Functional inequalities associated with Jordan-von Neumann type additive functional equations*, J. Inequal. Appl. **2007** (2007), Article ID 41820, 13 pages.
19. C. Park, J. Hou and S. Oh, *Homomorphisms between JC^*-algebras and between Lie C^*-algebras*, Acta Math. Sinica **21** (2005), 1391–1398.

20. J. M. Rassias, *On approximation of approximately linear mappings by linear mappings*, J. Funct. Anal. **46** (1982), 126–130.

21. J. M. Rassias, *On approximation of approximately linear mappings by linear mappings*, Bull. Sci. Math. **108** (1984), 445–446.

22. J. M. Rassias, *Solution of a problem of Ulam*, J. Approx. Theory **57** (1989), 268–273.

23. Th. M. Rassias, *On the stability of the linear mapping in Banach spaces*, Proc. Amer. Math. Soc. **72** (1978), 297–300.

24. Th. M. Rassias, *Problem 16; 2*, Report of the 27th International Symp. on Functional Equations, Aequationes Math. **39** (1990), 292–293; 309.

25. Th. M. Rassias, *The problem of S.M. Ulam for approximately multiplicative mappings*, J. Math. Anal. Appl. **246** (2000), 352–378.

26. Th. M. Rassias, *On the stability of functional equations in Banach spaces*, J. Math. Anal. Appl. **251** (2000), 264–284.

27. Th. M. Rassias, *On the stability of functional equations and a problem of Ulam*, Acta Appl. Math. **62** (2000), 23–130.

28. Th. M. Rassias, *Functional Equations, Inequalities and Applications*, Kluwer Academic Publishers, Dordrecht, Boston and London, 2003.

29. Th. M. Rassias and P. Šemrl, *On the Hyers-Ulam stability of linear mappings*, J. Math. Anal. Appl. **173** (1993), 325–338.

30. J. Rätz, *On inequalities associated with the Jordan-von Neumann functional equation*, Aequationes Math. **66** (2003), 191–200.

31. F. Skof, *Proprietà locali e approssimazione di operatori*, Rend. Sem. Mat. Fis. Milano **53** (1983), 113–129.

32. S. M. Ulam, *Problems in Modern Mathematics*, Wiley, New York, 1960.

Nonlinear Functional Analysis and Applications, Volume 1, 119–131

ON POSITIVE SOLUTIONS OF NONLINEAR n^{th}-ORDER SINGULAR BOUNDARY VALUE PROBLEM WITH NONLOCAL CONDITIONS

Xinan Hao[1], Lishan Liu[1,2,*] and Yonghong Wu[2]

Abstract. This paper consider the existence, nonexistence and multiplicity of positive solutions for the nonlinear n^{th}-order singular boundary value problem

$$u^{(n)}(t) + \lambda a(t) f(t, u) = 0, \quad t \in (0, 1);$$

$$u(0) = 0, \quad u'(0) = 0, \quad \cdots, \quad u^{(n-2)}(0) = 0, \quad \alpha u(\eta) = u(1),$$

where $0 < \eta < 1$, $0 < \alpha \eta^{n-1} < 1$, $\lambda > 0$. a may be singular at $t = 0$ and/or 1, $f(t, u)$ may also have singularity at $u = 0$. The proof of our results is based on the fixed point theory on cones. The main results improve and generalize many related results.

1. Introduction

The study of nonlocal boundary value problems (BVPs) for nonlinear differential equations was initiated by Il'in and Moiseev [1, 2] and Gupta [3]. Since then, more general nonlinear nonlocal BVPs have been studied extensively. Many authors refer to such problems as multipoint problems. They studied the existence and multiplicity of positive solutions by the lower and upper solutions method, the coincidence degree theory and the fixed point theorem in cones, see [5-7] and references therein. However, research for

Received September 10, 2007.   * Corresponding author.

2000 *Mathematics Subject Classification*. 45J05, 34k30.

Key words and phrases. Positive solution, singular boundary value problem, nonlocal conditions, fixed point theory.

The first and second authors were supported by the National Natural Science Foundation of China (10771117) and the State Ministry of Education Doctoral Foundation of China (20060446001). The third author was supported financially by the Australia Research Council through an ARC Discovery Project Grant.

[1] Department of Mathematics, Qufu Normal University, Qufu, 273165, Shandong, People's Republic of China (*E-mails*: haoxinan2004@163.com (X. Hao) and lls@mail.qfnu.edu.cn (L. Liu))

[2] Department of Mathematics and Statistics, Curtin University of Technology, Perth, WA 6845, Australia (*E-mail*: yhwu@maths.curtin.edu.au)

singular nonlocal BVPs have proceed vary slowly and the related results are few. Recently, Eloe and Ahmad [8] studied the positive solutions for the nth-order differential equation

$$(1.1) \qquad u^{(n)}(t) + a(t)f(u) = 0, \quad t \in (0,1),$$

subject to the nonlocal boundary conditions

$$(1.2) \qquad u(0) = 0, \quad u'(0) = 0, \quad \cdots, \quad u^{(n-2)}(0) = 0, \quad \alpha u(\eta) = u(1),$$

where $0 < \eta < 1$, $0 < \alpha \eta^{n-1} < 1$. For nonsingular case, Eloe and Ahmad established the existence theorem of positive solution of BVP(1.1) and (1.2) by applying the fixed point theorem in cones duo to Krasnosel'skii and Guo.

Motivated by the above works, we will consider the following nonlocal BVP for n^{th}-order singular differential equations

$$(1.3) \qquad \begin{cases} u^{(n)}(t) + \lambda a(t)f(t,u) = 0, \quad t \in (0,1), \\ u(0) = 0, \quad u'(0) = 0, \quad \cdots, \quad u^{(n-2)}(0) = 0, \quad \alpha u(\eta) = u(1), \end{cases}$$

where $0 < \eta < 1$, $0 < \alpha \eta^{n-1} < 1$, $\lambda > 0$. The function a may be singular at $t = 0$ and/or $t = 1$, $f(t,u)$ may also have singularity at $u = 0$. This paper consider the existence, nonexistence and multiplicity of positive solutions for the singular BVP(1.3). The results obtained improve and generalize essentially the results of [8]. We would like to point out that the existence results on infinitely many positive solutions of the BVP we established has been given rather less attention in the existed literature. The main results generalize and include the results corresponding to those obtained by many others.

A function u is said to be a positive solution of boundary value problem (1.3) if $u \in C^{n-2}([0,1],[0,+\infty)) \cap C^n((0,1),[0,+\infty))$ satisfies BVP (1.3) and $u(t) > 0$ for $t \in (0,1)$.

Let E be a Banach space and K be a cone in E. For $0 < r < R < +\infty$, let $K_r = \{u \in K : \|u\| < r\}$, $\partial K_r = \{u \in K : \|u\| = r\}$ and $\overline{K}_R = \{u \in K : \|u\| \le R\}$. The following lemma is needed in our argument.

Lemma 1.1. ([11]) *Let K be a positive cone in real Banach space E, $0 < r < R < +\infty$, and let $A : K \to K$ be a completely continuous operator and such that either*
(a) $\|Ax\| \le \|x\|$, $x \in \partial K_r$, $\|Ax\| \ge \|x\|$, $x \in \partial K_R$, or
(b) $\|Ax\| \ge \|x\|$, $x \in \partial K_r$, $\|Ax\| \le \|x\|$, $x \in \partial K_R$.
Then A has a fixed point in $\overline{K}_R \backslash K_r$.

2. Preliminaries and Several Lemmas

Let G be Green's function for the $u^{(n)}(t) = 0$ coupled with the boundary condition (1.2), then

$$(2.1) \qquad G(t,s) = \begin{cases} \dfrac{\phi(s)t^{n-1}}{(n-1)!}, & 0 \le t \le s \le 1, \\ \dfrac{\phi(s)t^{n-1} + (t-s)^{n-1}}{(n-1)!}, & 0 \le s \le t \le 1, \end{cases}$$

where

$$(2.1) \qquad \phi(s) = \begin{cases} -\dfrac{(1-s)^{n-1}}{1-\alpha\eta^{n-1}}, & \eta \le s, \\ -\dfrac{(1-s)^{n-1} - (\eta-s)^{n-1}}{1-\alpha\eta^{n-1}}, & s \le \eta. \end{cases}$$

It is easy to see that

(2.2) $$G(t,s) < 0, \quad t \in (0,1), \quad s \in (0,1).$$

Lemma 2.1. ([8]) *Let* $0 < \alpha\eta^{n-1} < 1$ *and* u *satisfies* $u^{(n)}(t) \leq 0$ *for* $0 < t < 1$, *with the nonlocal conditions* (1.2), *then*

$$\inf_{t \in [\eta,1]} u(t) \geq \gamma\|u\|,$$

where $\gamma = \min\{\alpha\eta^{n-1}, \ \alpha(1-\eta)(1-\alpha\eta)^{-1}, \ \eta^{n-1}\}$.

Defining $g(s) = \max_{t\in[0,1]} |G(t,s)|$. From the proof of Lemma 1.2, we know

(2.3) $$|G(t,s)| \geq \gamma g(s), \quad t \in [\eta,1], \ s \in [0,1].$$

Throughout the paper we assume that:

(H_1) $a(t) : (0,1) \to [0,\infty)$ is continuous and there exists $t_0 \in [\eta,1)$ such that $a(t_0) > 0$ and $\int_0^1 g(s)a(s)\,ds < +\infty$.

(H_2) $f(t,x) : [0,1] \times (0,\infty) \to [0,\infty)$ is continuous and singular at $x = 0$, for any $0 < r < R < +\infty$,

$$\lim_{n\to\infty} \sup_{u\in\overline{K}_R\setminus K_r} \int_{H(n)} g(s)a(s)f(s,u(s))\,ds = 0,$$

where $H(n) = [0,\frac{1}{n}]\bigcup[\frac{n-1}{n},1]$.

Remark 2.1. From (H_1), we can choose $\delta \in (\eta,1)$ such that $a(t) > 0$ for any $t \in (\eta,\delta)$ and $0 < \int_\eta^\delta g(s)a(s)\,ds < +\infty$. Under the conditions of Lemma 2.1, we also have $\inf_{t\in[\eta,\ \delta]} u(t) \geq \gamma\|u\|$.

For convenience in the following discussion, we use the notations:

$$f^\alpha := \limsup_{x\to\alpha} \max_{t\in[0,1]} \frac{f(t,x)}{x}, \quad \alpha = 0, +\infty,$$

$$f_\beta := \liminf_{x\to\beta} \min_{t\in[\eta,\ \delta]} \frac{f(t,x)}{x}, \quad \beta = 0, +\infty,$$

$$L := \int_0^1 g(s)a(s)\,ds, \quad l := \gamma^2 \int_\eta^\delta g(s)a(s)\,ds.$$

Set $E = C[0,1]$. It is easy to testify that E is a Banach space with the norm $\|u\| = \sup_{t\in[0,1]} |u(t)|$. We define a cone P as follows:

$$K = \left\{u \in E : u \geq 0, \ \min_{t\in[\eta,\ \delta]} u(t) \geq \gamma\|u\|\right\},$$

where γ is given in Lemma 2.1. Define an operator $A : \overline{K}_R\setminus K_r \to E$ by

(2.4) $$Au(t) = -\lambda \int_0^1 G(t,s)a(s)f(s,u(s))\,ds, \ t \in [0,1].$$

It is well known that the BVP (1.3) has a positive solution if and only if A has a fixed point in $\overline{K}_R\setminus K_r$.

Next, We establish some lemmas that provide us with some useful information concerning the behavior of solutions for BVP (1.3).

Lemma 2.2. $A(\overline{K}_R \setminus K_r) \subset K.$

Proof. For any $r > 0$, we first show that

$$(2.5) \qquad \sup_{u \in \partial K_r} \lambda \int_0^1 g(s)a(s)f(s, u(s))ds < +\infty.$$

By (H_2), there exists a natural number l such that

$$\sup_{u \in \partial K_r} \lambda \int_{H(l)} g(s)a(s)f(s, u(s))ds < 1.$$

If $u \in \partial K_r$, it is easy to see that

$$\gamma_1 r \leq u(t) \leq r, \quad \frac{1}{l} \leq t \leq \frac{l-1}{l},$$

where $\gamma_1 = \min\{\frac{1}{r} \min_{t \in [\frac{1}{l}, \frac{l-1}{l}]} u(t), \gamma\}$. Let $M_1 = \max\{f(t, x) : \frac{1}{l} \leq t \leq \frac{l-1}{l}, \; \gamma_1 r \leq x \leq r\}$. By (H_1) and (H_2), we have

$$\sup_{u \in \partial K_r} \lambda \int_0^1 g(s)a(s)f(s, u(s))ds$$

$$\leq \sup_{u \in \partial K_r} \lambda \int_{H(l)} g(s)a(s)f(s, u(s))ds + \sup_{u \in \partial K_r} \lambda \int_{\frac{1}{l}}^{\frac{l-1}{l}} g(s)a(s)f(s, u(s))ds$$

$$\leq 1 + M_1 \lambda \int_0^1 g(s)a(s)ds < +\infty.$$

Then $A : \overline{K}_R \setminus K_r \to E$ is well defined.

By (H_1), (H_2), (2.2) and (2.4) we know

$$Au(t) \geq 0, \quad t \in [0, 1].$$

For $u \in \overline{K}_R \setminus K_r$, $t \in [0, 1]$,

$$Au(t) = -\lambda \int_0^1 G(t, s)a(s)f(s, u(s))\, ds$$

$$= \lambda \int_0^1 |G(t, s)|a(s)f(s, u(s))\, ds$$

$$\leq \lambda \int_0^1 g(s)a(s)f(s, u(s))\, ds.$$

Thus,

$$(2.6) \qquad \|Au\| \leq \lambda \int_0^1 g(s)a(s)f(s, u(s))\, ds.$$

Then

$$\min_{t \in [\eta, \delta]} Au(t) = \min_{t \in [\eta, \delta]} \left[-\lambda \int_0^1 G(t, s) a(s) f(s, u(s)) \, ds \right]$$

$$= \min_{t \in [\eta, \delta]} \lambda \int_0^1 |G(t, s)| a(s) f(s, u(s)) \, ds$$

(2.7)

$$\geq \lambda \int_0^1 \min_{t \in [\eta, \delta]} |G(t, s)| a(s) f(s, u(s)) \, ds$$

$$\geq \gamma \lambda \int_0^1 g(s) a(s) f(s, u(s)) \, ds$$

$$\geq \gamma \|Au\|.$$

Hence, $A(\overline{K}_R \backslash K_r) \subset K$. □

Lemma 2.3. $A : \overline{K}_R \backslash K_r \to K$ *is completely continuous.*

Proof. Let $u_n, u_0 \in \overline{K}_R \backslash K_r$ and $\|u_n - u_0\| \to 0$ $(n \to \infty)$. For any $\varepsilon > 0$, by (H_2), there exists a natural number $m > 0$ such that

(2.8)
$$\sup_{u \in \overline{K}_R \backslash K_r} \lambda \int_{H(m)} g(s) a(s) f(s, u(s)) ds < \frac{\varepsilon}{4}.$$

Obviously, $r \leq \|u_n\| \leq R$, $r \leq \|u_0\| \leq R$, $u_n(t) \geq \gamma_2 r$, $u_0(t) \geq \gamma_2 r$, $t \in [\frac{1}{m}, \frac{m-1}{m}]$, where $\gamma_2 = \min\{\frac{1}{r} \min_{t \in [\frac{1}{m}, \frac{m-1}{m}]} u_n(t), \gamma\}$, $(n = 0, 1, 2, \cdots)$, then $\gamma_2 r \leq u_n(t) \leq R$, $t \in [\frac{1}{m}, \frac{m-1}{m}]$, $(n = 0, 1, 2, \cdots)$. Since $f(t, u)$ is uniformly continuous in $[\frac{1}{m}, \frac{m-1}{m}] \times [\gamma_2 r, R]$, we have

$$\lim_{n \to +\infty} |f(t, u_n(t)) - f(t, u_0(t))| = 0$$

holds uniformly on $t \in [\frac{1}{m}, \frac{m-1}{m}]$. Then the Lebesgue dominated convergence theorem yields that

$$\lambda \int_{\frac{1}{m}}^{\frac{m-1}{m}} g(s) a(s) \Big| f(s, u_n(s)) - f(s, u_0(s)) \Big| ds \to 0, \quad (n \to \infty).$$

Thus for above $\varepsilon > 0$, there exists a natural number N, for $n > N$, we have

(2.9)
$$\lambda \int_{\frac{1}{m}}^{\frac{m-1}{m}} g(s) a(s) \big| f(s, u_n(s)) - f(s, u_0(s)) \big| ds < \frac{\varepsilon}{2}.$$

It follows from (2.8) and (2.9) that, for $n > N$,

$$\|Au_n - Au_0\| \leq 2 \sup_{u \in \overline{K}_R \backslash K_r} \lambda \int_{H(m)} g(s) a(s) f(s, u_n(s)) ds$$

$$+ \lambda \int_{\frac{1}{m}}^{\frac{m-1}{m}} g(s) a(s) \mid f(s, u_n(s)) - f(s, u_0(s))) \mid ds$$

$$< \frac{\varepsilon}{2} + \frac{\varepsilon}{2} = \varepsilon.$$

Therefore, $A : \overline{K}_R \backslash K_r \to K$ is continuous.

Assume that $B \subset \overline{K}_R \backslash K_r$ is a bounded set, for any $u \in B$, $r \leq \|u\| \leq R$, from (H_2) and discussion above, $A(B)$ is uniform bounded. By the Arzela-Ascoli Theorem we need to show $A(B)$ is equicontinuous.

For any $\varepsilon > 0$, from (H_2) there exists a nature number k such that

$$\sup_{u \in \overline{K}_R \backslash K_r} \lambda \int_{H(k)} g(s)a(s)f(s, u(s))ds < \frac{\varepsilon}{4}.$$

Let $\gamma_3 = \min\{\frac{1}{r} \min_{t \in [\frac{1}{k}, \frac{k-1}{k}]} u(t), \gamma\}$, $M_2 = \max\{f(t, x) : \frac{1}{k} \leq t \leq \frac{k-1}{k}, \gamma_3 r \leq x \leq R\}$. By $G(t, s)$ is uniform continuous on $[0, 1] \times [0, 1]$, there exists $\sigma > 0$ such that

$$|G(t, s) - G(t', s)| \leq \frac{1}{2}\varepsilon(\lambda M_2 L)^{-1}g(s)$$

for $|t - t'| < \sigma$, $t, t' \in [0, 1]$ and $s \in [\frac{1}{k}, \frac{k-1}{k}]$. Hence, for $|t - t'| < \sigma$, $t, t' \in [0, 1]$ and $u \in B$, we have

$$|Au(t) - Au(t')| \leq 2 \sup_{u \in \overline{K}_R \backslash K_r} \lambda \int_{H(k)} g(s)a(s)f(s, u(s))ds$$

$$+ \sup_{u \in \overline{K}_R \backslash K_r} \lambda \int_{\frac{1}{k}}^{\frac{k-1}{k}} |G(t, s) - G(t', s)| \, a(s)f(s, u(s))ds$$

$$< \frac{\varepsilon}{2} + \frac{\varepsilon}{2} = \varepsilon.$$

Therefore, $A(B)$ is equicontinuous. The proof of Lemma 2.3 is completed.   □

3. Existence and Nonexistence of Positive Solution

Theorem 3.1. *Assume that conditions (H_1) and (H_2) hold and $0 \leq f^0 < L^{-1}$, $l^{-1} < f_\infty \leq +\infty$. Then BVP(1.3) has at least one positive solutions for any*

$$(3.1) \qquad \lambda \in \left(\frac{1}{lf_\infty}, \frac{1}{Lf^0} \right).$$

Proof. By (3.1), there exists $\varepsilon_0 > 0$ such that

$$(3.2) \qquad 0 < \frac{1}{(f_\infty - \varepsilon_0)l} \leq \lambda \leq \frac{1}{(f^0 + \varepsilon_0)L}.$$

By $0 \leq f^0 < L^{-1}$, there exists $r_1 > 0$ such that

$$(3.3) \qquad f(t, x) \leq (f^0 + \varepsilon_0)x \leq (f^0 + \varepsilon_0)r_1, \quad 0 \leq t \leq 1, \ 0 < x \leq r_1.$$

Then for $t \in [0, 1]$, $\gamma r_1 \leq u(t) \leq r_1$, by (3.2) and (3.3) we have

$$\|Au\| \leq \lambda \int_0^1 g(s)a(s)f(s, u(s))ds$$

$$\leq (f^0 + \varepsilon_0)r_1\lambda \int_0^1 g(s)a(s)ds \leq r_1 = \|u\|.$$

Thus,

$$(3.4) \qquad \|Au\| \leq \|u\|, \quad u \in \partial K_{r_1}.$$

Next, by $0 < l^{-1} < f_\infty \leq +\infty$, there exists an $\bar{r}_2 > 0$ such that

$$f(t,u) \geq (f_\infty - \varepsilon_0)u, \quad u \geq \bar{r}_2, \ \eta \leq t \leq \delta.$$

Let $r_2 = \max\{2r_1, \frac{\bar{r}_2}{\gamma}\}$, then $u \in \partial K_{r_2}$ implies that $\min_{t \in [\eta,\delta]} u(t) \geq \gamma \|u\| \geq \bar{r}_2$. So for $u \in \partial K_{r_2}$,

$$Au(\eta) = \lambda \int_0^1 |G(\eta,s)|a(s)f(s,u(s))ds$$

$$\geq \lambda(f_\infty - \varepsilon_0) \int_\eta^\delta |G(\eta,s)|a(s)u(s)ds$$

$$\geq \lambda(f_\infty - \varepsilon_0)\gamma^2\|u\| \int_\eta^\delta g(s)a(s)ds$$

$$\geq \|u\|.$$

Thus,

$$(3.5) \qquad\qquad \|Au\| \geq \|u\|, \quad u \in \partial K_{r_2}.$$

It follows from (3.4), (3.5) and Lemma 1.1 that A has a fixed point $u^* \in \overline{K}_{r_2} \backslash K_{r_1}$. That is, u^* is a positive solution of BVP(1.3). $\qquad\square$

Remark 3.1. From the proof of Theorem 3.1, $f(t,u)$ is not required sublinear or superlinear. Theorem 3.1 is true if one of the following conditions satisfied:

(1) If $f_\infty = \infty$, $f^0 > 0$, then for any $\lambda \in (0, \frac{1}{Lf^0})$;

(2) If $f_\infty = \infty$, $f^0 = 0$, then for any $\lambda \in (0, +\infty)$;

(3) If $f_\infty > l^{-1}$, $f^0 = 0$, then for any $\lambda \in \left(\frac{1}{lf_\infty}, +\infty\right)$.

Theorem 3.2. *Assume that conditions* (H_1) *and* (H_2) *hold and* $0 \leq f^\infty < L^{-1}$, $l^{-1} < f_0 \leq +\infty$. *Then BVP (1.3) has at least one positive solutions for any*

$$(3.6) \qquad\qquad \lambda \in \left(\frac{1}{lf_0}, \frac{1}{Lf^\infty}\right).$$

Proof. Suppose that λ satisfies (3.6). There exists an $\varepsilon_0 > 0$ such that $0 < \lambda f^\infty < L^{-1} - \varepsilon_0$, $l^{-1} + \varepsilon_0 < \lambda f_0$. By $0 \leq f^\infty < \frac{1}{\lambda}(L^{-1} - \varepsilon_0)$, there exists an $r_0 > 0$ such that

$$f(t,x) \leq \frac{1}{\lambda}(L^{-1} - \varepsilon_0)x, \quad x \geq r_0, \ 0 \leq t \leq 1.$$

Let $M_0 = \sup_{u \in \partial K_{r_0}} \int_0^1 \lambda g(s)a(s)f(s,u(s))ds$. From (2.5), we know $M_0 < +\infty$. Let $r_3 > \max\{r_0, \frac{1}{\varepsilon_0 L}M_0\}$. For any $u \in \partial K_{r_3}$, set $D(u) = \{t \in (0,1) : u(t) > r_0\}$. For any $t \in D(u)$, we have $r_0 < u(t) \leq \|u\| = r_3$, furthermore, $f(t,u) \leq \frac{1}{\lambda}(L^{-1} - \varepsilon_0)r_3$ for any $t \in D(u)$. Then for any $u \in \partial K_{r_3}$, let $u_1(t) = \min\{u(t), r_0\}$, then $u_1 \in \partial K_{r_0}$. Thus, for any $u \in \partial K_{r_3}$, we have

$$\|Au\| = \max_{t \in [0,1]} \lambda \int_0^1 |G(t,s)| a(s) f(s, u(s)) ds$$

$$\leq \max_{t \in [0,1]} \lambda \int_{D(u)} |G(t,s)| a(s) f(s, u(s)) ds$$

$$+ \lambda \int_{[0,1] \backslash D(u)} g(s) a(s) f(s, u(s)) ds$$

$$\leq \frac{1}{\lambda} (L^{-1} - \varepsilon_0) r_3 \lambda \int_0^1 g(s) a(s) ds$$

$$+ \lambda \int_0^1 g(s) a(s) f(s, u_1(s)) ds$$

$$\leq (L^{-1} - \varepsilon_0) r_3 L + M_0 < r_3 = \|u\|.$$

Hence,

(3.7) $\|Au\| < \|u\|, \quad u \in \partial K_{r_3}.$

Next, by $0 < \frac{1}{\lambda}(l^{-1} + \varepsilon_0) < f_0 \leq +\infty$, there exists $0 < r_4 < r_3$ such that

$$f(t, x) \geq \frac{1}{\lambda}(l^{-1} + \varepsilon_0) x, \quad 0 < x \leq r_4, \ \eta \leq t \leq \delta.$$

Then for $u \in \partial K_{r_4}$,

$$Au(\eta) = \lambda \int_0^1 |G(\eta, s)| a(s) f(s, u(s)) ds$$

$$\geq \frac{1}{\lambda}(l^{-1} + \varepsilon_0) \lambda \int_\eta^\delta |G(\eta, s)| a(s) u(s) ds$$

$$\geq (l^{-1} + \varepsilon) \gamma^2 r_4 \int_\eta^\delta g(s) a(s) ds$$

$$\geq r_4 = \|u\|.$$

Thus,

(3.8) $\|Au\| \geq \|u\|, \quad u \in \partial K_{r_4}.$

Therefore, by Lemma 1.1, A has a fixed point $u^* \in \overline{K}_{r_3} \backslash K_{r_4}$, and so u^* is a positive solution of BVP(1.3).   □

Remark 3.2. From the proof of theorem 3.2, $f(t, u)$ is not required sublinear or superlinear. Theorem 3.2 is true if one of the following conditions satisfied:
  (1) If $f^\infty < L^{-1}$, $f^0 = \infty$, then for any $\lambda \in (0, \frac{1}{Lf^\infty})$;
  (2) If $f^\infty = 0$, $f_0 = \infty$, then for any $\lambda \in (0, +\infty)$;
  (3) If $f^\infty = 0$, $f_0 > l^{-1}$, then for any $\lambda \in (\frac{1}{lf_0}, +\infty)$.

Theorem 3.3. *Assume that (H_1)-(H_2) hold. In addition, suppose that either*

$$\lambda L f(t, x) < x, \quad x \in (0, \infty), \ t \in (0, 1); \quad or \quad \lambda l f(t, x) > x, \quad x \in (0, \infty), \ t \in (\eta, \delta).$$

Then BVP(1.3) *has no positive solutions.*

Proof. Assume that $\lambda L f(t,x) < x$ for any $x \in (0,\infty)$, $t \in (0,1)$ and $u(t)$ is a positive solution of BVP (1.3), then $u(t) > 0$ for any $t \in (0,1)$. So, by (2.4) and (2.3), we have

$$\|u\| \leq \lambda \int_0^1 g(s)a(s)f(s,u(s))ds$$

$$< L^{-1} \int_0^1 g(s)a(s)u(s)ds$$

$$\leq L^{-1}\|u\| \int_0^1 g(s)a(s)ds \leq \|u\|,$$

which is a contradiction. The proof of another case is similar. □

4. Existence of Two Positive Solutions

In this section, we study the existence of at least two positive solutions for BVP(1.3).

Theorem 4.1. *Assume that conditions (H_1) and (H_2) hold. In addition, suppose that $f_0 = f_\infty = \infty$. Then for each $\lambda \in (0, \lambda^*)$, BVP(1.3) has at least two positive solutions, where*

$$\lambda^* = \sup_{m>0} \frac{m}{L \max_{0 \leq t \leq 1,\, \gamma m \leq x \leq m} f(t,x)}.$$

Proof. For any $m > 0$, define

$$h(m) = \frac{m}{L \max_{0 \leq t \leq 1,\, \gamma m \leq x \leq m} f(t,x)}.$$

It is easy to show that $h : (0, \infty) \to (0, \infty)$ is continuous and $\lim\limits_{m \to +0} h(m) = \lim\limits_{m \to \infty} h(m) = 0$. So, there exists $m_0 \in (0, \infty)$ such that $h(m_0) = \sup_{m>0} h(m) = \lambda^*$ and $0 < \lambda^* < +\infty$. For $\lambda \in (0, \lambda^*)$, there exist m_1 and m_2 such that $0 < m_1 < m_0 < m_2$ with $h(m_1) = h(m_2) = \lambda$. Thus

$$(4.1) \qquad f(t,x) \leq \frac{m_1}{\lambda L}, \quad x \in [\gamma m_1, m_1],\ t \in [0,1],$$

$$(4.2) \qquad f(t,x) \leq \frac{m_2}{\lambda L}, \quad u \in [\gamma m_2, m_2],\ t \in [0,1].$$

Next, by $f_0 = f_\infty = \infty$, there exist d_1 and d_2 with $0 < d_1 < m_1 < m_2 < \gamma d_2 < \infty$ such that

$$\frac{f(t,x)}{x} \geq \frac{1}{\lambda l}, \quad x \in (0, d_1] \cup [\gamma d_2, \infty),\ t \in [\eta, \delta].$$

Thus

$$(4.3) \qquad f(t,x) \geq \frac{\gamma d_1}{\lambda l}, \quad x \in [\gamma d_1, d_1],\ t \in [\eta, \delta],$$

$$(4.4) \qquad f(t,x) \geq \frac{\gamma d_2}{\lambda l}, \quad x \in [\gamma d_2, d_2],\ t \in [\eta, \delta].$$

It follows from (4.1) that for any $u \in \partial K_{m_1}$,

$$\|Au\| \leq \lambda \int_0^1 g(s)a(s)f(s,u(s))ds$$

$$\leq \lambda \int_0^1 g(s)a(s)\frac{m_1}{\lambda L}ds \leq \frac{m_1}{L}\int_0^1 g(s)a(s)ds$$

$$= m_1 = \|u\|.$$

Thus,

(4.5) $$\|Au\| \leq \|u\|, \quad u \in \partial K_{m_1}.$$

On the other hand, for any $u \in \partial K_{d_1}$, by (4.3) we have

$$Au(\eta) = \lambda \int_0^1 |G(\eta,s)|a(s)f(s,u(s))ds$$

$$\geq \gamma\lambda \int_\eta^\delta g(s)a(s)f(s,u(s))ds \geq \gamma\lambda \int_\eta^\delta g(s)a(s)\frac{\gamma d_1}{\lambda l}ds$$

$$\geq \frac{\gamma^2 d_1}{l}\int_\eta^\delta g(s)a(s)ds = d_1 = \|u\|.$$

Thus,

(4.6) $$\|Au\| \geq \|u\|, \quad u \in \partial K_{d_1}.$$

Applying Lemma 1.1 to (4.5) and (4.6) yields that A has a fixed point $u^* \in \overline{K}_{m_1}\backslash K_{d_1}$, which is a positive solution of BVP(1.3). Similarly, by (4.2),(4.4) and following the procedure used in the Theorem 3.2 there exists another positive solution $u^{**} \in \overline{K}_{d_2}\backslash K_{m_2}$. This ends the proof of Theorem 4.1.   □

Theorem 4.2. *Assume that conditions* (H_1) *and* (H_2) *hold. In addition, suppose that* $f^0 = f^\infty = 0$. *Then for each* $\lambda \in (\lambda^{**}, \infty)$, *BVP (1.3) has at least two positive solutions, where*

$$\lambda^{**} = \inf_{m>0} \frac{\gamma m}{l \min_{\eta \leq t \leq \delta, \, \gamma m \leq x \leq m} f(t,x)}.$$

Proof. For $m > 0$, define

$$h(m) = \frac{\gamma m}{l \min_{\eta \leq t \leq \delta, \, \gamma m \leq x \leq m} f(t,x)},$$

then $h : (0,\infty) \to (0,\infty)$ is continuous and $\lim_{m\to+0} h(m) = \lim_{m\to\infty} h(m) = \infty$. So there exists $m_0 \in (0,\infty)$ such that $h(m_0) = \inf_{m>0} h(m) = \lambda^{**}$ and $0 < \lambda^{**} < +\infty$. For $\lambda \in (\lambda^{**}, \infty)$, there exist m_1 and m_2 such that $0 < m_1 < m_0 < m_2$ with $h(m_1) = h(m_2) = \lambda$. Thus

(4.7) $$f(t,x) \geq \frac{\gamma m_1}{\lambda l}, \quad x \in [\gamma m_1, m_1], \, t \in [\eta, \delta].$$

(4.8) $$f(t,x) \geq \frac{\gamma m_2}{\lambda l}, \quad x \in [\gamma m_2, m_2], \, t \in [\eta, \delta].$$

Next, since $f^0 = 0$, there exists $c_1 : 0 < c_1 < m_1$ such that

$$\frac{f(t,x)}{x} \leq \frac{1}{\lambda L}, \quad x \in (0, c_1], \ t \in [0,1].$$

Thus,

(4.9) $$f(t,x) \leq \frac{c_1}{\lambda L}, \quad x \in [\gamma c_1, c_1], \ t \in [0,1].$$

By $f^\infty = 0$, there exists $\bar{c}_2 : m_2 < \gamma \bar{c}_2 < \infty$ such that

$$\frac{f(t,x)}{x} \leq \frac{1}{\lambda L}, \quad x \in [\gamma \bar{c}_2, \infty), \ t \in [0,1].$$

Let $M = \sup_{0 \leq t \leq 1, \gamma \bar{c}_2 \leq x \leq \bar{c}_2} f(t,x)$, $c_2 \geq \max\{\lambda M L, 2\bar{c}_2\}$, it is easy to see that

(4.10) $$f(t,x) \leq \frac{c_2}{\lambda L} \quad x \in [\gamma c_2, , c_2], \ t \in [0,1].$$

The rest of the proof is similar to that of Theorem 4.1. $\qquad \square$

5. Infinitely Many Positive Solutions

Theorem 5.1. *Assume that conditions (H_1) and (H_2) hold. In addition, suppose that $f(t,u)$ is nondecreasing in u. Then for each $\lambda \in (\frac{1}{lf_\infty}, \frac{1}{Lf^\infty})$, BVP (1.3) has a row of positive solutions $u_n^* \in K$ such that $\|u_n^*\| \to +\infty$ $(n \to \infty)$.*

Proof. By $\lambda \in (\frac{1}{lf_\infty}, \frac{1}{Lf^\infty})$, we have $f^\infty < \frac{1}{\lambda L}$, $f_\infty > \frac{1}{\lambda l}$. By $f^\infty < \frac{1}{\lambda L}$ and the fact that $f(t,u)$ is nondecreasing in u, there exist positive number sequence $\{a_n\}$, $a_n \to \infty$ such that

(5.1) $$\max_{0 \leq t \leq 1, \ \gamma a_n \leq u \leq a_n} f(t,u) = \max_{0 \leq t \leq 1} f(t, a_n) < \frac{a_n}{\lambda L}.$$

Next, by $f_\infty > \frac{1}{\lambda l}$, there exist positive number sequence $\{b_n\}$, $b_n \to \infty$ such that

(5.2) $$\min_{\eta \leq t \leq \delta, \ \gamma b_n \leq u \leq b_n} f(t,u) = \min_{\eta \leq t \leq \delta} f(t, \gamma b_n) > \frac{\gamma b_n}{\lambda l}.$$

Without loss of generality, we can suppose that

$$a_1 < b_1 < a_2 < b_2 < \cdots < a_n < b_n < \cdots.$$

For each natural number n, for any $u \in \partial K_{a_n}$ and $0 \leq t \leq 1$, which imply that $f(t,u) < \frac{a_n}{\lambda L}$. Then

$$\|Au\| \leq \lambda \int_0^1 g(s)a(s)f(s,u(s))ds$$
$$< \lambda \int_0^1 g(s)a(s)\frac{a_n}{\lambda L}ds = \frac{a_n}{L}\int_0^1 g(s)a(s)ds$$
$$= a_n = \|Au\|.$$

Thus,

(5.3) $$\|Au\| \leq \|u\|, \quad u \in \partial K_{a_n}.$$

For $u \in \partial K_{b_n}$, $\eta \le t \le \delta$, which imply that $\gamma b_n \le u(t) \le b_n$, $f(t, u) > \frac{\gamma b_n}{\lambda l}$. Then

$$Au(\eta) = \lambda \int_0^1 |G(\eta, s)| a(s) f(s, u(s)) ds$$

$$\ge \gamma \lambda \int_\eta^\delta g(s) a(s) f(s, u(s)) ds > \gamma \lambda \int_\eta^\delta g(s) a(s) \frac{\gamma b_n}{\lambda l} ds$$

$$= \frac{\gamma^2 b_n}{l} \int_\eta^\delta g(s) a(s) ds = b_n = \|u\|.$$

Therefore,

(5.4) $$\|Au\| \ge \|u\|, \quad u \in \partial K_{b_n}.$$

Applying (5.3), (5.4) and Lemma 1.1 we know that there exists an positive solution $u_n^* \in \overline{K}_{b_n} \backslash K_{a_n} \subset K$ satisfying $a_n \le \|u_n^*\| \le b_n$. Let $n \to \infty$, we find a row of positive solutions $u_n^* \in K$ and $\|u_n^*\| \to +\infty$. This finishes the proof. □

Remark 5.1. Theorem 5.1 is true if one of the following conditions satisfied:
(1) If $f_\infty = \infty$, $0 < f^\infty < \infty$, then for any $\lambda \in (0, \frac{1}{Lf^\infty})$;
(2) If $f_\infty = \infty$, $f^\infty = 0$, then for any $\lambda \in (0, +\infty)$;
(3) If $0 < f_\infty < \infty$, $f^\infty = 0$, then for any $\lambda \in (\frac{1}{lf_0}, \infty)$.

Similarly, we get the following theorem.

Theorem 5.2. *Assume that conditions* (H_1) *and* (H_2) *hold. In addition, suppose that* $f(t, u)$ *is nondecreasing in* u. *Then for each* $\lambda \in (\frac{1}{lf_0}, \frac{1}{Lf^0})$, *BVP*(1.3) *has a row of positive solutions* $u_n^{**} \in K$ *such that* $\|u_n^{**}\| \to 0$ $(n \to \infty)$.

Remark 5.2. Theorem 5.2 is true if one of the following conditions satisfied:
(1) If $f_0 = \infty$, $0 < f^0 < \infty$, then for any $\lambda \in (0, \frac{1}{Lf^0})$;
(2) If $f_0 = \infty$, $f^0 = 0$, then for any $\lambda \in (0, +\infty)$;
(3) If $0 < f_0 < \infty$, $f^0 = 0$, then for any $\lambda \in (\frac{1}{lf_0}, +\infty)$.

Remark 5.3. We obtain an explicit interval for λ such that for any λ in this interval, the existence of positive solutions to BVP (1.3) is guaranteed. Our results extend and improve the results in [8] and many other known results including singular and non-singular cases.

REFERENCES

1. V. A. Il'in and E. I. Moiseev, *Nonlocal boundary value problem of the first kind for a Sturm-Liouville operator in its differential and finite difference aspects*, Diff. Eqs. **23** (1987), 803–810.

2. V. A. Il'in and E. I. Moiseev, *Nonlocal boundary value problem of the second kind for a Sturm-Liouville operator*, Diff. Eqs. **23** (1987), 979–987.

3. C. P. Gupta, *Solvability of a three-point nonlinear boundary value problem for a second order ordinary differential equation*, J. Math. Anal. Appl. **168** (1992), 540–551.

4. P. W. Eloe and J. Henderson, *Positive solutions for higher order ordinary differential equations*, Electron J. Diff. Eqs. **3** (1985), 1–8.

5. X. Y. Liu, *Nontrivial solutions of singular nonlinear m-point boundary value problems*, J. Math. Anal. Appl. **284** (2003) 576–590.

6. R. Y. Ma, *Positive solutions for a nonlinear three-point boundary value problem,* Electron J. Diff. Eqs. **34** (1998), 1–8.

7. R. Y. Ma and H. Y. Wang, *Positive of nonlinear three-point boundary-value problems,* J. Math. Anal. Appl. **279** (2003), 216–227.

8. P. W. Eloe and B. Ahmad, *Positive solutions of a nonlinear nth order boundary value problem with nonlocal conditions,* Appl. Math. Lett. **18** (2005), 521–527

9. P. W. Eloe, *Maximum principles for a family of nonlocal boundary value problems,* Adv. Diff. Eqs. **3** (2004), 201–210.

10. X. A. Hao, L. S. Liu and Y. H. Wu, *Positive solutions for nonlinear nth-order singular nonlocal boundary value problems,* Boundary Value Problems **2007** (2007), Article ID 74517, 10 pages.

11. D. J. Guo and V. Lakshmikantham, *Nonlinear Problems in Abstract Cones,* Academic Press, New York/San Diego, 1988.

12. D. J. Guo, V. Lakshmikantham and X. Z. Liu, *Nonlinear Integral Equations in Banach Spaces,* Dordrecht, Kluwer Academic Publishers, 1996.

13. K. Deimling, *Nonlinear Functional Analysis,* Springer-Verlag, Berlin, 1985.

14. Z. B. Bai and H. Y. Wang, *On positive solutions of some nonlinear fourth-order beam equations,* J. Math. Anal. Appl. **270** (2002), 357–368.

15. Y. P. Sun, *Positive solutions of nonlinear second-order m-point boundary value problems,* Nonlinear Anal. **61** (2005), 1283–1294.

Nonlinear Functional Analysis and Applications, Volume 1, 133–138
© 2012 Nova Science Publishers, Inc.

AN ELEMENTARY APPROACH TO FIXED POINT INDEX FOR COUNTABLY CONDENSING MAPS

Yu Qing Chen[1], Yeol Je Cho[2,*] and Donal O'Regan[3]

ABSTRACT. We present a simple approach to defining the fixed point index for countably condensing single and multivalued maps based on topological degree in $\mathbf{R}^n$ and Dugundji's extension theorem.

1. Introduction

In this paper, we present a simple approach to defining the fixed point index for countably condensing single and multivalued maps. Of course, this index has been presented in the literature for these maps (see [1, 5]). However, in 1972, Granas [2, pp. 272–287] presented a beautiful and simple approach to defining the fixed point index for compact single valued maps of ANR's. His approach was based on topological degree in $\mathbf{R}^n$, Schauder projections and the Arens-Eells embedding theorem. This elementary argument is also presented in Section 10 and 12 of the recent Granas and Dugundji book [3].

Let X be an ANR and U open in X. Let $C(\overline{U}, X)$ denote the set of continuous compact maps from $\overline{U}$ to X and $C_{\partial U}(\overline{U}, X)$ the set of maps $f \in C(\overline{U}, X)$ with no fixed point on ∂U.

Now, Granas [2] showed that there exists an integer valued (fixed point index) function $f \mapsto i(f)$ for $f \in C_{\partial U}(\overline{U}, X)$ with the following properties:

(I) If $i(f, U) \neq 0$, then $Fix(f) = \{x \in \overline{U} : x = f(x)\} \neq \emptyset$;

Received September 11, 2007.   *Corresponding author.

2000 *Mathematics Subject Classification.* 47H10, 54H25.

Key words and phrases. Fixed point index, countable condensing map, Dugundji's extension theorem.

[1]Department of Mathematics, Foshan University, Foshan, 528000, Guangdong, People's Republic of China (*E-mail*: ychen64@163.com)

[2]Department of Mathematics Education and the Research Institute of Natural Sciences, Gyeongsang National University, Jinju 660-701, Korea (*E-mail*: yjcho@gnu.ac.kr)

[3]Department of Mathematics, National University of Ireland, Galway, Ireland (*E-mail*: donal.oregan@nuigalway.ie)

(II) If $h_t : \overline{U} \to X$ is a continuous compact homotopy in $C_{\partial U}(\overline{U}, X)$ (i.e., a map $h : \overline{U} \times [0,1] \to X$ is defined by $h(x,t) = h_t(x)$ for all $x \in \overline{U} \times [0,1]$ is continuous and compact and $h_t \in C_{\partial U}(\overline{U}, X)$ for each $t \in [0,1]$), then $i(h_0, U) = i(h_1, U)$.

Of course, we also have additivity, normalization, multiplicity, commutativity anfd others. In Section 2, we extend this index to continuous countably condensing maps. The advantage of this approach is that the argument is elementary, so everything follows from the above and Dugundji's extension theorem.

Let C be a closed convex set in a normed linear space E and let $U \subseteq C$ be open in C. We say that $F \in K(\overline{U}, C)$ if $F : \overline{U} \to CC(C)$ is a upper semicontinuous compact map; here $CC(C)$ denotes the family of nonempty convex compact subsets of C. Let $K_{\partial U}(\overline{U}, C)$ denote the set of maps $F \in K(\overline{U}, C)$ which are fixed point free on ∂U. In [3, pp. 337], if $F \in K_{\partial U}(\overline{U}, C)$, we can select a map $f \in C_{\partial U}(\overline{U}, C)$ satisfying $[f] = [F]$ (here $[f]$ (respectively, $[F]$) denotes the equivalence class of f (respectively F) in $C_{\partial U}(\overline{U}, C)$ (respectively, $K_{\partial U}(\overline{U}, C)$)) and we define

$$index(F, U) = i(f, U).$$

Again we have the following properties:

(I) If $index(F, U) \neq 0$, then $Fix(F) = \{x \in \overline{U} : x \in F(x)\} \neq \emptyset$;

(II) If $H_t : \overline{U} \to CC(C)$ is a upper semicontinuous compact homotopy in $K_{\partial U}(\overline{U}, C)$ (i.e., $H : \overline{U} \times [0,1] \to CC(C)$ is defined by $H(x,t) = H_t(x)$ for all $x \in \overline{U} \times [0,1]$ is upper semicontinuous and compact and $H_t \in K_{\partial U}(\overline{U}, C)$ for each $t \in [0,1]$), then

$$index(H_0, U) = index(H_1, U).$$

In Section 3, we discuss briefly (since the arguments are the same as in Section 2) the fixed point index for countably condensing Kakutani maps. The theory will rely on the following generalization of Dugundji's extension theorem [4, pp. 7].

Theorem 1.1. *Let A be a closed subset of a metrizable space X, E a locally convex space and $F : A \to CC(E)$ a upper semicontinuous map. Then F has a upper semicontinuous extension $G : X \to CC(E)$ with $G(X) \subseteq co(F(A))$.*

2. Single-valued Maps

Let E be a Banach space, C a closed convex subset of E and U an open bounded subset of C. We say that $T \in KC(\overline{U}, C)$ if $T : \overline{U} \to C$ is a continuous countably condensing map (recall a map $T : \overline{U} \to C$ is said to be countably condensing if $\alpha(T(X)) \leq \alpha(X)$ for each countably bounded subset X of $\overline{U}$ and $\alpha(T(Y)) < \alpha(Y)$ for each countably bounded noncompact subset Y of $\overline{U}$, here α denotes the Kuratowski measure of noncompactness). We say that $T \in KC_{\partial U}(\overline{U}, C)$ if $T \in KC(\overline{U}, C)$ and $x \neq T(x)$ for $x \in \partial U$ (here ∂U denotes the boundary of U in C).

Let E, C and U be as above and let $T \in KC_{\partial U}(\overline{U}, C)$. Define

$$A_1 = \overline{co}(T(\overline{U})), \quad A_n = \overline{co}(T(\overline{U} \cap A_{n-1}))$$

for $n \in \{2, 3, \cdots\}$ and

$$A_\infty = \cap_{n=1}^\infty A_n.$$

Of course, A_∞ is convex. Also [5, Theorem 2.2] guarantees that A_∞ is compact. Notice $T : \overline{U} \cap A_\infty \to A_\infty$ since $T(\overline{U} \cap A_\infty) \subseteq T(\overline{U} \cap A_{n-1}) \subseteq A_n$ for $n \in \{2, 3, \cdots\}$.

If $\overline{U} \cap A_\infty = \emptyset$, we let

$$ind(T, U) = 0.$$

Next, consider the case when $\overline{U} \cap A_\infty \neq \emptyset$. Using Dugundji's extension theorem [3, pp. 163], we see that $T : \overline{U} \cap A_\infty \to A_\infty$ extends to a continuous map $\tilde{T} : \overline{U} \to A_\infty$, so $\tilde{T}$ is compact. It is also immediate that $Fix(\tilde{T}) = Fix(T) \subseteq A_\infty$. We let

$$ind(T, U) = i(\tilde{T}, U),$$

where i is defined in Section 1.

We claim the definition does not depend on $\tilde{T}$. Suppose $\tilde{T}_1 : \overline{U} \to A_\infty$ is another continuous extension of $T : \overline{U} \cap A_\infty \to A_\infty$. Let $H_t : \overline{U} \to A_\infty$ (note A_∞ is convex) be a map defined by

$$H_t(x) = (1 - t)\tilde{T}(x) + t\tilde{T}_1(x).$$

Clearly, H_t is a continuous compact homotopy. Now, suppose $x = H_t(x)$ for some $x \in \overline{U}$ and $t \in [0, 1]$. Then, since $H_t(x) \in A_\infty$, we have $x \in A_\infty \cap \overline{U}$, so

$$x = (1 - t)\tilde{T}(x) + t\tilde{T}_1(x) = (1 - t)T(x) + tT_1(x) = T(x).$$

Consequently,

$$Fix(H_t) = Fix(T) \quad \text{for all } t \in [0, 1].$$

Now, (II) in Section 1 implies

$$i(H_0, U) = i(H_1, U),$$

i.e.,

$$i(\tilde{T}, U) = i(\tilde{T}_1, U),$$

so our index is well defined.

Remark 2.1. Note E Banach could be replaced by E Fréchet provided T countably condensing is replaced by T countably P–concentrative [1] or countably condensing in the sense of Vath [5, pp. 353, 356].

Property 2.1. *Suppose that A is a compact convex set, $A_\infty \subseteq A$ with $T(\overline{U} \cap A) \subseteq A$ and $A \cap \overline{U} \neq \emptyset$. Then, by Dugundji's extension theorem, we can extend $T : \overline{U} \cap A \to A$ to a continuous map $T^\star : \overline{U} \to A$, so $T^\star$ is compact. We claim*

(2.1) $$ind(T, U) = i(T^\star, U).$$

Remark 2.2. If $A \cap \overline{U} = \emptyset$, then $A_\infty \cap \overline{U} = \emptyset$, so trivially (2.1) is true.

To show (2.1), we need to consider two cases, namely, $A_\infty \cap \overline{U} = \emptyset$ and $A_\infty \cap \overline{U} \neq \emptyset$.

Case (i) $A_\infty \cap \overline{U} = \emptyset$.

Now, $int(T, U) = 0$. If $i(T^\star, U) \neq 0$, then $Fix(T^\star) \neq \emptyset$ by (I) of Section 1. Suppose $x = T^\star(x)$ for $x \in \overline{U}$. Then $x \in \overline{U}$ and $x \in A$, so $x = Tx$ since $T^\star = T$ on $\overline{U} \cap A$. As a result, $x \in A_\infty$, so $A_\infty \cap \overline{U} \neq \emptyset$, which is a contradiction. Thus $i(T^\star, U) = 0$.

Case (ii) $A_\infty \cap \overline{U} \neq \emptyset$.

Let $G_t : \overline{U} \to A$ (note A is convex) be a map defined by

$$G_t(x) = (1-t)\tilde{T}(x) + tT^\star(x).$$

Clearly, G_t is a continuous compact homotopy. Suppose $x = G_t(x)$ for some $x \in \overline{U}$ and $t \in [0,1]$. Then $G_t(x) \in A$, so $x \in \overline{U} \cap A$. Thus $T^\star(x) = T(x)$, so

$$G_t(x) = (1-t)\tilde{T}(x) + tT(x).$$

As a result, $x \in co(\{T(x)\} \cup A_\infty)$, which implies $x \in A_\infty$ (note trivially $T(x) \in \overline{co}(T(\overline{U})) = A_1$, so $x \in co(A_1 \cup A_\infty) = A_1$ and, so then $T(x) \in \overline{co}(T(\overline{U} \cap A_1)) = A_2$, so $x \in co(A_2 \cup A_\infty) = A_2$ etc.). Thus $x = (1-t)T(x) + tT(x) = T(x)$, so $Fix(G_t) = Fix(T)$. Now, (II) in Section 1 implies $i(G_0, U) = i(G_1, U)$, i.e., $i(\tilde{T}, U) = i(T^\star, U)$, so (2.1) holds.

Existence Property. *Let E, C, U be as above and $T \in KC_{\partial U}(\overline{U}, C)$. Then*

$$ind(T, U) \neq 0 \Rightarrow T \quad \text{has a fixed point in } U.$$

Proof. Now, $ind(T, U) \neq 0$ implies $\overline{U} \cap A_\infty \neq \emptyset$, so $i(\tilde{T}, U) \neq 0$. Now, (I) of Section 1 implies that there exists $x \in \overline{U}$ with $x = \tilde{T}(x)$. This immediately implies $x \in A_\infty$, so $x = T(x)$ since $\tilde{T} = T$ on $\overline{U} \cap A_\infty$. The result follows since T is fixed point free on ∂U. $\square$

Homotopy Property. *Let E, C, U be as above with $H : [0,1] \times \overline{U} \to C$ a continuous countably condensing map together with $H_t(.) = H(t,.) : \overline{U} \to C$ belonging to $KC_{\partial U}(\overline{U}, C)$ for each $t \in [0,1]$. Then*

(2.2) $$ind(H_0, U) = ind(H_1, U).$$

Proof. Let

$$A_1(H) = \overline{co}(H([0,1] \times \overline{U})), A_n(H) = \overline{co}(H([0,1] \times (\overline{U} \cap A_{n-1}(H))))$$

for $n \in \{2, 3, \cdots\}$ and

$$A_\infty(H) = \cap_{n=1}^\infty A_n(H).$$

We consider two cases, namely, $A_\infty(H) \cap \overline{U} = \emptyset$ and $A_\infty(H) \cap \overline{U} \neq \emptyset$.

Case (i): $A_\infty(H) \cap \overline{U} \neq \emptyset$.

For each $i \in [0,1]$, let

$$A_1(H_i) = \overline{co}(H_i(\overline{U})), \quad A_n(H_i) = \overline{co}(H_i(\overline{U} \cap A_{n-1}(H_i)))$$

for $n \in \{2, 3, \cdots\}$ and

$$A_\infty(H_i) = \cap_{n=1}^\infty A_n(H_i).$$

Notice that $A_\infty(H_i) \subseteq A_\infty(H)$ for each $i \in [0,1]$. Also, it is easy to see that $H_i : \overline{U} \cap A_\infty(H) \to A_\infty(H)$ for each $i \in [0,1]$. Using Dugundji's extension theorem, we see that $H_i : \overline{U} \cap A_\infty(H) \to A_\infty(H)$ extends to a continuous map $\tilde{H}_i : \overline{U} \to A_\infty(H)$ (so $\tilde{H}_i$ is compact) for each $i \in [0,1]$. This yields a continuous compact homotopy $\tilde{H} : [0,1] \times \overline{U} \to A_\infty(H)$ given by $\tilde{H} = \{\tilde{H}_i : i \in [0,1]\}$.

Note also that $Fix(\tilde{H}) = Fix(H)$.

Subcase (a): Assume $\overline{U} \cap A_\infty(H_i) \neq \emptyset$ for $i \in \{0,1\}$. Then, by Property 2.1,

$$ind(H_0, U) = i(\tilde{H}_0, U)$$

and

$$ind(H_1, U) = i(\tilde{H}_1, U).$$

Now, (II) in Section 1 implies

$$i(\tilde{H}_0, U) = i(\tilde{H}_1, U),$$

so (2.2) holds.

Subcase (b): Assume $\overline{U} \cap A_\infty(H_i) = \emptyset$ for $i \in \{0, 1\}$. Then (2.2) is trivially true since both sides are zero.

Subcase (c) Assume $\overline{U} \cap A_\infty(H_0) = \emptyset$ and $\overline{U} \cap A_\infty(H_1) \neq \emptyset$ (the other case $\overline{U} \cap A_\infty(H_0) \neq \emptyset$ and $\overline{U} \cap A_\infty(H_1) = \emptyset$ is similar, so omitted).

Suppose $ind(H_1, U) \neq 0$. Then, by Property 2.1,

$$i(\tilde{H}_1, U) = ind(H_1, U) \neq 0.$$

Also, (II) in Section 1 implies $i(\tilde{H}_1, U) = i(\tilde{H}_0, U)$, so $i(\tilde{H}_0, U) \neq 0$. Now, (I) in Section 1 implies there exists $x \in \overline{U}$ with $x = \tilde{H}_0(x)$. Thus $x \in \overline{U}$ and $x \in A_\infty(H)$, so $x = H_0(x)$ since $\tilde{H} = H$ on $\overline{U} \cap A_\infty(H)$. As a result, $x \in A_\infty(H_0)$, so $\overline{U} \cap A_\infty(H_0) \neq \emptyset$, which is a contradiction. Consequently, $ind(H_1, U) = 0$, so (2.2) holds.

Case (ii): $A_\infty(H) \cap \overline{U} = \emptyset$. Then $A_\infty(H_i) \cap \overline{U} = \emptyset$ for $i \in \{0, 1\}$, so (2.2) holds.   □

3. Multi-valued Maps

Let E be a Banach space, C a closed convex subset of E and U an open bounded subset of C. We say that $T \in KK(\overline{U}, C)$ if $T : \overline{U} \to CC(C)$ is a upper semicontinuous countably condensing map. We say that $T \in KK_{\partial U}(\overline{U}, C)$ if $T \in KK(\overline{U}, C)$ and $x \notin T(x)$ for $x \in \partial U$. Let

$$A_1 = \overline{co}(T(\overline{U})), \quad A_n = \overline{co}(T(\overline{U} \cap A_{n-1}))$$

for $n \in \{2, 3, \cdots\}$ and

$$A_\infty = \cap_{n=1}^{\infty} A_n.$$

If $\overline{U} \cap A_\infty = \emptyset$, we let

$$ind(T, U) = 0.$$

Next, consider the case when $\overline{U} \cap A_\infty \neq \emptyset$. Using Theorem 1.1, we see that $T : \overline{U} \cap A_\infty \to CC(A_\infty)$ extends to a upper semicontinuous map $\tilde{T} : \overline{U} \to CC(A_\infty)$, so $\tilde{T}$ is compact. It is also immediate that $Fix(\tilde{T}) = Fix(T) \subseteq A_\infty$. We let

$$ind(T, U) = index(\tilde{T}, U),$$

where *index* is defined in Section 1.

Notice that the definition does not depend on $\tilde{T}$. Suppose that $\tilde{T}_1 : \overline{U} \to CC(A_\infty)$ is another upper semicontinuous extension of $T : \overline{U} \cap A_\infty \to CC(A_\infty)$. Let $H_t : \overline{U} \to CC(A_\infty)$ be a map defined by

$$H_t(x) = (1 - t)\tilde{T}(x) + t\tilde{T}_1(x).$$

Clearly, H_t is a upper semicontinuous compact homotopy and as in Section 2, $Fix(H_t) = Fix(T)$ for all $t \in [0, 1]$ with $index(H_0, U) = index(H_1, U)$. Thus $index(\tilde{T}, U) = index(\tilde{T}_1, U)$.

Remark 3.1. Note that a Banach space E could be replaced by a Fréchet space E provided T countably condensing is replaced by T countably P-concentrative [1] or countably condensing in the sense of Vath [5, pp. 353, 356].

Property 3.1. *Suppose that A is a compact convex set, $A_\infty \subseteq A$ with $T(\overline{U} \cap A) \subseteq A$ and $A \cap \overline{U} \neq \emptyset$. Then, by Theorem 1.1, we can extend $T : \overline{U} \cap A \to CC(A)$ to a upper semicontinuous map $T^\star : \overline{U} \to CC(A)$, so $T^\star$ is compact. Essentially, the same reasoning as in Section 2 yields*

$$ind(T, U) = index(T^\star, U).$$

Essentially, the same reasoning as in Section 2 also yields the following:

Existence Property. *Let E, C, U be as above and $T \in KK_{\partial U}(\overline{U}, C)$. Then*

$$ind(T, U) \neq 0 \Rightarrow T \quad has \ a \ fixed \ point \ in \ U.$$

Homotopy Property. *Let E, C, U be as above with $H : [0,1] \times \overline{U} \to CC(C)$ a upper semicontinuous countably condensing map together with $H_t(.) = H(t,.) : \overline{U} \to CC(C)$ belonging to $KK_{\partial U}(\overline{U}, C)$ for each $t \in [0,1]$. Then*

$$ind(H_0, U) = ind(H_1, U).$$

REFERENCES

1. R. P. Agarwal and D. O'Regan, *An index theory for countably P-concentrative J maps*, Appl. Math. Letts. **16**(2003), 1265–1271.

2. A. Granas, *The Leray-Schauder index and the fixed point theory for arbitrary ANR's*, Bull. Soc. Math. France **100**(1972), 209–228.

3. A. Granas and J. Dugundji, *Fixed Point Theory*, Springer Verlag, New York, 2003.

4. T. W. Ma, *Topological degrees of set-valued compact fields in locally convex spaces*, Dissert. Math. **92** (1972), 1–43.

5. M. Vath, *Fixed point theorems and fixed point index for countably condensing maps*, Topological Methods in Nonlinear Analysis **13**(1999), 341–363.

Nonlinear Functional Analysis and Applications, Volume 1, 139–145

STRONG CONVERGENCE THEOREMS OF
THE CQ METHOD FOR SOLUTION TO
VARIATIONAL INEQUALITIES IN HILBERT SPACE

RUDONG CHEN[1,*] AND HUIMIN HE[2]

ABSTRACT. In this paper, we show that strongly convergence theorems of the CQ method for nonlinear variational inequality by projection methods in Hilbert spaces. The results presented in this paper extend and improve the corresponding results of Ram U. Verma[General convergence analysis for two-step projection methods and applications to variational problems, Appl. Math. Lett. 18 (2005), 1286–1292]and some others.

1. Introduction and Preliminaries

Throughout this paper, let H be a real Hilbert space with inner product $\langle \cdot, \cdot \rangle$ and norm $\| \cdot \|$. We write $x_n \rightharpoonup x$ to indicate that the sequence $\{x_n\}$ converges weakly to x. Similarly, $x_n \to x$ will symbolize strong convergence. we denote by N the sets of nonnegative integers, respectively. Let K be a closed convex subset of a Hilbert space H, and let $T : K \to H$ be a any mapping on K.

Considering the following nonlinear variational inequality: determine elements $x^*, y^* \in K$ such that

$$(1.1) \quad \begin{aligned} \langle \rho T(y^*) + x^* - y^*, x - x^* \rangle \geq 0, & \quad \forall x \in K, \, \rho > 0, \\ \langle \eta T(x^*) + y^* - x^*, x - y^* \rangle \geq 0, & \quad \forall x \in K, \, \eta > 0. \end{aligned}$$

The (1.1) is equivalent to the following projection formulas

$$(1.2) \quad \begin{aligned} x^* = P_K[y^* - \rho T(y^*)], & \quad \forall \rho > 0, \\ y^* = P_K[x^* - \eta T(x^*)], & \quad \forall \eta > 0. \end{aligned}$$

Received September 15, 2007.   * Corresponding author.

2000 *Mathematics Subject Classification.* 47J20, 49J40.

Key words and phrases. Variational inequalities, nonexpansive mapping, strong convergence, metric projection.

This work is supported by the National Science Foundation of China, Grant 10771050.

[1] Department of Mathematics, Tianjin Polytechnic University, Tianjin 300160, People's Republic of China (*E-mails:* chenrd@tjpu.edu.cn(R. Chen) and hehuimin20012000@yahoo.com.cn(H. Min)

where P_K is the projection of H onto K.

We note that for $\eta = 0$ the (1.1) reduces to the following problem: determine an element $x^* \in K$ such that

(1.3) $\langle T(x^*), x - x^* \rangle \geq 0, \quad \forall x \in K.$

The (1.3) is equivalent to the following projection formula

(1.4) $x^* = P_K[x^* - \rho T(x^*)], \quad \forall \rho > 0.$

We use F to denote the solution set of the nonlinear variational inequality(1.3) or (1.4); i.e., $F = \{x^* \in K : x^* = P_K[x^* - \rho T(x^*)]\}$

Projection type methods have played a significant role in the numerical resolution of variational inequalities based on their convergence analysis. However, the convergence analysis does require some sort of strong monotonicity besides the Lipschitz continuity. There have been some recent developments where convergence analysis for projection type methods under somewhat weak conditions such as cocoercivity [2,5] and partial relaxed monotonicity[8] is achieved. Recently, the author [7] introduced a two-step model for nonlinear variational inequalities and discussed the approximation solvability of this model based on the convergence analysis of a two-type projection method in a Hilbert space setting. The two-type projection methods contain several known as well as new projection methods as special cases,while some have have been applied to problems arising,especially from complementarity, computational mathematics, convex quadratic programming, and other variational problems. In 2005 R. U.Verma[9] introduced algorithms as follows:

(1.5)
$$x_{n+1} = (1 - \alpha_n)x_n + \alpha_n P_k[y_n - \rho T(y_n)],$$
$$y_n = (1 - \beta_n)x_n + \beta_n P_k[x_n - \eta T(x_n)],$$

where $0 \leq \alpha_n, \beta_n \leq 1$ and $\sum_{n=0}^{\infty} \alpha_n \beta_n = \infty$, Then the sequence $\{x_n\}$ and $\{y_n\}$, respectively, converge to x^* and y^*. And for $\eta = 0$ and $\beta_n = 1$ in (1.5), he arrived at

(1.6) $x_{n+1} = (1 - \alpha_n)x_n + \alpha_n P_k[x_n - \rho T(x_n)]$

where $0 \leq \alpha_n \leq 1$, Then the sequence $\{x_n\}$ converge to $x^* \in K$.

In this paper, we consider another algorithm for nonlinear variational inequality on K as follows:

(1.7)
$$\begin{cases} x_0 = x \in K, \\ y_n = \alpha_n x_n + (1 - \alpha_n)P_K[x_n - \rho T(x_n)], \\ C_n = \{z \in K; \|y_n - z\| \leq \|x_n - z\|\}, \\ Q_n = \{z \in K; \langle x_n - z, x_0 - x_n \rangle \geq 0\}, \\ x_{n+1} = P_{C_n \cap Q_n}(x_0) \end{cases}$$

for each $n \in N$, where $\alpha_n \in [0, a]$ for some $a \in [0, 1)$, then $\{x_n\}$ converges strongly to $z_0 \in F$ for $\rho > 0$ and $z_0 = P_F(x_0)$ i.e. z_0 satisfying the following variational inequality:

$$\langle z_0 - x_0, z_0 - z \rangle \leq 0 \quad \forall z \in F.$$

In the sequel, we shall need the following definitions and results.

Definition 1.1. A Banach space E is said to satisfy *Opial's condition* [1], if whenever $\{x_n\}$ is a sequence in E which converge weakly to x, as $n \to \infty$, then

$$\limsup_{n\to\infty} \|x_n - x\| < \limsup_{n\to\infty} \|x_n - y\|, \quad \forall y \in E \text{ with } x \neq y.$$

It is well known that Hilbert space and $l^p (1 < l < \infty)$ space satisfy Opial's condition[3].

Lemma 1.2. ([4]) *Let K be a nonempty closed convex subset of a Hilbert space H. Given $x \in H$ and $y \in K$. Then $y = P_K x$ if and only if there satisfies $\langle x - y, y - z \rangle \geq 0$, $\forall z \in K$.*

Lemma 1.3. ([4,6]) *Every Hilbert space H has Radon-Riesz Property or Kadets-Klee Property, i.e., for a sequence $\{x_n\} \subset H$ with $x_n \rightharpoonup x$ and $\|x_n\| \to \|x\|$, then there holds $x_n \to x$.*

Lemma 1.4. ([6]) *Let K be a nonempty closed convex subset of a Hilbert space H. and P_K the projection of H onto K. Then $P_K(x) : H \to K$ is nonexpansive mapping.*

A mapping $T : K \to H$ is called *monotonic* if for each $x, y \in K$, we have

$$\langle T(x) - T(y), x - y \rangle \geq 0.$$

A mapping $T : K \to H$ is called *r-strongly monotonic* if for each $x, y \in K$, we have

$$\langle T(x) - T(y), x - y \rangle \geq r \|x - y\|^2 \quad \text{for a constant } r > 0.$$

This implies that

$$\|T(x) - T(y)\| \geq r \|x - y\|,$$

that is, T is r-expansive, and when $r = 1$, it is expansive.

The mapping T is called *μ-Lipschitz continuous* (or *Lipschizian*) if there exists a constant $\mu \geq 0$ such that

$$\|T(x) - T(y)\| \leq \mu \|x - y\|, \quad \forall x, y \in K.$$

2. Main Results

Lemma 2.1. *Let K be a closed convex subset of a Hilbert space H. Let $T : K \to H$ be strongly r-monotonic and μ-Lipschitz continuous such that $F \neq \emptyset$, and the sequence $\{x_n\}$ generated by (1.7). where $\alpha_n \in [0, a]$ for some $a \in [0, 1)$ and $0 < \rho < 2r/\mu^2$, Then $\{x_n\}$ is well defined and $F \subset C_n \cap Q_n$ for every $n \in N$.*

Proof. It is obvious that C_n is closed and Q_n is closed and convex for every $n \in N$. It follows from that C_n is convex for every $n \in N$ because $\|y_n - z\| \leq \|x_n - z\|$ is equivalent to

$$\|y_n - x_n\|^2 + 2\langle y_n - x_n, x_n - z \rangle \leq 0.$$

So, $C_n \cap Q_n$ is closed and convex for every $n \in N$. Let $u \in F$. Then from

$$\|y_n - u\| = \|\alpha_n x_n + (1 - \alpha_n) P_K[x_n - \rho T(x_n)] - u\|$$
$$\leq \alpha_n \|x_n - u\| + (1 - \alpha_n) \|P_K[x_n - \rho T(x_n)] - P_K[u - \rho T(u)]\|$$
$$\leq \alpha_n \|x_n - u\| + (1 - \alpha_n) \|x_n - u - \rho[T(x_n) - T(u)]\|$$

Since T be strongly r-monotonic and μ-Lipschitz continuous, we have

(2.1)
$$
\begin{aligned}
&\|x_n - u - \rho[T(x_n) - T(u)]\|^2 \\
&= \|x_n - u\|^2 - 2\rho\langle T(x_n) - T(u), x_n - u\rangle + \rho^2 \|T(x_n) - T(u)\|^2 \\
&\leq \|x_n - u\|^2 - 2\rho r \|x_n - u\|^2 + (\rho^2\mu^2) \|x_n - u\|^2 \\
&= [1 - 2\rho r + \rho^2\mu^2] \|x_n - u\|^2 .
\end{aligned}
$$

As a result, we have

$$
\begin{aligned}
\|y_n - u\| &\leq \alpha_n \|x_n - u\| + (1 - \alpha_n)\theta \|x_n - u\| \\
&\leq \alpha_n \|x_n - u\| + (1 - \alpha_n) \|x_n - u\| \\
&= \|x_n - u\| .
\end{aligned}
$$

where $\theta = [1 - 2\rho r + \rho^2\mu^2]^{1/2} < 1$. Hence we have $u \in C_n$ for each $n \in N$. So, we have $F \subset C_n$ for all $n \in N$.

Next, we show by mathematical induction that $\{x_n\}$ is well defined and $F \subset C_n \cap Q_n$ for every $n \in N$. For $n = 0$, we have $x_0 = x \in K$ and $Q_0 = K$, and hence $F \subset C_0 \cap Q_0$. Suppose that x_k is given and $F \subset C_k \cap Q_k$ for some $k \in N$. There exists a unique element $x_{k+1} \in C_k \cap Q_k$ such that $x_{k+1} = P_{C_k \cap Q_k}(x_0)$. From $x_{k+1} = P_{C_k \cap Q_k}(x_0)$ there holds

$$
\langle x_{k+1} - z, x_0 - x_{k+1}\rangle \geq 0
$$

for each $z \in C_k \cap Q_k$. Since $F \subset C_k \cap Q_k$, we get $F \subset Q_{k+1}$, Therefore we have $F \subset C_{k+1} \cap Q_{k+1}$. The proof is completed. $\qquad \square$

Lemma 2.2. *Let K be a closed convex subset of a Hilbert space H. Let $T : K \to H$ be strongly r-monotonic and μ-Lipschitz continuous such that $F \neq \emptyset$, and the sequence $\{x_n\}$ generated by (1.7). where $\alpha_n \in [0, a]$ for some $a \in [0, 1)$ and $0 < \rho < 2r/\mu^2$, Then $\{x_n\}$ is bounded.*

Proof. Now, we show F is a closed convex subset of K. Firstly, we claim that F is closed. In fact, if $\{p_n\} \subset F, n \in N$, such that $\lim_{n\to\infty} p_n = p$, then we have

$$
\begin{aligned}
P_K[p - \rho T(p)] &= \lim_{n\to\infty} P_K[p_n - \rho T(p_n)] \\
&= \lim_{n\to\infty} p_n = p
\end{aligned}
$$

Thus $p \in F$.

Secondly, we show that F is convex, we shall use the following identity in Hilbert space.

(2.2)
$$
\|tx + (1 - t)y\|^2 = t \|x\|^2 + (1 - t) \|y\|^2 - t(1 - t) \|x - y\|^2 ,
$$

which holds $\forall x, y \in H$ and $\forall t \in [0,1]$ indeed,

$$\|tx + (1-t)y\|^2$$
$$= t^2 \|x\|^2 + (1-t)^2 \|y\|^2 + 2t(1-t)\langle x, y\rangle$$
$$= t \|x\|^2 + (1-t) \|y\|^2 + 2t(1-t)\langle x, y\rangle - t(1-t) \|x\|^2 - t(1-t) \|y\|^2$$
$$= t \|x\|^2 + (1-t) \|y\|^2 - t(1-t)(\|x\|^2 + \|y\|^2 - 2\langle x, y\rangle)$$
$$= t \|x\|^2 + (1-t) \|y\|^2 - t(1-t) \|x-y\|^2.$$

Let $p_1, p_2 \in F$ and $\forall t \in [0,1]$, $p = tp_1 + (1-t)p_2$, then

$$(2.3) \qquad p - p_1 = (1-t)(p_2 - p_1), \ \ p - p_2 = t(p_1 - p_2).$$

From (2.1), (2.2) and (2.3), we have

$$\|p - P_K[p - \rho T(p)]\|^2$$
$$= \|t(p_1 - P_K[p - \rho T(p)]) + (1-t)(p_2 - P_K[p - \rho T(p)])\|^2$$
$$= t \|p_1 - P_K[p - \rho T(p)]\|^2 + (1-t) \|p_2 - P_K[p - \rho T(p)]\|^2$$
$$\quad - t(1-t) \|p_1 - p_2\|^2$$
$$\leq t \|p_1 - p - \rho[T(p_1) - T(p)]\|^2 + (1-t) \|p_2 - p - \rho[T(p_2) - T(p)]\|^2$$
$$\quad - t(1-t) \|p_1 - p_2\|^2$$
$$\leq t\theta \|p_1 - p\|^2 + (1-t)\theta \|p_2 - p\|^2 - t(1-t) \|p_1 - p_2\|^2$$
$$\leq t \|p_1 - p\|^2 + (1-t) \|p_2 - p\|^2 - t(1-t) \|p_1 - p_2\|^2$$
$$= t(1-t)^2 \|p_1 - p_2\|^2 + t^2(1-t) \|p_1 - p_2\|^2 - t(1-t) \|p_1 - p_2\|^2$$
$$= t(1-t)(1-t+t-1) \|p_1 - p_2\|^2$$
$$= 0$$

Thus $p = P_K[p - \rho T(p)]$, i.e., $p \in F$.

Next, we show $\{x_n\}$ is bounded, Since F is a nonempty closed convex subset of K, there exists an unique element $z_0 \in F$ such that $z_0 = P_F(x_0)$. From $x_{n+1} = P_{C_n \cap Q_n}(x_0)$, we have

$$\|x_{n+1} - x_0\| \leq \|z - x_0\|, \quad \forall z \in C_n \cap Q_n.$$

It follows from lemma2.1 that $F \subset C_n \cap Q_n$ for every $n \in N$, together with $z_0 \in F$, we have

$$(2.4) \qquad \|x_{n+1} - x_0\| \leq \|z_0 - x_0\|, \quad \forall n \in N.$$

This implies that $\{x_n\}$ is bounded. This proof is completed. $\qquad \square$

Lemma 2.3. *Let K be a closed convex subset of a Hilbert space H. Let $T : K \to H$ be strongly r-monotonic and μ-Lipschitz continuous such that $F \neq \emptyset$, and the sequence $\{x_n\}$ generated by (1.7). where $\alpha_n \in [0, a]$ for some $a \in [0, 1)$ and $0 < \rho < 2r/\mu^2$, Then $\lim_{n \to \infty} \|x_{n+1} - x_n\| = 0.$*

Proof. Since $Q_n = \{z \in C; \langle x_n - z, x_0 - x_n \rangle \geq 0\}$, $x_n = P_{Q_n}(x_0)$. As $x_{n+1} \in C_n \cap Q_n \subset Q_n$, we obtain

$$\|x_{n+1} - x_0\| \geq \|x_n - x_0\|, \quad \forall z \in C_n \cap Q_n.$$

Therefore, by Lemma 2.2 the sequence $\{\|x_n - x_0\|\}$ is bounded and nondecreasing. So $\lim_{n \to \infty} \|x_n - x_0\|$ exists. On the other hand, from $x_{n+1} \in Q_n$, we get $\langle x_n - x_{n+1}, x_0 - x_n \rangle \geq 0$ and hence

$$\|x_n - x_{n+1}\|^2$$
$$= \|(x_n - x_0) - (x_{n+1} - x_0)\|^2$$
$$= \|x_n - x_0\|^2 - 2\langle x_n - x_0, x_{n+1} - x_0 \rangle + \|x_{n+1} - x_0\|^2$$
$$= \|x_n - x_0\|^2 + \|x_{n+1} - x_0\|^2 - 2\langle x_n - x_0, x_{n+1} - x_n + x_n - x_0 \rangle$$
$$= \|x_{n+1} - x_0\|^2 - \|x_n - x_0\|^2 - 2\langle x_n - x_{n+1}, x_0 - x_n \rangle$$
$$\leq \|x_{n+1} - x_0\|^2 - \|x_n - x_0\|^2.$$

So we have

$$\lim_{n \to \infty} \|x_{n+1} - x_n\| = 0.$$

This proof is completed. □

Theorem 2.4. *Let K be a closed convex subset of a Hilbert space H. Let $T : K \to H$ be strongly r-monotonic and μ-Lipschitz continuous such that $F \neq \emptyset$, and the sequence $\{x_n\}$ generated by (1.7), where $\alpha_n \in [0, a]$ for some $a \in [0, 1)$ and $0 < \rho < 2r/\mu^2$, Then the sequence $\{x_n\}$ converges strongly to $P_F x_0$.*

Proof. Since $\{x_n\}$ is bounded, we assume that a subsequence $\{x_{n_i}\}$ of $\{x_n\}$ converges weakly to w_0. It follows from $x_{n+1} \in C_n$ that

$$\|P_K[x_n - \rho T(x_n)] - x_n\| = \frac{1}{1 - \alpha_n}\|y_n - x_n\|$$
$$\leq \frac{1}{1 - \alpha_n}(\|y_n - x_{n+1}\| + \|x_{n+1} - x_n\|)$$
$$\leq \frac{2}{1 - \alpha_n}\|x_{n+1} - x_n\|, \quad \forall n \in N.$$

By lemma 2.3, we get

(2.5) $$\|P_K[x_n - \rho T(x_n)] - x_n\| \to 0.$$

Suppose that $w_0 \neq P_K[w_0 + \rho T(w_0)]$. From Opial's condition and (2.1), (2.5), we have

$$\liminf_{i \to \infty} \|x_{n_i} - w_0\| < \liminf_{i \to \infty} \|x_{n_i} - P_K[w_0 - \rho T(w_0)]\|$$
$$\leq \liminf_{i \to \infty}(\|x_{n_i} - P_K[x_{n_i} - \rho T(x_{n_i})]\|$$
$$\|P_K[x_{n_i} - \rho T(x_{n_i})] - P_K[w_0 - \rho T(w_0)]\|)$$
$$\leq \liminf_{i \to \infty} \|x_{n_i} - w_0 - \rho[Tx_{n_i} - Tw_0]\|$$
$$\leq \liminf_{i \to \infty} \theta \|x_{n_i} - w_0\|$$
$$\leq \liminf_{i \to \infty} \|x_{n_i} - w_0\|.$$

This is a contradiction. Hence we get

(2.6) $w_0 \in F.$

If $z_0 = P_F(x_0)$, it follows from (2.4), (2.6) and the lower semicontinuity of the norm that

$$\|x_0 - z_0\| \leq \|x_0 - w_0\| \leq \liminf_{i \to \infty} \|x_0 - x_{n_i}\|$$

$$\leq \limsup_{i \to \infty} \|x_0 - x_{n_i}\| \leq \|x_0 - z_0\|.$$

Thus, we obtain

$$\lim_{i \to \infty} \|x_{n_i} - x_0\| = \|x_0 - w_0\| = \|x_0 - z_0\|.$$

This implies

$$x_{n_i} \to w_0 = z_0.$$

Therefore, we have $x_n \to z_0$. The proof is completed.   □

References

1. Z. Opial, *Weak convergence of the sequence of sucessive approximations for nonexpansive mappings*, Bull. Amer. Math. Soc. **73** (1967), 591–597.

2. J. C. Dunn, *Convexity, monotonicity and gradient processes in Hilbert spaces*, J. Math. Anal. Appl. **53** (1976), 145–158.

3. K. Yanagi, *On some fixed point theorems for multivalued mappings*, Pacific J. Math. **87(1)** (1980), 233–240.

4. R. E. Megginson, *An Introduction to Banach Space Theory*, Springer-Verlag, New York, 1998.

5. R. U. Verma, *A class of quasivariational inequalities cocoecive mappings*, Adv. Nonlinear Var. Inequal. **2(2)** (1999), 1–12.

6. Wataru Takahashi, *Nonlinear Functional Analysis-Fixed point Theory and its Applications*, Yokohama Publishers inc., 2000.

7. R. U. Verma, *Projection methods,algorithms and a new system of nonlinear variational inequalities*, Comput. Math. Appl. **41** (2001), 1025–1031.

8. R. U. Verma, *Generalized class of partial relaxed monotonicity and its connections*, Adv. Nonlinear Var. Inequal. **7(2)** (2004), 155–164.

9. R. U. Verma, *General convergence analysis for two-step projection methods and applications to variational problems*, Appl. Math. Lett. **18** (2005), 1286–1292.

Nonlinear Functional Analysis and Applications, Volume 1, 147–156

ON THE NEAREST COMMON FIXED POINT PROBLEM FOR AN INFINITE FAMILY OF NON-SELF NONEXPANSIVE MAPPINGS

SHIH-SEN CHANG[1,*], R. F. RAO[1] AND JING AI LIU[3]

ABSTRACT. Strong convergence problem of the iteration $x_{n+1} = P(\alpha_{n+1}u + (1 - \alpha_{n+1})T_{n+1}x_n)$ for a family of infinite nonexpansive mappings $\{T_1, T_2, \cdots\}$ was studied in a Hilbert space H, where P is a sunny nonexpansive retraction of H onto $C \subset H$. It is proved that under suitable conditions the iteration scheme converges strongly to the nearest common fixed point of the infinite family of nonexpansive mappings. As application we utilize our results to study nonexpansive semigroups and other problems. The results presented in this paper expand and improve the corresponding results of [S. S. Chang, On the problem of nearest common fixed point of nonexpansive mappings, Acta Mathematica Sinica, Chinese Series 49(6)(2006), 1297–1302], [J. G. O'Hara, P. O'Hara Pillay and H. K. Xu, Iterative approaches to finding nearest common fixed points of nonexpansive mappings in Hilbert spaces, Nonlinear Anal.TMA 54 (2003), 1417–1426] and [T. Shimizu and W.Takahashi, Strong convergence to common fixed points of families of nonexpansive mappings, J. Math. Anal.Appl. 211 (1997), 71–83].

1. Introduction and Preliminaries

Very recently S. S. Chang [5] has studied the following iteration with a finite family of nonexpansive mappings $\{T_n\}_{n=1}^N$:

$$(1.1) \qquad x_{n+1} = P(\alpha_{n+1}f(x_n) + (1 - \alpha_{n+1})T_{n+1}x_n), \quad T_n = T_{n(mod\ N)} \quad \forall n \geq 0,$$

Received October 5, 2007.   * Corresponding author.

2000 *Mathematics Subject Classification.* 47H09, 47H05.

Key words and phrases. Nonexpansive semigroup, demi-closed principle, sunny nonexpansive retraction, common fixed point, nearest point projection.

This paper was supported by the Natural Science Foundation of Yibin University (No.2007Z3).
[1] Department of Mathematics, Yibin University, Yibin, Sichuan 644007, People's Republic of China (*E-mails*: changss@yahoo.cn(S.S. Chang)) and rrf2006wp@163.com(R.F. Rao))

[2] Department of Mathematics and Physics, Beijing Petro-chemical Engineering Institute, Beijing 102617, People's Republic of China (*E-mail*: liujingai@bipt.edu.cn)

which involved the convex feasibility problem (CFP). It is well known that there are considerable investigations on CFP in the setting of Hilbert spaces which captures applications in various disciplines such as image restoration, computer tomography, and radiation therapy treatment planning, and so on.

Throughout this paper, we assume, H is a real Hilbert space endowed with inner product $\langle \cdot, \cdot \rangle$ and norm $\| \cdot \|$, C is a nonempty closed convex subset of H, $\{T_1, T_2, \cdots, T_n, \cdots\}$ is a family of infinitely many nonexpansive mappings from C into H. Assume that I is the identity mapping, $\mathbb{N}$ is the set of natural numbers, $F(T) = \{x \in C : x = Tx\}$ is the set of fixed points of mapping $T : C \to H$. Denote the strong convergence and weak convergence by $\to$ and $\rightharpoonup$, respectively.

Definition 1.1. Let $T : C \to H$ be a mapping. T is said to be a (*non-self*) *nonexpansive mapping*, if

$$(1.2) \qquad \|Tx - Ty\| \leq \|x - y\|, \quad \forall \, x, y \in C.$$

Especially, T is said to be a *nonexpansive mapping* on C if $T : C \to C$ is a mapping satisfying (1.2).

Definition 1.2. ([5]) Let $P : H \to C$ be a mapping. P is said to be
(1) *sunny*, if for each $x \in C$ and $t \in [0, 1]$ we have

$$P(tx + (1 - t)Px) = Px;$$

(2) a *retraction* of H onto C, if $Px = x$ for all $x \in C$;
(3) a *sunny nonexpansive retraction* if P is sunny, nonexpansive and retraction of H onto C.

Definition 1.3. $Q_C : H \to C$ is called the *metric projection* (or the *nearest point projection*) of H onto C, if for each $x \in H$, $Q_C x \in C$ is the unique element such that $\|Q_C x - x\| = \inf_{y \in C} \|y - x\|$.

Lemma 1.4.([6]) *Assume, H is a real Hilbert space, C is a nonempty closed convex subset of H, then for any $x \in H$ and $y \in C$ we have the following conclusions :*
(a) $\langle z - Q_C x, \, Q_C x - x \rangle \geq 0, \, \forall z \in C$;
(b) *if* $\langle z - y, y - x \rangle \geq 0$ *for all* $z \in C$, *then* $y = Q_C x$.

Lemma 1.5. [5, Lemm 2.3(2)] *Let E be a smooth Banach space and let C be a nonempty closed convex subset of E. If $P : E \to C$ is a retraction and J is the normalized duality mapping on E, then the following conclusions are equivalent (see, [1], [7], [8], [11]):*
(a) *P is sunny and nonexpansive;*
(b) $\|Px - Py\|^2 \leq \langle x - y, J(Px - Py) \rangle$ *for all* $x, y \in E$;
(c) $\langle x - Px, J(y - Px) \rangle \leq 0$ *for all* $x \in E$ *and* $y \in C$.

Since H is a Hilbert space, we know, $J = I$. Moreover, by Lemma 1.4 and Lemma 1.5 we can immediately infer the following conclusion:

Proposition 1.6. *Let H be a Hilbert space, C be a nonempty closed convex subset of H, P be a sunny nonexpansive retraction of H onto C, and Q_C be the nearest point projection of H onto C, then $Q_C x = Px$ for all $x \in H$.*

In (1.1), letting $f(x_n) = u \in H$ for all $n \geq 0$, but extending $\{T_n\}$ from a finite family of nonexpansive mappings $\{T_n\}_{n=1}^{N}$ to an infinite family of nonexpansive mappings $\{T_n\}_{n=1}^{\infty}$, we obtain the following iteration

$$(1.3) \qquad x_{n+1} = P(\alpha_{n+1}u + (1 - \alpha_{n+1})T_{n+1}x_n), \quad \forall n \geq 0.$$

In this paper, we will prove that $\{x_n\}$ defined by (1.3) converges strongly to $Q_F u$ under some suitable conditions, where $F = \bigcap_{n=1}^{\infty} F(T_n)$ is a nonempty closed convex subset of H, and Q_F is the nearest point projection of H onto F. For the convenience of proof, we may firstly recall the following Lemmas:

Lemma 1.7. ([9]) *Let $\{a_n\}$, $\{b_n\}$, $\{c_n\}$ be three nonnegative real sequences such that*

$$a_{n+1} \leq (1 - \lambda_n)a_n + b_n + c_n, \quad n \geq n_0,$$

where n_0 is some nonnegative integer, $\lambda_n \in (0,1)$, $\sum_{n=1}^{\infty} \lambda_n = \infty$, $b_n = o(\lambda_n)$ and $\sum_{n=1}^{\infty} c_n < \infty$, then $a_n \to 0\,(n \to \infty)$.

Lemma 1.8. ([2]) *Assume, E is real Banach space, E^* is the dual space of E, and $J : E \to 2^{E^*}$ is the normalized duality mapping defined by*

$$J(x) = \{f \in E^*, \langle x, f \rangle = \|x\| \cdot \|f\| = \|x\|^2 = \|f\|^2\}, \quad x \in E,$$

where $\langle \cdot, \cdot \rangle$ denotes the generalized duality pair. Then for any given $x, y \in E$,

$$\|x + y\|^2 \leq \|x\|^2 + 2\langle y, j(x + y) \rangle, \quad \forall j(x+y) \in J(x+y),$$

Especially, if E is a Hilbert space, then $J = I$, i.e.,

$$\|x + y\|^2 \leq \|x\|^2 + 2\langle y, x + y \rangle, \quad \forall x, y \in E.$$

Lemma 1.9. ([5, Lemma 2.6(2)] or [6]) *If E is a reflexive Banach space which admits a weakly sequentially continuous normalized duality mapping, and if $E \to E$ is a nonexpansive mapping, then the mapping T is demiclosed, that is, for any sequence $\{x_n\}$ in E, if $x_n \rightharpoonup x$ and $(x_n - Tx_n) \to y$, then $(I - T) = y$.*

Definition 1.10. ([3]) *Let E is a real Banach space, E is said to be strictly convex if, for all $x, y \in E$ with x and y be linearly independent,*

$$\|x + y\| < \|x\| + \|y\|.$$

2. Main Results

By the same method as given in Goebel and Reich [7], we can prove that the following lemma holds.

Lemma 2.1. *Let E be a strictly convex Banach space, C be a nonempty closed convex subset of E, $T : C \to E$ be a non-self nonexpansive mapping, then the fixed point set $F(T)$ of T is a closed convex subset of C.*

Now we give the main results of this paper.

Theorem 2.2. *Let H be a real Hilbert space, C be a nonempty closed convex subset of H, P be a sunny nonexpansive retraction of H onto C, and $\{T_n : C \to H\}_{n=1}^{\infty}$ be a* *family of infinitely many nonexpansive mappings with $F = \bigcap_{n=1}^{\infty} F(T_n) \neq \emptyset$. Assume that $\{\alpha_n\}$ is a real sequence in $(0,1)$ with $\lim_{n \to \infty} \alpha_n = 0$ and $\sum_{n=1}^{\infty} \alpha_n = \infty$. For any given $x_0 \in C$ and $u \in H$, $\{x_n\} \subset C$ is defined by the following iteration:*

$$(2.1) \qquad x_{n+1} = P(\alpha_{n+1}u + (1 - \alpha_{n+1})T_{n+1}x_n), \quad \forall n \geq 0,$$

If there exists a family of nonexpansive mappings $\{G_\gamma : H \to H\}_{\gamma \in \Gamma}$ such that
   (a) $\bigcap_{\gamma \in \Gamma} F(G_\gamma) \neq \emptyset$ and $\bigcap_{\gamma \in \Gamma} F(G_\gamma) \subset \bigcap_{n=1}^{\infty} F(T_n)$;
   (b) $\limsup_{n \to \infty} \|PT_{n+1}x_n - G_\gamma(PT_{n+1}x_n)\| = 0$, for each $\gamma \in \Gamma$,
   then $\{x_n\}$ defined by (2.1) converges strongly to $Q_F u \in F = \bigcap_{n=1}^{\infty} F(T_n)$, where Γ is a finite or infinite index set, Q_F is the metric projection of H onto F.

Proof. First, by Lemma 2.1 we know that $F(T_n)$ is closed and convex for each $n \in \mathbb{N}$. Therefore, $F = \bigcap_{n=1}^{\infty} F(T_n)$ is a closed and convex subset of C, which implies that the nearest point projection of $Q_F : H \to F$ is well defined, and that there exists unique element $Q_F u \in F$ for any given $u \in H$.

Next, we claim that the sequence $\{x_n\}$ defined by (2.1) is bounded.

Indeed, for any given $x \in F = \bigcap_{n=1}^{\infty} F(T_n) \subset C$, considering that P is a sunny nonexpansive mappings of H onto C, by (2.1) we have

$$
\begin{aligned}
\|x_{n+1} - x\| &= \|P(\alpha_{n+1}u + (1 - \alpha_{n+1})T_{n+1}x_n) - Px\| \\
&\leq \|\alpha_{n+1}u + (1 - \alpha_{n+1})T_{n+1}x_n - x\| \\
(2.2) \qquad &\leq \alpha_{n+1}\|u - x\| + (1 - \alpha_{n+1})\|T_{n+1}x_n - x\| \\
&\leq \alpha_{n+1}\|u - x\| + (1 - \alpha_{n+1})\|x_n - x\| \\
&\leq \max\{\|u - x\|, \|x_n - x\|\}.
\end{aligned}
$$

Then by induction we have

$$(2.3) \qquad \|x_{n+1} - x\| \leq \max\{\|u - x\|, \|x_0 - x\|\} \quad \text{for all } n \geq 0.$$

This implies that $\{x_n\}$ is bounded. Since T_n and PT_n both are nonexpansive, $\{T_{n+1}x_n\}$ and $\{PT_{n+1}x_n\}$ are also bounded. Therefore, there exists a constant $M > 0$ such that

$$(2.4) \qquad \|x_n\| + \|T_{n+1}x_n\| + \|PT_{n+1}x_n\| \leq M, \quad \text{for all } n \geq 0.$$

Considering that for all $n \geq 0$,

$$
\begin{aligned}
|\langle PT_{n+1}x_n - Q_F u, u - Q_F u \rangle| &\leq \|PT_{n+1}x_n - Q_F u\| \cdot \|u - Q_F u\| \\
&\leq (M + \|Q_F u\|) \cdot (\|u\| + \|Q_F u\|)
\end{aligned}
$$

and

$$
\begin{aligned}
|\langle x_{n+1} - Q_F u, u - Q_F u \rangle| &\leq \|x_{n+1} - Q_F u\| \cdot \|u - Q_F u\| \\
&\leq (M + \|Q_F u\|) \cdot (\|u\| + \|Q_F u\|).
\end{aligned}
$$

This implies that both $\{\langle PT_{n+1}x_n - Q_F u, u - Q_F u \rangle\}$ and $\{\langle x_{n+1} - Q_F u, u - Q_F u \rangle\}$ are bounded. Hence each of $\limsup_{n \to \infty} \langle PT_{n+1}x_n - Q_F u, u - Q_F u \rangle$ and $\limsup_{n \to \infty} \langle x_{n+1} - Q_F u, u - Q_F u \rangle$ exists. Thus, we can assume that, there exists a subsequence $\{x_j\} \subset \{x_n\}$ such that

$$(2.5) \qquad \limsup_{n \to \infty} \langle PT_{n+1}x_n - \overline{P}_F u, u - Q_F u \rangle = \lim_{j \to \infty} \langle PT_{n_j+1}x_{n_j} - Q_F u, u - \overline{P}_F u \rangle.$$

On the other hand, by the condition (b), we have

(2.6)
$$0 = \limsup_{n \to \infty} \|PT_{n+1}x_n - G_\gamma(PT_{n+1}x_n)\|$$
$$\geq \limsup_{j \to \infty} \|PT_{n_j+1}x_{n_j} - G_\gamma(PT_{n_j+1}x_{n_j})\|, \quad \forall \gamma \in \Gamma,$$

which implies

(2.7)
$$\lim_{j \to \infty} \|PT_{n_j+1}x_{n_j} - G_\gamma(PT_{n_j+1}x_{n_j})\| = 0, \quad \forall \gamma \in \Gamma.$$

By the boundedness of $\{PT_{n+1}x_n\}$, we know that $\{PT_{n_j+1}x_{n_j}\}$ is bounded, and that there exists a subsequence $\{x_l\} \subset \{x_j\}$ and $q \in H$ such that

(2.8)
$$PT_{n_l+1}x_{n_l} \rightharpoonup q, \quad \text{as } l \to \infty.$$

Now we claim $q \in F$.

Indeed, by $\{x_l\} \subset \{x_j\}$, (2.7), (2.8) and Lemma 1.9, we have

$$q \in F(G_\gamma), \quad \forall \gamma \in \Gamma.$$

Then by the condition (a) we know

(2.9)
$$q \in \bigcap_{\gamma \in \Gamma} F(G_\gamma) \subset \bigcap_{n=1}^{\infty} F(T_n).$$

Considering F being a nonempty closed convex subset of C, by $\{x_l\} \subset \{x_j\}$, (2.5), (2.8), (2.9) and Lemma 1.4 we have

(2.10)
$$\limsup_{n \to \infty} \langle PT_{n+1}x_n - Q_F u, u - Q_F u \rangle$$
$$= \lim_{j \to \infty} \langle PT_{n_j+1}x_{n_j} - Q_F u, u - Q_F u \rangle$$
$$= \lim_{l \to \infty} \langle PT_{n_l+1}x_{n_l} - Q_F u, u - Q_F u \rangle$$
$$= \langle q - Q_F u, u - Q_F u \rangle \leq 0.$$

On the other hand, by $\lim_{n \to \infty} \alpha_n = 0$, (2.1) and (2.4) we have

(2.11)
$$\|x_{n+1} - PT_{n+1}x_n\| = \|P(\alpha_{n+1}u + (1 - \alpha_{n+1})T_{n+1}x_n) - PT_{n+1}x_n\|$$
$$\leq \alpha_{n+1}\|u - T_{n+1}x_n\|$$
$$\leq \alpha_{n+1}(\|u\| + M) \to 0, \quad \text{as } n \to \infty.$$

By (2.10) and (2.11) we have

(2.12)
$$\limsup_{n \to \infty} \langle x_{n+1} - Q_F u, u - Q_F u \rangle \leq 0.$$

Defined

$$\beta_{n+1} = \max\{\langle x_{n+1} - Q_F u, u - Q_F u \rangle, 0\}, \quad \forall n \geq 0.$$

Then we know $\beta_n \geq 0$ for all $n \in \mathbb{N}$. Similarly as (2.11) in S. S. Chang [3], we can also prove and obtain

(2.13)
$$\lim_{n \to \infty} \beta_n = 0.$$

Finally, by (2.1) and Lemma 1.8 we have

(2.14)

$$
\begin{aligned}
&\|x_{n+1} - Q_F u\|^2 \\
&= \|P(\alpha_{n+1}u + (1 - \alpha_{n+1})T_{n+1}x_n) - PQ_F u\|^2 \\
&\leq \|\alpha_{n+1}u + (1 - \alpha_{n+1})T_{n+1}x_n - Q_F u\|^2 \\
&= \|(1 - \alpha_{n+1})(T_{n+1}x_n - Q_F u) + \alpha_{n+1}(u - Q_F u)\|^2 \\
&\leq (1 - \alpha_{n+1})^2\|T_{n+1}x_n - Q_F u\|^2 + 2\alpha_{n+1}\langle u - Q_F u, x_{n+1} - Q_F u\rangle \\
&\leq (1 - \alpha_{n+1})\|x_n - Q_F u\|^2 + 2\alpha_{n+1}\beta_{n+1}, \quad \forall n \in \mathbb{N}.
\end{aligned}
$$

Now, let $a_n = \|x_n - \overline{P}_F u\|^2$, $\lambda_n = \alpha_{n+1}$, $b_n = 2\alpha_{n+1}\beta_{n+1}$, $c_n = 0$ in (2.14), then $\{a_n\}$, $\{b_n\}$, $\{c_n\}$ are three nonnegative real sequences satisfying all assumptions of Lemma 1.7. Hence, by (2.14) and Lemma 1.7 we have $\|x_n - Q_F u\|^2 \to 0$, which completes the proof of Theorem 2.2. □

3. Some Applications of Theorem 2.2

The following results can be obtained from Theorem 2.2 immediately.

Theorem 3.1. *Let H be a real Hilbert space, C be a bounded nonempty closed and convex subset of H. Let S and T be nonexpansive mapping of C into itself such that $F = F(S) \cap F(T)$ is nonempty. Suppose that $\{\alpha_n\} \subset (0,1)$ satisfies*
(a) $\lim\limits_{n\to\infty} \alpha_n = 0$ *and*
(b) $\sum\limits_{n=1}^{\infty} \alpha_n = \infty$.
Let

$$
T_n x = \frac{2}{n(n+1)} \sum_{k=0}^{n-1} \sum_{i+j=k} S^i T^j x, \quad x \in C, \ n \in \mathbb{N},
$$

then, for an arbitrary $u \in C$, the sequence $\{x_n\}_{n=0}^{\infty}$ generated by $x = x_0 \in C$ and

(3.1) $\qquad x_{n+1} = \alpha_{n+1}u + (1 - \alpha_{n+1})\dfrac{2}{(n+1)(n+2)} \sum\limits_{k=0}^{n} \sum\limits_{i+j=k} S^i T^j x_n, \quad \forall n \geq 0,$

converges strongly to $Q_F u \in F$, where Q_F is the metric projection of H onto $F = \bigcap_{n=1}^{\infty} F(T_n)$.

Proof. First, it is obvious that T_n is a nonexpansive mapping for each $n \in \mathbb{N}$. Now we claim that $F(S) \cap F(T) \subset \bigcap\limits_{n=1}^{\infty} F(T_n) = F$.

Indeed, for each $x \in F(S) \cap F(T)$ and $n \in \mathbb{N}$, we have

$$
T_n x = \frac{2}{n(n+1)} \sum_{k=0}^{n-1} \sum_{i+j=k} S^i T^j x = x.
$$

Next, we claim, (3.1) is just the iteration (2.1) in the particular case.

Indeed, since $u \in C$, $x_0 \in C$ and $T_n : C \to C$, then $x_n \in C$ or $\alpha_{n+1}u + (1 - \alpha_{n+1})T_{n+1}x_n \in C$ for all $n \geq 0$. Thereby, for a sunny nonexpansive retraction P of H onto C, we have

$$P(\alpha_{n+1}u + (1 - \alpha_{n+1})T_{n+1}x_n) = \alpha_{n+1}u + (1 - \alpha_{n+1})T_{n+1}x_n, \quad \forall n \geq 0,$$

which implies that iteration (3.1) is just (2.1).

Finally, we only need to prove that the condition (b) of Theorem 2.1 is satisfied.

Indeed, considering $T_{n+1}x_n \in C$ and $PT_{n+1}x_n = T_{n+1}x_n$, and letting $\{G_\gamma\}_{\gamma \in \Gamma} = \{S, T\}$, by [12, Lemma 1] we have

$$0 = \lim_{n \to \infty} \sup_{x \in C} \|T_n x - S(T_n x)\| \geq \limsup_{n \to \infty} \|T_{n+1}x_n - S(T_{n+1}x_n)\|$$

$$\text{and} \quad 0 = \lim_{n \to \infty} \sup_{x \in C} \|T_n x - T(T_n x)\| \geq \limsup_{n \to \infty} \|T_{n+1}x_n - T(T_{n+1}x_n)\|,$$

which implies that all the conditions of Theorem 2.2 are satisfied in Lemma 3.1, and that the proof of Lemma 3.1 is completed by Theorem 2.1. $\qquad\square$

Definition 3.2. A family $\{S(t)\}_{t \in R^+}$ of C into itself is called a *nonexpansive semigroup* on C, if it satisfies the following conditions:

(1) $S(t_1 + t_2)x = S(t_1)S(t_2)x$ for each $t_1, t_2 \in R^+$ and $x \in C$;

(2) $S(0)x = x$ for each $x \in C$;

(3) for each $x \in C$, $t \to S(t)x$ is continuopus;

(4) $\|S(t)x - S(t)y\| \leq \|x - y\|$ for each $t \in R^+$ and $x, y \in C$,

where R^+ is denoted by the set of nonnegative real numbers.

Theorem 3.2. *Let H be a real Hilbert space, C be a bounded nonempty closed and convex subset of H. Let $\{S(t)\}_{t \in R^+}$ be a nonexpansive semigroup on C such that $\bigcap_{t \in R^+} F(S(t))$ is nonempty. Suppose that $\{\alpha_n\} \subset (0, 1)$ satisfies*

(a) $\lim_{n \to \infty} \alpha_n = 0$ and

(b) $\sum_{n=1}^{\infty} \alpha_n = \infty$.

Then, for an arbitrary $u \in C$, the sequence $\{x_n\}_{n=0}^{\infty}$ generated by $x = x_0 \in C$ and

$$(3.2) \qquad x_{n+1} = \alpha_{n+1}u + (1 - \alpha_{n+1})\frac{1}{t_n}\int_0^{t_n} S(v)x_n dv, \quad \forall n \geq 0$$

converges strongly to $Q_F u \in F$, where Q_F is the metric projection of H onto $F = \bigcap_{n=1}^{\infty} F(T_n)$, and $T_{n+1}x = \frac{1}{t_n}\int_0^{t_n} S(v)x dv$ for all $n \geq 0$, $x \in C$, $\{t_n\}_{t=0}^{\infty}$ is a positive real divergent sequence.

Proof. First, since

$$T_{n+1}x = \frac{1}{t_n}\int_0^{t_n} S(v)x dv \quad \text{for all} n \geq 0, \ x \in C,$$

we say, for each $n \in \mathbb{N}$, T_{n+1} is a nonexpansive mapping of C into itself.

Indeed, for any $n \in \mathbb{N}$ and $x, y \in C$, we have

$$
\begin{aligned}
\|T_{n+1}x - T_{n+1}y\| &= \left\| \frac{1}{t_n} \int_0^{t_n} (S(v)x - S(v)y)dv \right\| \\
&\leq \frac{1}{t_n} \int_0^{t_n} \|S(v)x - S(v)y\| dv \\
&\leq \frac{1}{t_n} \int_0^{t_n} \|x - y\| dv = \|x - y\|.
\end{aligned}
$$

Next, we claim, (3.2) is just the iteration (2.1) in the particular case.

Indeed, since $u \in C$, $x_0 \in C$ and $T_n : C \to C$, then $x_n \in C$ or $\alpha_{n+1}u + (1 - \alpha_{n+1})T_{n+1}x_n \in C$ for all $n \geq 0$. Thereby, for a sunny nonexpansive retraction P of H onto C, we have

$$
P(\alpha_{n+1}u + (1 - \alpha_{n+1})T_{n+1}x_n) = \alpha_{n+1}u + (1 - \alpha_{n+1})T_{n+1}x_n, \quad \forall n \geq 0,
$$

which implies that iteration (3.2) is just (2.1).

Now we claim $\bigcap_{s \in R^+} F(S(s)) \subset \bigcap_{n=1}^{\infty} F(T_n)$.

Indeed, for each $x \in \bigcap_{s \in R^+} F(S(s))$, we get $S(s)x = x$ for each $s \in R^+$. Thus, for each $n \geq 0$ we have

$$
T_{n+1}x = \frac{1}{t_n} \int_0^{t_n} S(v)x dv = \frac{1}{t_n} \int_0^{t_n} x dv = x, \quad \text{i.e., } x \in \bigcap_{n=1}^{\infty} F(T_n).
$$

Finally, we only need to prove that the condition (b) of Theorem 2.1 is satisfied.

Indeed, considering $T_{n+1}x_n \in C$ and $PT_{n+1}x_n = T_{n+1}x_n$, and letting $\{G_\gamma\}_{\gamma \in \Gamma} = \{S(s)\}_{s \in R^+}$, by [12, Lemma 2] we have

$$
0 = \lim_{n \to \infty} \sup_{x \in C} \|T_n x - S(s)(T_n x)\| \geq \limsup_{n \to \infty} \|T_{n+1}x_n - S(s)(T_{n+1}x_n)\|,
$$

which implies that all the conditions of Theorem 2.2 are satisfied. The conclusion of Theorem 3.2 is obtained.   □

Theorem 3.3.   Let $T_1, T_2, \cdots, T_N$ be N nonexpansive mappings of C into itself, satisfying $T_i T_j = T_j T_i$ for each $i, j = 1, 2, \cdots, N$, $i \neq j$ and $F = \bigcap_{n=1}^{N} F(T_n) \neq \emptyset$. Suppose that $\{\alpha_n\} \subset (0,1)$ satisfies

(a) $\lim_{n \to \infty} \alpha_n = 0$ and

(b) $\sum_{n=1}^{\infty} \alpha_n = \infty$.

Then, for an arbitrary $u \in C$, the sequence $\{x_n\}_{n=0}^{\infty}$ generated by $x = x_0 \in C$ and

(3.3) $x_{n+1} = \alpha_{n+1}u + (1 - \alpha_{n+1})T_{n+1}x_n, \quad T_n = T_{n(\mathrm{mod}\, N)}, \quad \forall n \geq 0.$

If for each $i = 1, 2, \cdots, N$,

(3.4) $\lim_{n \to \infty} \|x_n - T_i x_n\| = 0,$

then $\{x_n\}$ defined by (3.3) converges strongly to $Q_F u \in F$, where Q_F is the metric projection of H onto $F = \bigcap_{n=1}^{N} F(T_n)$.

Proof. Similarly as the proof of Lemma 3.1 and 3.3, we can also prove that (3.3) is just the iteration (2.1) in the particular case.

Now we only need to prove that the condition (b) is satisfied in Lemma 3.4.

Indeed, by (3.4) we have

$$\limsup_{n\to\infty} \|T_{n+1}x_n - T_iT_{n+1}x_n\| = \limsup_{n\to\infty} \|T_{n+1}x_n - T_{n+1}T_ix_n\|$$

$$\leq \limsup_{n\to\infty} \|x_n - T_ix_n\| = 0.$$

Letting $\{G_\gamma\}_{\gamma\in\Gamma} = \{T_1, T_2, \cdots, T_N\}$ in Theorem 2.2, and considering $PT_{n+1}x_n = T_{n+1}x_n$, we know that the condition (b) is satisfied in Lemma 3.4. Then we can easily prove that all the conditions of Theorem 2.2 are satisfied. The conclusion 0f Theorem 3.3 is proved. $\qquad\square$

Theorem 3.4. *Assume, H is a real Hilbert space, C is a nonempty closed convex subset of H, $\{T_n : C \to C, \ n = 1, 2, \cdots\}$ is an infinite family of nonexpansive mappings such that $F = \bigcap_{n=1}^\infty F(T_n) \neq \emptyset$. Assume, $\{\alpha_n\}$ is a real sequence in $(0,1)$ such that $\alpha_n \to 0$ and $\sum_{n=1}^\infty \alpha_n = \infty$. For any given $x_0, u \in C$, $\{x_n\} \subset C$ is defined by the following iteration*

$$(3.5) \qquad x_{n+1} = \alpha_{n+1}u + (1 - \alpha_{n+1})T_{n+1}x_n, \quad \forall n \geq 0,$$

If there exists a family of nonexpansive mappings $\{G_\gamma\}_{\gamma\in\Gamma}$ such that
(a) $\bigcap_{\gamma\in\Gamma} F(G_\gamma) \neq \emptyset$ and $\bigcap_{\gamma\in\Gamma} F(G_\gamma) \subset \bigcap_{n=1}^\infty F(T_n)$;
(b) $\limsup_{n\to\infty} \|T_{n+1}x_n - G_\gamma(T_{n+1}x_n)\| = 0$, for each $\gamma \in \Gamma$,
then $\{x_n\}$ converges strongly to $\overline{P}_Fu \in F$, where Γ is a finite or an infinite index set, $\overline{P}_F$ is the metric projection of H onto F.

Proof. It is easy to see that $P(\alpha_{n+1}u + (1-\alpha_{n+1})T_{n+1}x_n) = \alpha_{n+1}u + (1-\alpha_{n+1})T_{n+1}x_n$, and that $PT_{n+1}x_n = T_{n+1}x_n$, which implies that all conditions of Theorem 2.2 are satisfied. Hence the conclusion of Theorem 3.4 can be obtained from Theorem 2.2 immediately. $\qquad\square$

Remark. (1) Theorem 3.1 and Theorem 3.2 are the main results of Shimizu and Takahashi [12, Theorem 1, 2].

(2) Theorem 3.3 is Theorem 3.3 in O'Hara, Pillay and Xu [10].

(3) Theorem 3.4 is Theorem 2.1 in S. S. Chang [4].

REFERENCES

1. R. E. Bruck Jr., *Nonexpansive projections on subsets of Banach spaces,* Pacific J. Math. **47** (1973), 341–355.
2. S. S. Chang, *On Chidume's open questions and approximation solutions of multivalued strongly accretive mappings equations in Banach spaces,* J. Math. Anal. Appl. **216** (1997), 94–111.
3. S. S. Chang, *Viscosity approximation methods for a finite family of nonexpansive mappings in Banach spaces,* J. Math. Anal. Appl. **323** (2006), 1402–1416.
4. S. S. Chang, *On the problem of nearest common fixed point of nonexpansive mappings,* Acta Mathematica Sinica, Chinese Series **49(6)** (2006), 1297–1302.
5. S. S. Chang, J. C. Yao, J. K. Kim and L. Yang, *Iterative approximation to convex feasibility problems in Banach space,* to appear in Fixed Point Theory and Applications.

6. K. Goebel and W.A. Kirk, *Topics in Metric Fixed Point Theory*, Cambridge Studies in Advanced Mathematics, 1990.

7. K. Goebel and S. Reich, *Uniform Convexity,Hyperbolic Geometry,and Nonexpansive Mappings*, Monographs and Textbooks in Pure and Applied Mathematics, Vol. 83, Marcel Dekker, NewYork, 1984.

8. E. Kopecká and S. Reich, *Nonexpansives retractions in Banach spaces,* Erwin Schrodinger Institute (Preprint No. 1787), (2006).

9. L. S. Liu, *Ishikawa and Mann iterative processes with errors for nonlinear strongly accretive mappings in Banach space, J. Math. Anal. Appl.* **194** (1995), 114–125.

10. J. G. O'Hara, P. O'Hara Pillay and H. K. Xu, *Iterative approaches to finding nearest common fixed points of nonexpansive mappings in Hilbert spaces,* Nonlinear Anal., TMA **54** (2003), 1417–1426.

11. S. Reich, *Asymptotic behavior of contractions in Banach spaces, J. Math. Anal. Appl.* **44(1)** (1973), 57–70.

12. T. Shimizu and W.Takahashi, *Strong convergence to common fixed points of families of nonexpansive mappings, J. Math. Anal. Appl.* **211** (1997), 71–83.

Nonlinear Functional Analysis and Applications, Volume 1, 157–164

A NOTE ON UNIFORM CONVERGENCE
OF ITERATES OF ASYMPTOTIC CONTRACTIONS

SIMEON REICH[1,*] AND ALEXANDER J. ZASLAVSKI[2]

ABSTRACT. We provide sufficient conditions for the iterates of certain asymptotic contractions defined on a closed subset K of a complete metric space X to converge to their unique fixed points, uniformly on each bounded subset of K.

1. Introduction

Our main purpose in this paper, which may be considered a continuation of [2] and [3], is to establish an extension of the main result in [3] to nonself-mappings, that is, mappings which map a nonempty closed subset K of a complete metric space (X, d) into X. For the convenience of the reader, we begin by briefly recalling the setting of our results.

According to Banach's fixed point theorem, the iterates of any strict contraction on X converge to its unique fixed point. Since this classical theorem has found numerous important applications, it has also been extended in many directions. Perhaps the first such extension is due to Rakotch [11]. Several years later Boyd and Wong [4] obtained even a more general result. See [9] for a comprehensive survey of the results available for various types of contraction mappings up to 2001.

More recently, Kirk [10] has introduced the notion of an asymptotic contraction and proved the following fixed point theorem for such mappings by using ultrapower techniques. His theorem may be considered an asymptotic version of the Boyd-Wong fixed point theorem [4] mentioned above.

Received October 13, 2007.   * Corresponding author.

2000 *Mathematics Subject Classification.* 47H10, 54E50, 54H25.

Key words and phrases. Asymptotic contraction, complete metric space, fixed point, iteration.

This research was supported by the Israel Science Foundation (Grant No. 647/07), the Fund for the Promotion of Research at the Technion and by the Technion President's Research Fund.

[1] Department of Mathematics, The Technion-Israel Institute of Technology, 32000 Haifa, Israel (*E-mail*: sreich@tx.technion.ac.il)

[2] Department of Mathematics, The Technion-Israel Institute of Technology, 32000 Haifa, Israel (*E-mail*: ajzasl@tx.technion.ac.il)

Theorem 1.1. *Let $T : X \to X$ be a continuous mapping such that*

$$d(T^n x, T^n y) \leq \phi_n(d(x, y))$$

for all $x, y \in X$ and all natural numbers n, where $\phi_n : [0, \infty) \to [0, \infty)$ and $\lim_{n \to \infty} \phi_n = \phi$, uniformly on the range of d. Suppose that ϕ and all ϕ_n are continuous and that $\phi(t) < t$ for all $t > 0$. If there exists $x_0 \in X$ which has a bounded orbit $O(x_0) = \{x_0, Tx_0, T^2 x_0, \cdots\}$, then T has a unique fixed point $x_* \in X$ and $\lim_{n \to \infty} T^n x = x_*$ for all $x \in X$.

Arandelović [1] then provided a short and simple proof of Kirk's Theorem 1.1. Jachymski and Jóźwik [8] extended this result with a constructive proof and obtained a complete characterization of asymptotic contractions on a compact metric space.

The following theorem is the main result of Chen [7]. It improves upon Kirk's original theorem [10].

Theorem 1.2. *Let $T : X \to X$ be such that*

$$d(T^n x, T^n y) \leq \phi_n(d(x, y))$$

for all $x, y \in X$ and all natural numbers n, where $\phi_n : [0, \infty) \to [0, \infty)$ and $\lim_{n \to \infty} \phi_n = \phi$, uniformly on any bounded interval $[0, b]$. Suppose that ϕ is upper semicontinuous and that $\phi(t) < t$ for all $t > 0$. Furthermore, suppose that there exists a positive integer n_* such that ϕ_{n_*} is upper semicontinuous and $\phi_{n_*}(0) = 0$. If there exists $x_0 \in X$ which has a bounded orbit $O(x_0) = \{x_0, Tx_0, T^2 x_0, \cdots\}$, then T has a unique fixed point $x_* \in X$ and $\lim_{n \to \infty} T^n x = x_*$ for all $x \in X$.

Note that Theorem 1.2 does not provide us with uniform convergence of the iterates of T on bounded subsets of X, although this does hold for many classes of mappings of contractive type (e.g., those considered in [5] and [11]). This property is important because it yields stability of the convergence of iterates even in the presence of computational errors [6]. In [2] we show that this conclusion can be derived in the setting of Theorem 1.2. To this end, we first prove in [2] a somewhat more general result (Theorem 1.3 below) which, when combined with Theorem 1.2, yields our strengthening of Chen's result (Theorem 1.4).

Theorem 1.3. *Let $x_* \in X$ be a fixed point of $T : X \to X$. Assume that*

$$d(T^n x, x_*) \leq \phi_n(d(x, x_*))$$

for all $x \in X$ and all natural numbers n, where $\phi_n : [0, \infty) \to [0, \infty)$ and $\lim_{n \to \infty} \phi_n = \phi$, uniformly on any bounded interval $[0, b]$. Suppose that ϕ is upper semicontinuous and $\phi(t) < t$ for all $t > 0$. Then $T^n x \to x_*$ as $n \to \infty$, uniformly on each bounded subset of X.

Theorem 1.4. *Let $T : X \to X$ be such that*

$$d(T^n x, T^n y) \leq \phi_n(d(x, y))$$

for all $x, y \in X$ and all natural numbers n, where $\phi_n : [0, \infty) \to [0, \infty)$ and $\lim_{n \to \infty} \phi_n = \phi$, uniformly on any bounded interval $[0, b]$. Suppose that ϕ is upper semicontinuous and $\phi(t) < t$ for all $t > 0$. Furthermore, suppose that there exists a positive integer n_* such that ϕ_{n_*} is upper semicontinuous and $\phi_{n_*}(0) = 0$. If there exists $x_0 \in X$ which has a bounded orbit $O(x_0) = \{x_0, Tx_0, T^2x_0, \cdots\}$, then T has a unique fixed point $x_* \in X$ and $\lim_{n\to\infty} T^n x = x_*$, uniformly on each bounded subset of X.

We now state the main result of the paper [3]. In contrast with Theorem 1.3, here we only assume that a subsequence of $\{\phi_n\}_{n=1}^\infty$ converges to ϕ.

Theorem 1.5. *Let $x_* \in X$ be a fixed point of $T : X \to X$. Assume that*

$$d(T^n x, x_*) \le \phi_n(d(x, x_*))$$

for all $x \in X$ and all natural numbers n, where the functions $\phi_n : [0, \infty) \to [0, \infty)$, $n = 1, 2, \cdots$, satisfy the following conditions:
(a) *For each $b > 0$, there is a natural number n_b such that*

$$\sup\{\phi_n(t) : t \in [0, b] \text{ and all } n \ge n_b\} < \infty;$$

(b) *there exist an upper semicontinuous function $\phi : [0, \infty) \to [0, \infty)$ satisfying $\phi(t) < t$ for all $t > 0$ and a strictly increasing sequence of natural numbers $\{m_k\}_{k=1}^\infty$ such that $\lim_{k\to\infty} \phi_{m_k} = \phi$, uniformly on any bounded interval $[0, b]$.*
Then $T^n x \to x_*$ as $n \to \infty$, uniformly on any bounded subset of X.

As we have already mentioned, in the present paper we will prove a generalization of Theorem 1.5 for nonself-mappings, that is, mappings which map a nonempty closed subset K of (X, d) into X (see Section 2, Theorem 2.1). Section 3 contains an extension of Theorem 2.1. In Section 4 we present another version of Theorem 1.2 with a new proof.

2. A Generalization of Theorem 1.5

Let K be a nonempty closed subset of a complete metric space (X, d). We consider a mapping $T : K \to X$ and set $T^0 x = x$ for all $x \in K$. If $x \in K$ and $Tx \in K$, we set $T^2 x = T(Tx)$. More generally, if $x \in K$, $n \ge 0$ is an integer and $T^n x \in K$, we define $T^{n+1} x = T(T^n x)$.

Theorem 2.1. *Let $x_* \in K$ be a fixed point of $T : K \to X$. Assume that*

(2.1) $$d(T^n x, x_*) \le \phi_n(d(x, x_*))$$

for all $x \in K$ and all natural numbers n for which $T^n x \in X$ is defined, where the functions $\phi_n : [0, \infty) \to [0, \infty)$, $n = 1, 2, \cdots$, satisfy the following conditions:
(a) *For each $b > 0$, there is a natural number n_b such that*

$$\sup\{\phi_n(t) : t \in [0, b] \text{ and all } n \ge n_b\} < \infty;$$

(b) *there exist an upper semicontinuous function $\phi : [0, \infty) \to [0, \infty)$ satisfying $\phi(t) < t$ for all $t > 0$ and a strictly increasing sequence of natural numbers $\{m_k\}_{k=1}^\infty$ such that $\lim_{k\to\infty} \phi_{m_k} = \phi$, uniformly on any bounded interval $[0, b]$.*
Then for any nonempty bounded set $M \subset K$ and any $\epsilon > 0$, there is an integer $n_0 \ge 1$ such that if $x \in M$ and $T^n x$ is defined for an integer $n \ge n_0$, then $d(T^n x, x_*) < \epsilon$.

Proof. We may assume that $\phi(0) = 0$ and that $\phi_n(0) = 0$ for all integers $n \geq 1$. For each $x \in X$ and each $r > 0$, set

$$B(x, r) = \{z \in X : d(x, z) \leq r\}.$$

Let $M > 0$ and $\epsilon \in (0, 1)$. By (a), there are $M_1 > M$ and an integer $n_1 \geq 1$ such that

(2.2) $\qquad\qquad \phi_i(t) \leq M_1 \quad$ for all $t \in [0, M + 1]$ and all integers $i \geq n_1$.

In view of (2.1) and (2.2), for each $x \in B(x_*, M) \subset K$ and each integer $n \geq n_1$ for which $T^n x$ is defined,

(2.3) $\qquad\qquad\qquad\qquad d(T_n x, x_*) \leq \phi_n(d(x, x_*)) \leq M_1.$

Since the function $t - \phi(t)$ is lower semicontinuous, there is $\delta > 0$ such that

(2.4) $\qquad\qquad\qquad\qquad\qquad\qquad \delta < \epsilon/8$

and

(2.5) $\qquad\qquad\qquad\qquad t - \phi(t) \geq 2\delta, \quad t \in [\epsilon/8, 4M_1 + 4].$

By (b), there is an integer $n_2 \geq 2n_1 + 2$ such that

(2.6) $\qquad\qquad\qquad |\phi_{n_2}(t) - \phi(t)| \leq \delta, \quad t \in [0, 4M_1 + 4].$

Assume that

(2.7) $\qquad\qquad\qquad x \in B(x_*, M_1 + 4) \cap K$ and that $T^{n_2} x$ is defined.

If $d(x, x_*) \leq \epsilon/8$, then it follows from (2.1), (2.4), (2.6) and (2.7) that

$$d(T^{n_2} x, x_*) \leq \phi_{n_2}(d(x, x_*)) \leq \phi(d(x, x_*)) + \delta$$
$$\leq d(x, x_*) + \delta < \epsilon/4.$$

If $d(x, x_*) \geq \epsilon/8$, then relations (2.1), (2.5), (2.6) and (2.7) imply that

$$d(T^{n_2} x, x_*) \leq \phi_{n_2}(d(x, x_*)) \leq \phi(d(x, x_*)) + \delta$$
$$\leq d(x, x_*) - 2\delta + \delta = d(x, x_*) - \delta.$$

Thus in both cases we have

(2.8) $\qquad\qquad\qquad d(T^{n_2} x, x_*) \leq \max\{d(x, x_*) - \delta, \epsilon/4\}.$

Now choose a natural number $q > 2$ such that

(2.9) $\qquad\qquad\qquad\qquad\qquad q > (8 + 2M_1)\delta^{-1}.$

Assume that

$$x \in B(x_*, M_1 + 4),$$

and that

(2.10) $\quad T^{qn_2} x$ is defined and $T^{in_2} x$ belongs to $B(x_*, M_1 + 4)$ for each $i = 1, \cdots, q - 1$.

We claim that

(2.11) $\qquad\qquad\qquad \min\{d(T^{jn_2} x, x_*) : j = 1, \cdots, q\} \leq \epsilon/4.$

Assume the contrary. Then by (2.8) and (2.10), for each $j = 1, \cdots, q$, we have

$$d(T^{jn_2} x, x_*) \leq d(T^{(j-1)n_2} x, x_*) - \delta$$

and
$$d(T^{qn_2}x, x_*) \le d(T^{(q-1)n_2}x, x_*) - \delta \le \cdots \le d(x, x_*) - q\delta$$
$$\le M_1 + 4 - q\delta.$$

This contradicts (2.9). The contradiction we have reached proves (2.11).

Assume that an integer j satisfies $1 \le j \le q-1$ and
$$d(T^{jn_2}x, x_*) \le \epsilon/4.$$

When combined with (2.8) and (2.10), this implies that
$$d(T^{(j+1)n_2}x, x_*) \le \max\{d(T^{jn_2}x, x_*) - \delta, \epsilon/4\} \le \epsilon/4.$$

It follows from this inequality and (2.11) that

(2.12)
$$d(T^{qn_2}x, x_*) \le \epsilon/4$$

for all x satisfying (2.10).

Assume now that $x \in B(x_*, M) \cap K$ and let an integer s be such that $s \ge n_1 + qn_2$ and $T^s x$ is defined. By (2.3),
$$T^i x \in B(x_*, M_1) \text{ for all integers } i \ge n_1 \text{ such that } i \le s$$

and

(2.13)
$$T^{s-qn_2}x \in B(x_*, M_1).$$

Since $T^s x = T^{qn_2}(T^{s-qn_2}x)$, it follows from (2.12) and (2.13) that
$$d(T^s x, x_*) = d(T^{qn_2}(T^{s-qn_2}x), x_*) < \epsilon/4.$$

This completes the proof of Theorem 2.1. $\square$

3. An Extension of Theorem 2.1

In this section we present an extension of Theorem 2.1 (Theorem 3.1 below) and point out that Theorem 2.1 also has a converse.

Theorem 3.1. *Let $x_* \in K$ be a fixed point of $T : K \to X$. Assume that $\{m_k\}_{k=1}^{\infty}$ is a strictly increasing sequence of natural numbers such that*
$$d(T^{m_k}x, x_*) \le \phi_{m_k}(d(x, x_*))$$

for all $x \in K$ such that $T^{m_k}x$ is defined and all natural numbers k, where T and the functions $\phi_{m_k} : [0, \infty) \to [0, \infty)$, $k = 1, 2, \cdots$, satisfy the following conditions:

(a) For each $M > 0$, there is $M_1 > 0$ such that
$$\{T^i x : x \in B(x_*, M) \text{ and } T^i x \text{ is defined}\} \subset B(x_*, M_1) \text{ for each integer } i \ge 0;$$

(b) there exists an upper semicontinuous function $\phi : [0, \infty) \to [0, \infty)$ satisfying $\phi(t) < t$ for all $t > 0$ such that $\lim_{k \to \infty} \phi_{m_k} = \phi$, uniformly on any bounded interval $[0, b]$. Then for any nonempty bounded set $M \subset K$ and any $\epsilon > 0$, there is an integer $n_0 \ge 1$ such that if $x \in M$ and $T^n x$ is defined for an integer $n \ge n_0$, then $d(T^n x, x_*) < \epsilon$.

Proof. Let i be a natural number such that $i \ne m_k$ for all natural numbers k. For each $t \ge 0$, set
$$\phi_i(t) = \sup\{d(T^i x, x_*) : x \in B(x_*, t) \cap K \text{ and } T^i x \text{ is defined}\}.$$

Clearly, $\phi_i(t)$ is finite for all $t \geq 0$. Since it is not difficult to check that all the assumptions of Theorem 2.1 hold, Theorem 3.1 is now seen to be a consequence of Theorem 2.1.

Assume now that $T : K \to X$, $T(C)$ is bounded for any bounded subset $C \subset K$, $x_* \in K$ is a fixed point of T, and that the following property holds:

For any nonempty bounded set $M \subset K$ and any $\epsilon > 0$, there is an integer $n_0 \geq 1$ such that if $x \in M$ and $T^n x$ is defined for an integer $n \geq n_0$, then $d(T^n x, x_*) < \epsilon$.

We claim that T necessarily satisfies all the hypotheses of Theorem 2.1 with an appropriate sequence $\{\phi_n\}_{n=1}^{\infty}$. Indeed, fix a natural number n and for all $t \geq 0$, set

$$\phi_n(t) = \sup\{d(T^n x, x_*) : x \in B(x_*, t) \cap K \text{ such that } T^n x \text{ is defined}\}.$$

Clearly, $\phi_n(t)$ is finite for all $t \geq 0$ and all natural numbers n and

$$d(T^n x, x_*) \leq \phi_n(d(x, x_*))$$

for all $x \in K$ and all natural numbers n such that $T^n x$ is defined. It is also obvious that $\phi_n \to 0$ as $n \to \infty$, uniformly on any bounded subinterval of $[0, \infty)$, and that, for any $b > 0$,

$$\sup\{\phi_n(t) : t \in [0, b], \ n \geq 1\} < \infty.$$

Thus all the assumptions of Theorem 2.1 hold with $\phi(t) = 0$ identically, as claimed. $\square$

4. Another Version of Theorem 1.2

In this section we present a version of Chen's result [7] (see Theorem 1.2) with a different proof.

Theorem 4.1. *Let $T : X \to X$ be such that*

$$d(T^n x, T^n y) \leq \phi_n(d(x, y))$$

for all $x, y \in X$ and all natural numbers n, where $\phi_n : [0, \infty) \to [0, \infty)$ and $\lim_{n \to \infty} \phi_n = \phi$, uniformly on any bounded interval $[0, b]$. Suppose that ϕ is upper semicontinuous and that $\phi(t) < t$ for all $t > 0$. Assume that there exists $x_0 \in X$ which has a bounded orbit $O(x_0) = \{x_0, Tx_0, T^2 x_0, \cdots\}$. Then there exists $x_* = \lim_{n \to \infty} T^n x_0$ in (X, d).

Moreover, if there exists a positive integer n_* such that ϕ_{n_*} is upper semicontinuous and $\phi_{n_*}(0) = 0$, then x_* is the unique fixed point of T and $\lim_{n \to \infty} T^n x = x_*$ for all $x \in X$.

Proof. Set $T^0 x = x$, $x \in X$. We may assume that $\phi(0) = 0$ and $\phi_n(0) = 0$ for all integers $n \geq 1$. Choose a positive number b such that

(4.1) $E := \{T^i x_0 : i = 1, 2, \cdots\} \subset B(x_0, b).$

We will show that $\{T^i x_0\}_{i=1}^{\infty}$ is a Cauchy sequence. To this end, let $\epsilon \in (0, 1)$ be given. Since the function $t - \phi(t)$ is lower semicontinuous, there is $\delta > 0$ such that

(4.2) $\delta < \epsilon/8$

and

$$t - \phi(t) \geq 2\delta, \quad \forall t \in [\epsilon/8, 4b + 4].$$

There is an integer n_1 such that

(4.3) $|\phi_{n_1}(t) - \phi(t)| \leq \delta, \quad \forall t \in [0, 4b + 4].$

Assume that

$$y, z \in E.$$

If $d(y,z) \leq \epsilon/8$, then by (4.2) and (4.3),

$$d(T^{n_1}y, T^{n_1}z) \leq \phi_{n_1}(d(y,z)) \leq \phi(d(y,z)) + \delta \leq \epsilon/8 + \delta < \epsilon/4.$$

If $d(y,z) > \epsilon/8$, then by (4.1)-(4.3),

$$d(T^{n_1}y, T^{n_1}z) \leq \phi_{n_1}(d(y,z)) \leq \phi(d(y,z)) + \delta \leq d(y,z) - \delta.$$

Therefore in both cases we have

(4.4) $$d(T^{n_1}y, T^{n_1}z) \leq \max\{d(y,z) - \delta, \epsilon/4\}$$

for all $y, z \in E$. Choose a natural number

(4.5) $$q > (2 + 2b)\delta^{-1}$$

and assume that $y, z \in E$. If, for all $j = 1, \cdots, q$,

$$d(T^{jn_1}y, \ T^{jn_1}z) > \epsilon/4,$$

then, by (4.4) and (4.1),

$$d(T^{qn_1}y, T^{qn_1}z) \leq d(T^{(q-1)n_1}y, T^{(q-1)n_1}z) - \delta \leq d(y,z) - q\delta \leq 2b - q\delta.$$

This contradicts (4.5). The contradiction we have reached proves that there is $j \in \{1, \cdots, q\}$ such that

(4.6) $$d(T^{jn_1}y, T^{jn_1}z) \leq \epsilon/4.$$

If $j < q$, then, by (4.1) and (4.4),

$$d(T^{(j+1)n_1}y, T^{(j+1)n_1}z) \leq \epsilon/4.$$

Thus (4.6) holds with $j = q$ for all $y, z \in E$. This implies that $\{T^i x_0\}_{i=1}^\infty$ is indeed a Cauchy sequence and there exists $x_* = \lim_{i \to \infty} T^i x_0$ in (X,d). The rest of the proof is analogous to that of Chen's. Indeed, let $n_* \geq 1$ be an integer such that ϕ_{n_*} is upper semicontinuous with $\phi_{n_*}(0) = 0$. Then for any integer $n > 0$,

$$d(T^{n_*+n}x_0, T^{n_*}x_*) \leq \phi_{n_*}(d(T^n x_0, x_*)).$$

Since

$$\limsup_{n \to \infty} \phi_{n_*}(d(T^n x_0, x_*)) \leq \phi_{n_*}(0) = 0,$$

we have $\lim_{n \to \infty} T^{n_*+n}x_0 = T^{n_*}x_*$, so that $T^{n_*}x_* = x_*$.

Note that $T^{n_*}(Tx_*) = T(T^{n_*}x_*) = Tx_*$. Therefore, x_* and Tx_* are fixed points of T^{n_*}. But it is clear that T^{n_*} has no more than one fixed point. Therefore, $x_* = Tx_*$ and x_* is the unique fixed point of T. Now it is easy to see that $\{T^i x : i = 1, 2, \cdots\}$ is bounded for any $x \in X$. Therefore $\lim_{i \to \infty} T^i x = x_*$ for any $x \in X$. Theorem 4.1 is proved. $\qquad \square$

REFERENCES

1. I. D. Aranđelović, *On a fixed point theorem of Kirk,* J. Math. Anal. Appl. **301** (2005), 384–385.
2. M. Arav, F. E. Castillo Santos, S. Reich and A. J. Zaslavski, *A note on asymptotic contractions,* Fixed Point Theory Appl. **2007** (Article ID 39465), 1–6.
3. M. Arav, S. Reich and A. J. Zaslavski, *Uniform convergence of iterates for a class of asymptotic contractions,* Fixed Point Theory **8** (2007), 3–9.
4. D. W. Boyd and J. S. W. Wong, *On nonlinear contractions,* Proc. Amer. Math. Soc. **20** (1969), 458–464.
5. F. E. Browder, *On the convergence of successive approximations for nonlinear functional equations,* Indag. Math. **30** (1968), 27–35.
6. D. Butnariu, S. Reich and A. J. Zaslavski, *Asymptotic behavior of inexact orbits for a class of operators in complete metric spaces,* J. Appl. Anal. **13** (2007), 1–11.
7. Y.-Z. Chen, *Asymptotic fixed points for nonlinear contractions,* Fixed Point Theory Appl. **2005** (2005), 213–217.
8. J. Jachymski and I. Jóźwik, *On Kirk's asymptotic contractions,* J. Math. Anal. Appl. **300** (2004), 147–159.
9. W. A. Kirk, *Contraction mappings and extensions,* Handbook of Metric Fixed Point Theory, Kluwer, Dordrecht, 2001, 1–34.
10. W. A. Kirk, *Fixed points of asymptotic contractions,* J. Math. Anal. Appl. **277** (2003), 645–650.
11. E. Rakotch, *A note on contractive mappings,* Proc. Amer. Math. Soc. **13** (1962), 459–465.

Nonlinear Functional Analysis and Applications, Volume 1, 165–173

COMPARATIVE STATIONARITY OF STOCHASTIC EXPONENTIAL AND MONOMIAL DENSITIES AND RELATED NONLINEAR APPROXIMATIONS

Nassar H. S. Haidar[1,*], Adnan M. Hamzeh[1] and Soumaya M. Hamzeh[1]

ABSTRACT. Stochastic exponential and monomial densities can serve as important sets of base functions both in linear and nonlinear functional analysis. The present paper is an extension of a work [1] of the first author to stochastic exponential densities that are compactly supported over $[a, b]$ in R. In addition to establishing certain stationarity properties for these densities, we announce a new stochastic operator which generates, via a nonlinear technique, unique approximants, over $[a, b]$, to functions $\mathfrak{g}(x) \in$ II (a certain noncommutative inner product space).

1. Introduction and Preliminaries

Random variables (RV's) like Y_λ, of a density $\rho_\lambda(x)$, $[Y_\lambda$: $\rho_\lambda(x) = \lambda^{-1} e^{-x/\lambda}$, $0 \leq x \leq \infty$ (the exponential density)], are clearly parameterized by λ. Other examples of such RV's can be, $\yen_\theta$: $\rho_\theta(x) = \frac{1}{\Gamma(\theta-1)} e^{-x} x^\theta$, $0 \leq x \leq \infty$, (the Gamma density) or H_n : $\rho_n(x) = \frac{1}{n!} e^{-x} x^n$, $0 \leq x \leq \infty$ (a Poisson like density), etc. Equivalently, Y_λ may also be conceived with a different parameterization in the form of Y_α : $u_\alpha(x) = \left(\ln \frac{1}{\alpha}\right) \alpha^x$, $0 \leq x \leq \infty$. Furthermore, during the process of transmission of random information, and for purposes of data contraction, densities of Y_λ, $\yen_\theta$ or H_n, which could be defined over $[0, \infty)$, may always be approximated by a monomial density, say, that is compactly

Received October 13, 2007.   * Corresponding author.
2000 *Mathematics Subject Classification.* 60H10, 93E15.
Key words and phrases. Stochastic exponential densities, monomial densities, variance stationarity, noncommutative inner product spaces, nonlinear approximation.
Partially supported by the Lebanese National Council for Scientific research, grant for basic research 53-08-03.
[1] Center for Research in Applied Mathematics and Statistics (CRAMS), Department of Basic Sciences, Arts, Science of Technology University in Lebanon (*E-mails:* nhaidar@suffolk.edu and nassar.haidar@aul.edu.lb(N.H.S. Haidar), adnan.hamzeh@aul.edu.lb(A.M. Hamzeh) and soumaya.hamzeh@aul.edu.lb(S.M. Hamzeh))

supported over a suitably chosen $[a, b]$ interval of the real line, i.e. by

$$(1.1) \qquad f_\alpha(x) = (\alpha + 1)x^\alpha / (b^{\alpha+1} - a^{\alpha+1}), \quad a \leq x \leq b.$$

This density invokes $(X_\alpha = x : [0, b] \to [a, b])$ on the probability space $\Im = ([a, b], \text{Œ}, F_\alpha)$, where Œ is the Borel σ-field, and F_α is the distribution whose density is $f_\alpha(x)$.

However, in addition to various monomial-type (Laurent, Taylor or Frobenius) series or power function representations [2], solutions to many problems in mathematics take the form of series of exponential functions of various types. These range inexhaustibly from the classical complex Dirichlet series to the real truncated exponential series solution for spectrometric integral equations, that was advanced by Haidar [3] in 2000. So in a similar fashion, after redefining $u_\alpha(x)$ over $[a, b]$ as

$$(1.2) \qquad h_\alpha(x) = (\ln \alpha)\alpha^x / (\alpha^b - \alpha^a), \quad a \leq x \leq b,$$

we invoke another real RV $(\gamma_\alpha = x : [0, b] \to [a, b])$ on the probability space $\aleph = ([a, b], \text{Æ}, B_\alpha)$, where Æ is the Borel σ-field, and B_α is the distribution whose density is $h_\alpha(x)$.

The expectations of X_α and γ_α are respectively given by

$$(1.3) \qquad E[X_\alpha] = G_\alpha\{x\} = \int_a^b x f_\alpha(x) dx = \mu^f(\alpha),$$

$$(1.4) \qquad E[\gamma_\alpha] = Q_\alpha\{x\} = \int_a^b x h_\alpha(x) dx = \mu^h(\alpha),$$

Moreover, with the RV's

$$(1.5) \qquad U_\alpha = g(f_\alpha(x)) = g_\alpha(x); \quad V_\alpha = q(h_\alpha(x)) = q_\alpha(x)$$

it is possible to associate the respective generalized moments

$$(1.6) \qquad G_\beta\{g_\alpha(x)\} = \int_a^b g_\alpha(x) f_\beta(x) dx; \quad Q_\beta\{q_\alpha(x)\} = \int_a^b q_\alpha(x) h_\beta(x) dx$$

As for $g_\alpha(x)$ (or $q_\alpha(x)$) we may consider now the negative logarithmic function of the density of the RV's X_α (or γ_α), defined viz,

$$g_\alpha(x) = \varphi_\alpha(x) = -\ln f_\alpha(x), \quad q_\alpha(x) = \psi_\alpha(x) = -\ln h_\alpha(x)$$

with $\varphi_\alpha(x)$ ($\psi_\alpha(x)$) as a convex functions defined on the convex set $[a, b]$.

The generalized moments

$$(1.7) \qquad G_\alpha\{\varphi_\alpha(x)\} = \int_a^b g_\alpha(x) f_\alpha(x) dx; \quad Q_\alpha\{\psi_\alpha(x)\} = -\int_a^b h_\alpha(x) \ln h_\alpha(x) dx$$

are clearly the Boltzmann S-entropies

$$(1.8) \qquad S[f_\alpha] = -\langle f_\alpha, \ln f_\alpha \rangle = G_\alpha\{\varphi_\alpha(x)\}; \quad S[h_\alpha] = -\langle h_\alpha, \ln h_\alpha \rangle = Q_\alpha\{\psi_\alpha(x)\}.$$

which can be employed as a measure of some uncertainty in various stochastic contexts [4].

In section 2 of this communication we study the stationarity of the X- and γ- and the U- and V- sequences. Section 3 reports on the asymptotic variance stationarity of the X- and γ-sequences. In section 4 we, 4.1, study Gram-Schmidt orthogonalization of the sets

$\{h_n(x)\}_{n=2}^{\infty}$ and $\{f_n(x)\}_{n=1}^{\infty}$ over $[a, b]$, 4.2 announce an asymptotic result on the cross correlation functions for both the $h_\alpha(x)$ and the $f_\alpha(x)$ functions, 4.3 introduce, a new kind of "correlative-convolutive" multiplication over a noncommutative inner product space II and develop an associated stochastic operator which generates, via an essentially nonlinear technique, unique low order spline approximants over $[a, b]$ to functions $\mathfrak{g}(x) \in$ II.

2. Substationarity of the X- and γ- and the U- and V-sequences

It is well known that an RV sequence $J = (J_\alpha; \alpha \in Z^+)$ is strictly stationary when the density $j_\alpha(x)$ of J_α satisfies

$$j_\alpha(x) \equiv j_{\alpha-1}(x), \quad \forall \alpha.$$

Moreover there exists a possibility for weakening the above condition for certain classes of semi-stationary random processes which happen to involve random sequences that are stationary at large, i.e. only in their various - order moments and with respect to the same densities. Consider then the sequence X (or γ) and the sequence $U = (U_\alpha = \varphi_\alpha(x);$ $\alpha \in Z^+)$, (or $V = (V_\alpha = \psi_\alpha(x); \alpha \in Z^+))$, where $U_\alpha : [a, b] \to R$, $(V_\alpha : [a, b] \to R)$, is defined on the same probability space Π_α of X_α, (or Φ_α of γ_α), in light of the definition that follows.

Definition 2.1. ([1]) A sequence $Y = (Y_\alpha = g_\alpha(x); \alpha \in Z^+)$ of real-valued RV's is said to be *uniformly stationary* if

$$G_\alpha\{g_\alpha(x)\} = G_{\alpha-1}\{g_{\alpha-1}(x)\}, \quad \forall \alpha.$$

Moreover if

$$G_\alpha\{g_\alpha(x)\} \geq (\leq)G_{\alpha-1}\{g_{\alpha-1}(x)\}, \quad \forall \alpha,$$

then Y is called *uniformly sub-(super-)stationary;* while if

$$G_\alpha\{g_\alpha(x)\} > (<)G_{\alpha-1}\{g_{\alpha-1}(x)\}, \forall \alpha,$$

then Y is called *uniformly strictly sub-(super-)stationary.*

Similarly, the stationarity of the sequence $\Upsilon = (\Upsilon_\alpha = q_\alpha(x) ; \alpha \in Z^+)$ is defined via the satisfaction of

$$Q_\alpha\{q_\alpha(x)\} \geq (\leq)Q_{\alpha-1}\{q_{\alpha-1}(x)\}, \quad \forall \alpha.$$

Definition 2.2. We say $b \gg a$ if $\lim_{b\to\infty}(b - a) = \infty$ for any fixed $a \in [0, \infty)$.

Consequently, since $\frac{a}{b} = \frac{a}{b-a}/[1 + \frac{a}{b-a}]$, then the following remark naturally follows.

Remark 2.1. If $b \gg a$ then $\lim_{b\to\infty}\frac{a}{b} = 0$.

2.1. The X- and γ-sequences

Theorem 2.1. *The sequences $X = (X_\alpha; \alpha \in Z^+)$ and $\gamma = (\gamma_\alpha; \alpha \in Z^+)$ are both uniformly substationary when $b \gg a > 0$.*

Proof. It is straightforward to show, when $b \gg a > 0$, that

$$E[X_\alpha] = \frac{(\alpha + 1)}{(\alpha + 2)}b, \quad \left(E[\gamma_\alpha] = b - \frac{1}{\ln \alpha}\right),$$

and

$$E[X_{\alpha-1}] = \frac{\alpha}{(\alpha+1)}, \quad \left(E[\gamma_{\alpha-1}] = b - \frac{1}{\ln(\alpha-1)}\right).$$

Therefore:

$$E[X_\alpha] = \left[1 + \frac{1}{\alpha(\alpha+2)}\right] E[X_{\alpha-1}], \quad \left(E[\gamma_\alpha] = \frac{1 - 1/b\ln\alpha}{1 - 1/b\ln(\alpha-1)} E[\gamma_{\alpha-1}]\right)$$

or

$$E[X_\alpha] \geq E[X_{\alpha-1}], \quad \forall\alpha, \quad (E[\gamma_\alpha] \geq E[\gamma_{\alpha-1}], \quad \forall\alpha),$$

i.e., X (or γ) is uniformly strictly substationary and the equality holds only when $\alpha \to \infty$; so

$$\lim_{\alpha\to\infty} E[X_\alpha] - E[X_{\alpha-1}]\} = 0, \quad \left(\lim_{\alpha\to\infty} E[\gamma_\alpha] - E[\gamma_{\alpha-1}]\} = 0\right).$$

The proof completes by taking into consideration that

$$E[X_\alpha] = G_\alpha\{x\}, \quad (E[\gamma_\alpha] = Q_\alpha\{x\}).$$

$\square$

2.2. U- and V-sequences

Let us introduce now the coefficient

(2.1) $$A_\alpha = \frac{1}{(\alpha+1)} - \frac{1}{\alpha}\ln\left(\frac{\alpha+1}{b^{\alpha+1} - a^{\alpha+1}}\right)$$

into (1.7) to enable writing

(2.2) $$G_\alpha\{\varphi_\alpha(x)\} = \alpha \left[A_\alpha(b^{\alpha+1} - a^{\alpha+1}) - (b^{\alpha+1}\ln b - a^{\alpha+1}\ln a)\right]/(b^{\alpha+1} - a^{\alpha+1}).$$

Further substitution of

(2.3) $$f_{\alpha+1}(b) = (\alpha+2)b^{\alpha+1}/(b^{\alpha+2} - a^{\alpha+2}),$$

and consideration of the notation

$$\Gamma_\alpha = \frac{\alpha}{(\alpha+2)} \frac{(b^{\alpha+2} - a^{\alpha+2})}{(b^{\alpha+1} - a^{\alpha+1})}$$

in (2.2) lead to

$$G_\alpha\{\varphi_\alpha(x)\} = S[f_\alpha] = \Gamma_\alpha\{f_{\alpha+1}(b)[A_\alpha - \ln b] - f_{\alpha+1}(a)[A_\alpha - \ln a]\}.$$

Following a similar procedure with

$$\Gamma_{\alpha-1} = \frac{(\alpha-1)}{(\alpha+1)} \frac{(b^{\alpha+1} - a^{\alpha+1})}{(b^\alpha - a^\alpha)}$$

yields

$$G_{\alpha-1}\{\varphi_{\alpha-1}(x)\} = S[f_{\alpha-1}] = \Gamma_{\alpha-1}\{f_\alpha(b)[A_{\alpha-1} - \ln b] - f_\alpha(a)[A_{\alpha-1} - \ln a]\}.$$

Theorem 2.2. *The sequences $U = (U_\alpha; \alpha \in Z^+)$ and $V = (V_\alpha; \alpha \in Z^+)$ are both uniformly strictly substationary when $b \gg a > 0$.*

Proof. Jensen's inequality is, strictly speaking, inapplicable here. Indeed, a formal analysis based on the Jensen's inequality leads to the uncertain results

$$G_\alpha\{\varphi_\alpha(x)\} \gtreqless G_{\alpha-1}\{\varphi_{\alpha-1}(x)\}; \quad Q_\alpha\{\psi_\alpha(x)\} \gtreqless Q_{\alpha-1}\{\psi_{\alpha-1}(x)\}.$$

Alternatively, consider $b \gg a$, first in

$$G_\alpha\{\varphi_\alpha(x) - G_{\alpha-1}\{\varphi_{\alpha-1}(x)\}$$
$$= S[f_\alpha] - S[f_{\alpha-1}]$$
$$= [\Gamma_\alpha A_\alpha f_{\alpha+1}(b) - \Gamma_{\alpha-1} A_{\alpha-1} f_\alpha(b)] - [\Gamma_\alpha f_{\alpha+1}(b) - \Gamma_{\alpha-1} f_\alpha(b)] \ln b$$
$$- [\Gamma_\alpha A_\alpha f_{\alpha+1}(a) - \Gamma_{\alpha-1} A_{\alpha-1} f_\alpha(a)] + [\Gamma_\alpha f_{\alpha+1}(a) - \Gamma_{\alpha-1} f_\alpha(a)] \ln a$$

to eliminate the last two terms in the relation above and to establish that

$$[\Gamma_\alpha f_{\alpha+1}(b) - \Gamma_{\alpha-1} f_\alpha(b)] \ln b \approx \ln b.$$

Under the same assumption of $b \gg a$, the first term in the first relation becomes

$$\Gamma_\alpha A_\alpha f_{\alpha+1}(b) - \Gamma_{\alpha-1} A_{\alpha-1} f_\alpha(b) = \frac{1}{\alpha(\alpha+1)} + \ln\left(\frac{\alpha}{\alpha+1}\right) + \ln b.$$

Substitution of

$$\varepsilon_\alpha = \frac{1}{\alpha(\alpha+1)} + \ln\left(\frac{\alpha}{\alpha+1}\right),$$

for which

$$\lim_{\alpha \to \infty} \varepsilon_\alpha = 0,$$

in the earlier expression to rewrite it as

$$S[f_\alpha] - S[f_{\alpha-1}] = \varepsilon_\alpha + \ln b > 0$$

Consequently, as $\ln b > 0$, we have, on one hand, that

$$G_\alpha\{\varphi_\alpha(x)\} > G_{\alpha-1}\{\varphi_{\alpha-1}(x)\}$$

On the other hand, we note that for sufficiently large $\alpha > 2$

$$Q_\alpha\{\psi_\alpha(x)\} = S[h_\alpha] = -\langle h_\alpha, \ln h_\alpha\rangle$$
$$= 1 - bh_\alpha(b) + ah_\alpha(a) + \ln(\alpha^b - \alpha^a) - \ln(\ln \alpha)$$

becomes negative definite. Consideration, secondly, of $b \gg a$ in $S[h_\alpha]$ and $S[h_{\alpha-1}]$ allows us to establish that

$$\frac{S[h_\alpha]}{S[h_{\alpha-1}]} = \frac{1 - \ln(\ln \alpha)}{1 - \ln[\ln(\alpha-1)]} \geqslant 1,$$

or

$$Q_\alpha\{\psi_\alpha(x)\} > Q_{\alpha-1}\{\psi_{\alpha-1}(x)\}.$$

i.e., both U and V-sequences are uniformly strictly substationary. $\qquad \square$

3. Asymptotic Variance Stationarity of the X- and γ-sequences

The existing correlation between $f_\alpha(x)$ and $f_{\alpha-1}(x)$, [or $h_\alpha(x)$ and $h_{\alpha-1}(x)$], with a continuous $\alpha \in T = Z^+$, as indicated by Theorem 2.1, may further be quantified by considering two different RV's $X_\alpha = X(\alpha)$ and $X_{\alpha+\tau} = X(\alpha + \tau)$, $[\gamma_\alpha = \gamma(\alpha)$ and $\gamma_{\alpha+\tau} = \gamma(\alpha + \tau)]$, on the same probability space with α and $\tau \in T = Z^+$ and having the density functions $f_\alpha(x)$ and $f_{\alpha+\tau}(x)$, $[h_\alpha(x)$ and $h_{\alpha+\tau}(x)]$, respectively.

By defining

$$\mu^f(\alpha + \tau) = E[X(\alpha + \tau) = \int_a^b x f_{\alpha+\tau}(x) dx;$$

$$\mu^h(\alpha + \tau) = E[\gamma(\alpha + \tau) = \int_a^b x h_{\alpha+\tau}(x) dx,$$

it can be easily shown, when $b \gg a > 0$, that

$$\mu^f(0) = E[X(0)] = b/2; \quad \mu^h(0) = E[\gamma(0)] = b,$$

while

$$\mu^f(\infty) = E[X(\infty)] = b; \quad \mu^h(\infty) = E[\gamma(\infty)] = b.$$

Clearly then the stochastic process $\{X(\alpha); \alpha \in Z^+\}$ cannot represent a process that could be weakly stationary, and not even in the mean. This fact alone cannot rule out, however, the possibility for this process to become stationary in some weak sense. Unlike $X(\alpha)$, the stochastic process $\{\gamma(\alpha); \alpha \in Z^+\}$ may, however, distinctively be weakly stationary in the mean. This, of course, shall not rule out the possibility for $\gamma(\alpha)$ to become also stationary in some other weak sense. In order to analyze such situations, the following additional notation [4] is required.

$$[\sigma^f(\alpha)]^2 = \int_a^b [x - \mu^f(\alpha)]^2 f_\alpha(x) dx;$$

$$[\sigma^h(\alpha)]^2 = \int_a^b [x - \mu^h(\alpha)]^2 h_\alpha(x) dx$$

and

$$[\sigma^f(\alpha + \tau)]^2 = \int_a^b [x - \mu^f(\alpha + \tau)]^2 f_{\alpha+\tau}(x) dx;$$

$$[\sigma^h(\alpha + \tau)]^2 = \int_a^b [x - \mu^h(\alpha + \tau)]^2 h_{\alpha+\tau}(x) dx$$

are respectively the squared standard deviations of $X(\alpha)$ and $X(\alpha+\tau)$, $[\gamma(\alpha)$ and $\gamma(\alpha+\tau)]$, and

$$\eta^f(\alpha, \tau) = \int_a^b [x - \mu^f(\alpha + \tau)][x - \mu^f(\alpha)] f_\alpha(x) dx;$$

$$\eta^h(\alpha, \tau) = \int_a^b [x - \mu^h(\alpha + \tau)][x - \mu^h(\alpha)] h_\alpha(x) dx$$

is formally similar to the covariance of $X(\alpha + \tau)$ and $X(\alpha)$, $[\gamma(\alpha)$ and $\gamma(\alpha + \tau)]$, while

$$\Re^f(\alpha, \tau) = \eta^f(\alpha, \tau)/\sigma^f(\alpha + \tau)\sigma^f(\alpha) = \sigma^f(\alpha)/\sigma^f(\alpha + \tau);$$

$$\Re^h(\alpha, \tau) = \eta^h(\alpha, \tau)/\sigma^h(\alpha + \tau)\sigma^h(\alpha) = \sigma^h(\alpha)/\sigma^h(\alpha + \tau)$$

is clearly a relative variance.

Definition 3.1. ([1]) A stochastic process $(Y(\alpha); \alpha \in Z^+)$ is *variance stationary* if its associated $\Re(\alpha, \tau)$ is independent of α.

Theorem 3.1. *For $b \gg a > 0$, the stochastic process $(X(\alpha); \alpha \in Z^+)$, $[$or $\gamma(\alpha); \alpha \in Z^+)]$, is characterized by*

$$(3.1) \qquad \Re^f(\alpha, \tau) = 1 + \frac{\tau}{\alpha}; \quad \Re^h(\alpha, \tau) = \frac{\ln(\alpha + \tau)}{\ln \alpha}$$

and is therefore asymptotically variance stationary.

Corollary 3.1. *Given the asymptotically variance stationary process of Theorem 3.1. Two different $X(\alpha)$ and $X(\alpha + \tau)$, $[$or $\gamma(\alpha)$ and $\gamma(\alpha + \tau)]$, RV's of this process can be correlated only when*

$$(3.2) \qquad \tau \in [-2\alpha, 0]\backslash\{\alpha\}; \quad \left[\tau \in \left[\frac{1}{\alpha} - \alpha, 0\right]\backslash\{\alpha\}\right]$$

Proof. The general domain $[-2\alpha, 0]$, $\left[$or $\left[\frac{1}{\alpha} - \alpha, 0\right]\right]$, in (3.2) follows from considering (3.1) in the well-established fact that

$$(3.3) \qquad -1 \le \Re^f(\alpha, \tau) \le 1; \quad -1 \le \Re^h(\alpha, \tau) \le 1.$$

Elimination of $\{\alpha\}$ from this domain is an obvious consequence of

$$\Re^f(\alpha, -\alpha) = 0; \quad \left[\Re^h(\alpha, \alpha) = 1 + \frac{\ln 2}{\ln \alpha}\right].$$

$\square$

Remark 3.1. The RV's X_α and $X_{\alpha-1}$, $[$or γ_α and $\gamma_{\alpha-1}]$, have when $b \gg a > 0$ and $\alpha \gg 3$, $[$or $\alpha \gg 1]$, the relative variance

$$\Re^f(\alpha, -1) = 1 - \frac{1}{\alpha}, \quad \left[\Re^h(\alpha, -1) = \frac{\ln(\alpha - 1)}{\ln \alpha}\right].$$

Observe also here that although the above $\Re^f(\alpha, -1)$, $[\Re^h(\alpha, -1)]$, is not a maximum for $\Re^f(\alpha, \tau)$, $[\Re^h(\alpha, \tau)]$, over $[-2\alpha, 0]\backslash\{\alpha\}$, $\left[$or $\left[\frac{1}{\alpha} - \alpha, 0\right]\backslash\{\alpha\}\right]$, it is however quite close to it, especially when $\alpha \to \infty$.

4. Low-Order Functional Splines over $[a, b]$

4.1. Orthogonalized Densities

Our interest here is in indexed sets of stochastic exponential and monomial densities. Clearly, the linearly independent sets $\{h_n(x)\}_{n=2}^\infty$ and $\{f_n(x)\}_{n=1}^\infty$ are both not orthogonal over $[a, b]$. Gram-Schmidt orthogonalization over an inner product space $\mho$ leads straightforwardly, however, to the associated orthogonal sets $\{\hbar_n(x)\}_{n=2}^\infty$, $\{\mathcal{L}_n(x)\}_{n=1}^\infty$, with the respective elements

$$\hbar_n(x) = h_n(x) - \sum_{j=2}^{n-1}(h_n, \hbar_j)\hbar_j(x); \quad \hbar_2(x) = h_2(x),$$

$$\mathcal{L}_n(x) = f_n(x) - \sum_{j=1}^{n-1}(f_n, \mathcal{L}_j)\mathcal{L}_j(x); \quad \mathcal{L}_1(x) = f_1(x).$$

These, no more density-like functions, can be utilized in the representation of functions $\mathfrak{g}(x)$ defined over $[a, b]$ viz.

$$\mathfrak{g}(x) = \sum_{j=2}^{\infty} \frac{(\mathfrak{g}, \hbar_j)}{(\hbar_j, \hbar_j)} \hbar_j(x) \quad \text{or} \quad \mathfrak{g}(x) = \sum_{j=1}^{\infty} \frac{(\mathfrak{g}, \mathcal{L}_j)}{(\mathcal{L}_j, \mathcal{L}_j)} \mathcal{L}_j(x).$$

4.2. Cross Correlations

Let us adopt for the cross correlation functions, over [a,b], of the monomial and exponential densities the respective notation,

$$\chi_{\alpha,\alpha-1}^{f}(x) = \int_a^b f_\alpha(x+\tau) f_{\alpha-1}(\tau) d\tau;$$

$$\chi_{\alpha,\alpha-1}^{h}(x) = \int_a^b h_\alpha(x+\tau) h_{\alpha-1}(\tau) d\tau,$$

in order to state the following result.

Theorem 4.1.

$$\lim_{\alpha \to \infty} \frac{\chi_{\alpha,\alpha-1}^{h}(x)}{\chi_{\alpha-1,\alpha}^{h}(x)} = 1^x = 1$$

and

$$\lim_{\alpha \to \infty} \frac{\chi_{\alpha,\alpha-1}^{f}(x)}{\chi_{\alpha-1,\alpha}^{f}(x)} = x^1 = x.$$

Proof. Details of the proof of this result may be found in [5].   □

4.3. Functional Splines

The previous facts serve as motivation for the implementation of a noncommutative orthogonality, over $[a, b]$, under the following correlative-convolutive product, $-\!\circ$, in a generalized inner product space, equipped with

$$(\mathfrak{g} -\!\circ y) = \int_a^b \int_a^b y(x-\nu) \int_a^b \mathfrak{g}(\tau) y(\nu+\tau) d\tau d\nu dx$$
$$= (y * \mathfrak{g} \leftrightarrow y) \in C; \quad y, \mathfrak{g} \in \mathrm{II},$$

with C: the complex field, II: is the space of y and $\mathfrak{g}$ with finite $(\mathfrak{g} -\!\circ y)$, and

$$\mathfrak{g} -\!\circ y = \int_a^b y(x-\nu) \int_a^b \mathfrak{g}(\tau) y(\nu+\tau) d\tau d\nu = y * \mathfrak{g} \leftrightarrow y.$$

Unlike the convolutive product $y * \mathfrak{g} = \mathfrak{g} * y$, the cross correlative product $\mathfrak{g} \leftrightarrow y = \chi_{\mathfrak{g},y}(x) = \chi_{y,\mathfrak{g}}(-x)$ is in general not commutative. For this reason $\mathfrak{g} -\!\circ y$ is conceived as $y * [\mathfrak{g} \leftrightarrow y] = [\mathfrak{g} \leftrightarrow y] * y$, i.e., $(\mathfrak{g} -\!\circ y) = (y * [\mathfrak{g} \leftrightarrow y]) = ([\mathfrak{g} \leftrightarrow y] * y)$, and in the notation of the stochastic operator

$$\wp^{y_\alpha}[] = ([] -\!\circ y_\alpha) : \mathrm{II} \longrightarrow C,$$

$([] -\!\circ y_\alpha)$ is in general noncommutative.

Definition 4.1. Two functions $z(x)$ and $y(x) \in \mathrm{II}$ are said to be *orthogonal*, under the $-\circ$ multiplication, over $[a, b]$ if

$$(z -\circ y) = \begin{cases} 0, & y \neq z, \\ (z -\circ z), & y = z. \end{cases}$$

After adopting the notation

$$\varepsilon_{\mathfrak{g}}^h(\sigma, \rho) = \frac{(\mathfrak{g} -\circ h_\sigma)^2}{(h_\sigma -\circ h_\sigma)^2}(h_\sigma^2 -\circ \mathfrak{g}^2) + \frac{(\mathfrak{g} -\circ h_\sigma)(\mathfrak{g} -\circ h_\rho)}{(h_\sigma -\circ h_\sigma)(h_\rho -\circ h_\rho)}(h_\sigma h_\rho -\circ \mathfrak{g}^2)$$

$$+ \frac{(\mathfrak{g} -\circ h_\rho)^2}{(h_\rho -\circ h_\rho)^2}(h_\rho^2 -\circ \mathfrak{g}^2) - 2\frac{(\mathfrak{g} -\circ h_\sigma)}{(h_\sigma -\circ h_\sigma)}(\mathfrak{g}h_\sigma -\circ \mathfrak{g}^2)$$

$$- 2\frac{(\mathfrak{g} -\circ h_\rho)}{(h_\rho -\circ h_\rho)}(\mathfrak{g}h_\rho -\circ \mathfrak{g}^2) + (\mathfrak{g}^2 -\circ \mathfrak{g}^2),$$

the following theory can then be implemented in II for the representation of $\mathfrak{g}(x)$ functions, defined over $[a, b]$.

Theorem 4.1. *For any $\mathfrak{g}(x) \in \mathrm{II}$, if σ and ρ are both positive roots of the simultaneous system of transcendental functional equations*

$$\frac{\partial}{\partial \sigma}\varepsilon_{\mathfrak{g}}^h(\sigma, \rho)d\sigma + \frac{\partial}{\partial \rho}\varepsilon_{\mathfrak{g}}^h(\sigma, \rho)d\rho = 0,$$

$$\varphi^h(\sigma, \rho) = (\sigma^{b-a} + \sigma^{a-b}) - (\rho^{b-a} + \rho^{a-b}) = 0,$$

then the exponential spline

$$\breve{\mathfrak{g}}(x) = \frac{(\mathfrak{g} -\circ h_\sigma)}{(h_\sigma -\circ h_\sigma)}h_\sigma(x) + \frac{(\mathfrak{g} -\circ h_\rho)}{(h_\rho -\circ h_\rho)}h_\rho(x) \approx \mathfrak{g}(x), \quad \forall x \in [a, b]$$

is a unique minimizer of the correlative-convolutive mean square error,

$$([\breve{\mathfrak{g}} - g]^2 -\circ \mathfrak{g}^2).$$

REFERENCES

1. N. S. H. Haidar, *On stochastic monomial densities and their entropy functions,* Random Oper. and Stoch. Equ. **9(3)** (2001), 219–234.
2. S. Talwong, V. Laohakosol and S. S. Cheng, *Power function solutions of iterative functional differential equations,* Appl. Math. E-notes **4** (2004), 160–163.
3. N. S. H. Haidar, *WKB theory for sampled unfolding,* Commun. Appl. Nonlinear Anal. **7(3)** (2000), 31–46.
4. R. S. Lipster and A. N. Shiryayev, *Statistics of Random Processes,* Springer-Verlag, New York, 1977.
5. N. S. H. Haidar, A. M. Hamzeh, S. M. Hamzeh and E. El-Nakat, *Stochastic exponential & monomial splines,* Random Oper. and Stoch. Equ. **12(4)**, (2004), 349–360.

Nonlinear Functional Analysis and Applications, Volume 1, 175–192
© 2012 Nova Science Publishers, Inc.

OPTIMAL CONTROL PROBLEMS FOR SEMILINEAR SECOND ORDER VOLTERRA INTEGRO-DIFFERENTIAL EQUATIONS IN HILBERT SPACE

Shin-ichi Nakagiri[1] and Jin-Soo Hwang[2,*]

This paper is dedicated to the memory of the late Prof. Taro Yoshizawa

ABSTRACT. We study the quadratic optimal control problems for the system described by the semilinear second order Volterra integro-differential equations in Hilbert space. Based on the variational method we prove the Gâteaux differentiability of the solution mapping from the space of controls to the space of solutions. For the quadratic cost problems involving distributive and terminal value observations we establish the necessary conditions of optimality for the costs in terms of the proper adjoint systems according to the types of nonlinear terms. An application to the semilinear wave equation with a fading memory is also given.

1. Introduction

In the present paper we study the quadratic optimal control problems for the systems described by the semilinear second order Volterra integro-differential evolution equations. The equations give the abstract models of nonlinear viscoelastic matrials with long memory. The second order Volterra integro-differential evolution equations has been studied extensively by many authors, among them we refer recent papers by Cavalcanti and Oquendo [2], Giorgi, Rivera and Pata [5], Rivera, Naso and Vegni [13], Tiehu and Guoxi

Received October 13, 2007.   * Corresponding author.
2000 *Mathematics Subject Classification.* 93C20, 35L70, 45K05.
Key words and phrases. Integro-differential equation, optimal control, transposition method, necessary condition of optimality.
The first author is supported by Grant-in-Aid for Scientific Research (C)(2) 16540194.
[1] Department of Applied Mathematics, Faculty of Engineering, Kobe University Nada, Kobe 657-8501, Japan (*E-mail*: nakagiri@kobe-u.ac.jp)
[2] The Graduate School of Science and Technology, Kobe University, Nada, Kobe 657-8501, Japan (*E-mail*: hwang@cs.kobe-u.ac.jp)

[15] and a book by Renardy, Hrusa and Nohel [12] and a monograph edited by Da Prato and Iannelli [3]. The above works are mainly focused on the operator theoretical treatment on existence, uniqueness and regularity of solutions. However, there are few reseaches on the variational treatment of second order nonautonomous Volterra equations (cf. Dautray and Lions [4]) and the related optimal control problems (cf. Seidman and Antman [14] and Hwang and Nakagiri [17]).

In [9] Lions has developed the quadratic optimal control theory for distributed parameter systems in full extent. The theory of Lions [9] covers a wide variety of distributed parameter systems, and especially he studied the optimal control problems for linear undamped second order control systems. Recently based on [9], Ha and Nakagiri [6] have extended his results on the optimal control problems for the semilinear damped second order systems

$$(1.1) \qquad\qquad y'' + A_2(t)y' + A_1(t)y = f(t,y) + Bv,$$

where $f(t,y)$ is a nonlinear forcing function, B is a controller and v is a control variable.

In this paper, we study quadratic optimal control problems as in [??] for the control systems described by the semilinear Volterra integro-differential equations of the form

$$(1.2) \qquad \begin{cases} y'' + A(t)y + \int_0^t K(t,s)y(s)ds = f(t,y,y') + Bv & \text{in } (0,T) \\ y(0) = y_0 \in V, \quad y'(0) = y_1 \in H, \end{cases}$$

where $A(t), K(t,\cdot)$ are time varying operators on appropriate Hilbert spaces embedded in a pivot Hilbert space H and $f(t,y,y')$ is a nonlinear forcing function. Here we note that the nonlinear term $f(t,y,y')$ contains the velocity term y', but the term is not considered in [6].

We have no general optimal control theory for (1.2) as of Lions [9] and the main aim here is to give new necessary conditions of optimality for the cost problem by using the method of transposition. The treatment is different from [6] becasue of the existence of y' term in $f(t,y,y')$.

We shall explain the main feature of this paper. Let $\mathcal{U}$ be a Hilbert space of control variables and $\mathcal{U}_{ad} \subset \mathcal{U}$ be an admissible set. We study the optimal control problem:

$$(1.3) \qquad\qquad \text{Minimize} \quad J(v) \text{ over } \mathcal{U}_{ad}$$

with the quadratic cost function

$$(1.4) \qquad\qquad J(v) = \|Cy(v) - z_d\|_M^2 + (Rv,v)_{\mathcal{U}}, \quad v \in \mathcal{U},$$

where $y = y(v)$ is the solution of (1.2). Here in (1.4), M is a Hilbert space of observation variables, z_d is a desired element in M and C is an observer, and R is a regulator on $\mathcal{U}$.

Our main concern for the control problem (1.3 is to characterize the necessary optimality conditions for nonlinear system (1.2) on the optimal control u, according to the specific types of observations C in time and space. For this purpose we have to show the Gâteaux differentiability of the nonlinear map $v \to y(v)$, which is used to define the associate adjoint system. The differentiability is proved by assuming the Fréchet differentiability of $f(t,y,y')$ in y and y'. Using the system and adjoint system equations, the necessary optimality conditions are completely characterized. These conditions extend the former results in [1, 9, 11] and others to two directions, one is nonlinear equations and the other is Volterra integro-differential equations.

This paper is composed of four sections. In Section 2, after giving the assumptions on operators $A(t), K(t, \cdot)$ and nonlinear Lipschitz continuous term $f(t, y, y')$, we state the existence, uniqueness and regularity results to the weak solutions of the system (1.2). Based on the results of Section 2 we study in detail the optimal control problems for the nonlinear system (1.2) with the quadratic cost (1.4) to be minimized on any admissible set $\mathcal{U}_{ad}$ in Section 3. That is, in Section 3 the necessary conditions of optimality are established for the distributive and terminal values observations by using the method of transposition (cf. [9], [10]). For the specific nonlinear term $f = f(t, y)$ we also give the necessary conditions without using the transposition method. In Section 4 we give an application to the optimal control problem for a semilinear wave equation with fading memory.

2. Volterra Integro-Differential Equations

Let H be a real pivot Hilbert space, and the inner product and the norm are denoted by $(\cdot, \cdot)$ and $|\cdot|$, respectively. Let V be a real separable Hilbert space with the norm $\|\cdot\|$. Assume that each pair (V, H) is a Gelfand triple space and that V is continuously embedded in H. We are given a family of symmmetric bilinear forms on $V \times V$. Let $a(t; \phi, \varphi), t \in [0, T]$ be a bilinear form on $V \times V$ satisfying

$$(2.1) \quad \begin{cases} (a) \ a(t; \phi, \psi) = a(t; \psi, \phi) \text{ for all } \phi, \psi \in V \text{ and } t \in [0, T], \\[4pt] (b) \ \text{there exists } c_1 > 0 \text{ such that } |a(t; \phi, \varphi)| \leq c_1 \|\phi\| \|\varphi\| \\ \quad \text{for all } \phi, \psi \in V \text{ and } t \in [0, T], \text{ and there exists } \alpha > 0 \text{ such that} \\ \quad a(t; \phi, \phi) \geq \alpha \|\phi\|^2 \text{ for all } \phi \in V \text{ and } t \in [0, T], \\[4pt] (c) \ \text{the function } t \to a(t; \phi, \varphi) \text{ is continuously differentiable} \\ \quad \text{in } [0, T] \text{ and there exists } c_2 > 0 \text{ such that} \\ \quad |a'(t; \phi, \varphi)| \leq c_2 \|\phi\| \|\varphi\| \text{ for all } \phi, \psi \in V \text{ and } t \in [0, T], \end{cases}$$

where $' = \frac{d}{dt}$. Then we can define the operator $A(t) \in \mathcal{L}(V, V')$ for $t \in [0, T]$ deduced by the relation

$$a(t; \phi, \varphi) = \langle A(t)\phi, \varphi \rangle \quad \text{for all } \phi, \varphi \in V,$$

where $\langle \cdot, \cdot \rangle$ denoted the duality pairing between V and V'.

Next, we consider a family of bilinear forms $k(t, s; \phi, \varphi)$ over $V \times V$ defined over $[0, T] \times [0, T]$ with

$$(2.2) \quad \begin{cases} (a) \ \text{there exists } k_0 > 0 \text{ such that } |k(t, s; \phi, \varphi)| \leq k_0 \|\phi\| \|\varphi\| \\ \quad \text{for all } \phi, \varphi \in V, \text{ and for all } (t, s) \in [0, T] \times [0, T], r \\[4pt] (b) \ \text{The function } (t, s) \to k(t, s; \phi, \varphi) \text{ is differentiable for all } \phi, \varphi \in V \\ \quad \text{and there exists } k_1 > 0 \text{ with } |\partial_t k(t, s; \phi, \varphi)| \leq k_1 \|\phi\| \|\varphi\| \\ \quad \text{for all } \phi, \varphi \in V, \text{ and for all } (t, s) \in [0, T] \times [0, T], \end{cases}$$

where $\partial_t k = \frac{\partial k}{\partial t}$. This family $k(t, s; \phi, \varphi)$ defines a family of operators $K(t, s) \in \mathcal{L}(V, V')$ by

$$(2.3) \qquad k(t, s; \phi, \varphi) = \langle K(t, s)\phi, \varphi \rangle.$$

We impose the following assumptions on the nonlinear term $f(t, y, z) : [0, T] \times V \times H \to H$ in (1.2):

(A1) $t \to f(t, y, z)$ is strongly measurable in H for all $(y, z) \in V \times H$;

(A2) there exists a $\beta \in L^2(0, T; \mathbf{R}^+)$ such that

$$|f(t, \xi_1, \eta_1) - f(t, \xi_2, \eta_2)| \leq \beta(t)(\|\xi_1 - \xi_2\| + |\eta_1 - \eta_2|) \quad \text{a.e. } t \in [0, T]$$

for all $\xi_i \in V$ and $\eta_i \in H$, $i = 1, 2$;

(A3) there exists a $\gamma \in L^2(0, T; \mathbf{R}^+)$ such that $|f(t, 0, 0)| \leq \gamma(t)$ a.e. $t \in [0, T]$.

We consider the following semilinear Volterra integro-differential equation

$$(2.4) \qquad \begin{cases} y'' + A(t)y + \int_0^t K(t, s)y(s)ds = f(t, y, y') + g & \text{in } (0, T), \\ y(0) = y_0 \in V, \quad y'(0) = y_1 \in H, \end{cases}$$

where f satisfies (A1)-(A3) and $g \in L^2(0, T; H)$. The solution Hilbert space $W(0, T)$ of (2.4) is defined by

$$W(0, T) = \{g | g \in L^2(0, T; V), g' \in L^2(0, T; H), g'' \in L^2(0, T; V')\},$$

endowed with the norm

$$\|g\|_{W(0,T)} = \left(\|g\|_{L^2(0,T;V)}^2 + \|g'\|_{L^2(0,T;H)}^2 + \|g''\|_{L^2(0,T;V')}^2 \right)^{\frac{1}{2}}.$$

We denote by $\mathcal{D}'(0, T)$ the space of distributions on $(0, T)$. A function $y = y(\cdot)$ is said to be a weak solution of (2.4) if $y \in W(0, T)$ and y satisfies

$$(2.5) \qquad \begin{cases} \langle y''(\cdot), \phi \rangle + a(\cdot; y(\cdot), \phi) + \int_0^{\cdot} k(\cdot, s; y(s), \phi)ds = (f(\cdot, y(\cdot), y'(\cdot)) + g(\cdot), \phi) \\ \quad \text{for all } \phi \in V \text{ in the sense of } \mathcal{D}'(0, T) \\ y(0) = y_0 \in V, \quad y'(0) = y_1 \in H. \end{cases}$$

The following theorem on the existence, uniqueness, regularity and energy equality of a weak solution of (2.4) is proved in Hwang and Nakagiri [8].

Theorem 2.1. *Assume that a and k satisfy (2.1) and (2.2), respectively and f satisfy (A1)-(A3). Then the equation (2.4) has a unique weak solution y in $W(0, T)$. Moreover, the solution $y \in C([0, T]; V)$, $y' \in C([0, T]; H)$ and the following energy equality holds for all $t \in [0, T]$:*

$$a(t; y(t), y(t)) + |y'(t)|^2$$

$$= a(0; y_0, y_0) + |y_1|^2 + \int_0^t a'(s; y(s), y(s))ds$$

$$(2.6)$$
$$+ 2\int_0^t (f(s, y(s), y'(s)) + g(s), y'(s))ds + 2\int_0^t k(s, s; y(s), y(s))ds$$

$$+ 2\int_0^t \left(\int_0^s \partial_t k(s, \sigma; y(\sigma), y(s))d\sigma - k(t, s; y(s), y(t)) \right)ds.$$

3. Quadratic Optimal Control Problems

In this section we study the quadratic cost optimal control problems for the semilinear Volterra integro-differential equations according to the framework of Lions [9]. Let $\mathcal{U}$ be a Hilbert space of control variables, and let B be an operator,

$$(3.1) \qquad\qquad B \in \mathcal{L}(\mathcal{U}, L^2(0, T; H)),$$

called a *controller* and $v \in \mathcal{U}$ be a control. We consider the following controlled semilinear Volterra integro-differential system:

$$(3.2) \quad \begin{cases} y''(v) + A(t)y(v) + \int_0^t K(t,s)y(v;s)ds = f(t, y(v), y'(v)) + Bv & \text{in } (0, T), \\ y(v; 0) = y_0 \in V, \quad y'(v; 0) = y_1 \in H. \end{cases}$$

Here in (3.2), $A(t)$, $K(t, \cdot)$ and $f(t, y, y')$ are differential operators and the nonlinear function satisfying the assumptions given in Section 2. By virtue of Theorem 2.1 and (3.1), we can define uniquely the solution map $v \to y(v)$ of $\mathcal{U}$ into $W(0, T)$, because $\tilde{f}(t, \xi, \eta) \equiv f(t, \xi, \eta) + Bv(t)$ satisfies the assumptions (A1)-(A3). We shall call the weak solution $y(v)$ of (3.2) the state of the control system (3.2). The observation of the state is assumed to be given by

$$(3.3) \qquad\qquad z(v) = Cy(v), \quad C \in \mathcal{L}(W(0, T), M),$$

where C is an operator called the *observer*, and M is a Hilbert space of observation variables. The quadratic cost function associated with the control system (3.2) is given by

$$(3.4) \qquad\qquad J(v) = \|Cy(v) - z_d\|_M^2 + (Rv, v)_\mathcal{U} \quad \text{for all} v \in \mathcal{U},$$

where $z_d \in M$ is a desired value of $z(v)$ and $R \in \mathcal{L}(\mathcal{U}, \mathcal{U})$ is symmetric and positive, i.e.,

$$(3.5) \qquad\qquad (Rv, v)_\mathcal{U} = (v, Rv)_\mathcal{U} \geq d\|v\|_\mathcal{U}^2$$

for some $d > 0$. Let $\mathcal{U}_{ad}$ be a closed convex subset of $\mathcal{U}$, which is called the admissible set. An element $u \in \mathcal{U}_{ad}$ which attains the minimum of $J(v)$ over $\mathcal{U}_{ad}$ is called an optimal control for the cost (3.4). The first one of nonlinear optimal control problems is the existence problem of optimal controls. For the problem we have no general answer under any natural condition, because the nonlinear term in (3.2) requires strong convergence of minimizing sequences. The strong convergence can not be derived under natural conditions on $\mathcal{U}_{ad}$ and B. In fact, we can show that if $\mathcal{U}_{ad}$ is compact or B is a compact operator, then there exists at least one optimal control u for (3.4). Since the compactness conditions seem to be severe, we omit to give the proof in this paper.

The second important one of the optimal control problems is to characterize the optimal controls by giving necessary conditions for optimality. For this it is necessary to write down the necessary condition for the cost (3.4)

$$(3.6) \qquad\qquad DJ(u)(v - u) \geq 0 \quad \text{for all } v \in \mathcal{U}_{ad}$$

and to analyze this inequality in view of the proper adjoint state system, where $DJ(u)$ denotes the Gâteaux derivative of $J(v)$ at $v = u$. In the calculation of $DJ(u)$, we need to calculate the Gâteaux derivative of $y(v)$ at $v = u$. That is, we have to prove that the mapping $v \to y(v)$ of $\mathcal{U}$ into $W(0, T)$ is Gâteaux differentiable at $v = u$.

The solution mapping $v \to y(v)$ of $\mathcal{U}$ into $W(0,T)$ is said to be Gâteaux differentiable at $v = u$ if for any $w \in \mathcal{U}$ there exists a $Dy(u) \in \mathcal{L}(\mathcal{U}, W(0,T))$ such that

$$\left\| \frac{1}{\lambda}(y(u + \lambda w) - y(u)) - Dy(u)w \right\|_{W(0,T)} \to 0 \quad \text{as } \lambda \to 0.$$

The operator $Dy(u)$ is called the *Gâteaux derivative* of $y(u)$ at $v = u$ and the function $Dy(u)w \in W(0,T)$ is called the Gâteaux derivative in the direction $w \in \mathcal{U}$.

Let us denote by $C^1(V,H)$ and $C^1(H)$ the sets of continuously Fréchet differentiable functions from V to H and H into H, respectively. Now, in order to obtain the Gâteaux differentiability of the solution map $v \to y(v)$, we impose the following further assumptions on the nonlinear term $f(t,y,z)$.

(A4) For each $t \in [0,T]$ and $z \in H$, $f(t,y,z) \in C^1(V,H)$, and for each $t \in [0,T]$, $f_y(t,y,z) \in C(V \times H, \mathcal{L}(V,H))$ and there is $\beta_1 \in L^2(0,T;\mathbf{R}^+)$ such that

$$\|f_y(t,y,z)\|_{\mathcal{L}(V,H)} \le \beta_1(t)(\|y\| + |z| + 1) \quad \text{a.e. } t \in [0,T].$$

(A5) For each $t \in [0,T]$ and $y \in V$, $f(t,y,z) \in C^1(H)$ and $f_z(t,y,z) \in C(V \times H, \mathcal{L}(H))$, and there is $\beta_2 \in L^2(0,T;\mathbf{R}^+)$ such that

$$\|f_z(t,y,z)\|_{\mathcal{L}(H)} \le \beta_2(t)(\|y\| + |z| + 1) \quad \text{a.e. } t \in [0,T].$$

Theorem 3.1. *Assume that (A4) and (A5) hold. Then the map $v \to y(v)$ of $\mathcal{U}$ into $W(0,T)$ is Gâteaux differentiable at $v = u$ and such the Gâteaux derivative of $y(v)$ at $v = u$ in the direction $w \in \mathcal{U}$, say $z = Dy(u)w$, is a unique weak solution satisfying the following equation*

(3.7)
$$\begin{cases} z'' + A(t)z + \int_0^t K(t,s)z(s)ds = f_y(t,y(u;t),y'(u;t))z \\ \quad + f_z(t,y(u;t),y'(u;t))z' + Bw \quad \text{in } (0,T), \\ z(0) = 0, \quad z'(0) = 0. \end{cases}$$

Proof. Let $\lambda \in (-1,1)$, and let $y_\lambda \equiv y(u + \lambda w)$ and $y \equiv y(u)$ be the weak solutions of (3.2) corresponding to $u + \lambda w$ and u, respectively. We set $z_\lambda = \lambda^{-1}(y_\lambda - y)$, $\lambda \ne 0$. Then z_λ is a unique weak solution of

(3.8)
$$\begin{cases} z_\lambda'' + A(t)z_\lambda + \int_0^t K(t,s)z_\lambda(s)ds \\ \quad = \frac{1}{\lambda}\{f(t,y_\lambda,y_\lambda') - f(t,y,y')\} + Bw \quad \text{in } (0,T), \\ z_\lambda(0) = 0, \quad z_\lambda'(0) = 0. \end{cases}$$

By the energy equality (2.6), z_λ satisfies

$$
\begin{aligned}
a(t; & z_\lambda(t), z_\lambda(t)) + |z'_\lambda(t)|^2 \\
&= \int_0^t a'(s; z_\lambda(s), z_\lambda(s))ds + 2\int_0^t (Bw(s), z'_\lambda(s))ds \\
&\quad + 2\int_0^t \int_0^s \partial_t k(s, \sigma; z_\lambda(\sigma), z_\lambda(s))d\sigma ds \\
&\quad - 2\int_0^t k(t, s; z_\lambda(s), z_\lambda(t))ds + 2\int_0^t k(s, s; z_\lambda(s), z_\lambda(s))ds \\
&\quad + 2\int_0^t (\lambda^{-1}[f(s, y_\lambda(s), y'_\lambda(s)) - f(s, y(s), y'(s))], z'_\lambda(s))ds.
\end{aligned}
$$
(3.9)

Note that the last term of (3.9) can be estimated as

$$
\begin{aligned}
&\left| 2\int_0^t (\lambda^{-1}[f(s, y_\lambda(s), y'_\lambda(s)) - f(s, y(s), y'(s))], z'_\lambda(s))ds \right| \\
&\quad \leq \frac{1}{\epsilon}\int_0^t \beta^2(s)(\|z_\lambda(s)\|^2 + |z'_\lambda(s)|^2)ds + 2\epsilon\int_0^t |z'_\lambda(s)|^2 ds
\end{aligned}
$$
(3.10)

for any $\epsilon > 0$. It is evident that

$$
2\left| \int_0^t (Bw(s), z'_\lambda(s))ds \right| \leq \|Bw\|^2_{L^2(0,T;H)} + \int_0^t |z'_\lambda(s)|^2 ds
$$
(3.11)

Also, from (A2) we have the estimate

$$
\begin{aligned}
2\left| \int_0^t k(t, s; z_\lambda(s), z_\lambda(t))ds \right| &\leq 2k_0\|z_\lambda(t)\|\int_0^t \|z_\lambda(s)\|ds \\
&\leq \epsilon\|z_\lambda(t)\|^2 + c_\epsilon\int_0^t \|z_\lambda(s)\|^2 ds
\end{aligned}
$$
(3.12)

for any $\epsilon > 0$ and some $c_\epsilon > 0$, and the estimates

$$
(3.13) \quad
\begin{cases}
2\left| \int_0^t \int_0^s \partial_t k(s, \sigma; z_\lambda(\sigma), z_\lambda(s))d\sigma ds \right| \leq 2k_1\left(\int_0^t \|z_\lambda(s)\|ds \right)^2 \\
2\left| \int_0^t k(s, s; z_\lambda(s), z_\lambda(s))ds \right| \leq 2k_0\int_0^t \|z_\lambda(s)\|^2 ds.
\end{cases}
$$

Therefore by using (A1), (3.10) and (3.12)-(3.13), we obtain the following inequality

$$
\begin{aligned}
|z'_\lambda(t)|^2 & + \alpha\|z_\lambda(t)\|^2 \\
&\leq \epsilon\|z_\lambda(t)\|^2 + (c_2 + 2k_0 + c_\epsilon)\int_0^t \|z_\lambda(s)\|^2 ds + 2k_1\left(\int_0^t \|z_\lambda(s)\|ds \right)^2 \\
&\quad + \|Bw\|^2_{L^2(0,T;H)} + (2\epsilon + 1)\int_0^t |z'_\lambda(s)|^2 ds \\
&\quad + \int_0^t \frac{\beta(s)^2}{\epsilon}(\|z_\lambda(s)\|^2 + |z'_\lambda(s)|^2)ds.
\end{aligned}
$$
(3.14)

Since

$$\left(\int_0^T \|z_\lambda(s)\| ds \right)^2 \leq T \int_0^T \|z_\lambda(s)\|^2 ds,$$

as shown in [8], (3.14) imlpies the inequality

(3.15)
$$|z_\lambda'(t)|^2 + (\alpha - \epsilon)\|z_\lambda(t)\|^2$$
$$\leq C_\epsilon(\|Bw\|_{L^2(0,T;H)}^2 + \int_0^t h(s)(\|z_\lambda(s)\|^2 + |z_\lambda'(s)|^2)ds)$$

for some constant $C_\epsilon > 0$ and some $h \in L^1(0,T;\mathbf{R}^+)$. Then, by taking $\epsilon = \frac{\alpha}{2} > 0$ and by applying the Gronwall's inequality to (3.15), we have

(3.16) $$\|z_\lambda(t)\|^2 + |z_\lambda'(t)|^2 \leq K\|Bw\|_{L^2(0,T;H)}^2 \exp(K\|h\|_{L^1(0,T;\mathbf{R}^+)}),$$

so that

(3.17) $$\|y_\lambda(t) - y(t)\|^2 + |y_\lambda'(t) - y'(t)|^2 \leq K\lambda^2\|Bw\|_{L^2(0,T;H)}^2 \exp(K\|h\|_{L^1(0,T;\mathbf{R}^+)})$$

for some $K > 0$. Therefore there exists a $z \in W(0,T)$ and a sequence $\{\lambda_k\} \subset (-1,1)$ tending to 0 such that

(3.18)
$$\begin{cases} z_{\lambda_k} \to z \text{ weak star in } L^\infty(0,T;V) \text{ and weakly in } L^2(0,T;V) \quad \text{as } k \to \infty, \\ z_{\lambda_k}' \to z' \text{ weak star in } L^\infty(0,T;H) \text{ and weakly in } L^2(0,T;H) \quad \text{as } k \to \infty, \\ z(0) = 0, \quad z'(0) = 0. \end{cases}$$

Let us prove that

(3.19) $$\frac{1}{\lambda_k}\{f(t,y_{\lambda_k},y_{\lambda_k}') - f(t,y,y'))\} \to f_y(t,y,y')z \quad \text{weakly in } L^2(0,T;H),$$

and

(3.20) $$\frac{1}{\lambda_k}\{f(t,y,y_{\lambda_k}') - f(t,y,y')\} \to f_z(t,y,y')z' \quad \text{weakly in } L^2(0,T;H),$$

as $k \to \infty$. From (A4), it is followed by the integral mean value theorem for Fréchet differentiable functions that

$$\frac{1}{\lambda_k}\{f(t,y_{\lambda_k},y_{\lambda_k}') - f(t,y,y_{\lambda_k}')\} = \int_0^1 f_y(t,\theta_1 y_{\lambda_k} + (1-\theta_1)y, y_{\lambda_k}')d\theta_1 \, z_{\lambda_k}.$$

Since $f_y(t,y,z)$ is continuous on y and z in $\mathcal{L}(V,H)$ by (A4), $y_{\lambda_k} \to y$ strongly in $C([0,T];V)$ and $y_{\lambda_k}' \to y'$ strongly in $C([0,T];H)$ as $k \to \infty$ by (3.17), we have

(3.21) $$f_y(t,\theta_1 y_{\lambda_k} + (1-\theta_1)y, y_{\lambda_k}') \to f_y(t,y,y') \quad \text{in } \mathcal{L}(V,H)$$

uniformly in $\theta_1 \in [0,1]$. Similarily by (A5), we verify that

(3.22) $$f_z(t,y,\theta_2 y_{\lambda_k}') + (1-\theta_2)y') \to f_z(t,y,y') \quad \text{in } \mathcal{L}(H)$$

uniformly in $\theta_2 \in [0,1]$. Then by (3.21) and (3.22), we have

(3.23) $$\int_0^1 f_y(t,\theta_1 y_{\lambda_k} + (1-\theta_1)y, y_{\lambda_k}')d\theta_1 \to f_y(t,y,y') \quad \text{a.e. } t \in [0,T],$$

and

$$(3.24) \qquad \int_0^1 f_z(t, y, \theta_2 y'_{\lambda_k} + (1 - \theta_2)y')d\theta_2 \to f_z(t, y, y') \quad \text{a.e. } t \in [0, T].$$

For simplicity of calculations we set

$$\Psi_\lambda(t) = \int_0^1 f_y(t, \theta_1 y_\lambda + (1 - \theta_1)y, y'_\lambda)d\theta_1,$$

$$\Phi_\lambda(t) = \int_0^1 f_z(t, y, \theta_2 y'_\lambda + (1 - \theta_2)y')d\theta_2.$$

Again by (3.17), there is a k_0 independent of t such that for all

$$(3.25) \qquad \|y_{\lambda_k}(t)\| \le (\|y(t)\| + 1) \quad \text{and} \quad |y'_{\lambda_k}(t)| \le (|y'(t)| + 1).$$

Then, by (3.25) and assumptions (A4) and (A5) we have for all $k \ge k_0$

$$(3.26) \qquad \|\Psi_{\lambda_k}(t)\|_{\mathcal{L}(V,H)} \le \beta_1(t)(\|\overline{y}\|_{L^\infty(0,T;V)} + \|y'\|_{L^\infty(0,T;H)} + 2),$$

$$(3.27) \qquad \|\Phi_{\lambda_k}(t)\|_{\mathcal{L}(H)} \le \beta_2(t)(\|y\|_{L^\infty(0,T;V)} + \|y'\|_{L^\infty(0,T;H)} + 2).$$

Hence, for each $\phi \in L^2(0, T; H)$, we have

$$(3.28) \qquad \begin{aligned} &\left| \int_0^T (\Psi_{\lambda_k}(t) - f_y(t, y, y'))z_{\lambda_k}, \phi)dt \right| \\ &\le \int_0^T \|\Psi_{\lambda_k}(t) - f_y(t, y, y')\|_{\mathcal{L}(V,H)} \|z_{\lambda_k}(t)\| |\phi(t)| dt \\ &\le \|z_{\lambda_k}\|_{L^\infty(0,T;V)} \int_0^T \|\Psi_{\lambda_k}(t) - f_y(t, y, y')\|_{\mathcal{L}(V,H)} |\phi(t)| dt \end{aligned}$$

and

$$(3.29) \qquad \begin{aligned} &\left| \int_0^T ((\Phi_{\lambda_k}(t) - f_z(t, y, y'))z'_{\lambda_k}(t), \phi(t))dt \right| \\ &\le \|z'_{\lambda_k}\|_{L^\infty(0,T;H)} \int_0^T \|\Phi_{\lambda_k}(t) - f_z(t, y, y')\|_{\mathcal{L}(H,V'_2)} |\phi(t)| dt. \end{aligned}$$

Here we note that the integrands in the last term of (3.28) and (3.29) are bounded by an L^1-integrable function due to (3.26) and (3.27) $\beta_1, \beta_2 \in L^2(0, T; \mathbf{R}^+) \subset L^1(0, T; \mathbf{R}^+)$. By using the Lebesgue dominated convergence theorem thanks to (3.26) and (3.27), we see that the both last terms of (3.28) and (3.29) converge to 0. On the other hand, by (3.18), (A4) and (A5) it is evident that

$$(3.30) \qquad \int_0^T (f_y(t, y, y')(z(t) - z_{\lambda_k}(t)), \phi(t))dt \to 0$$

and

$$(3.31) \qquad \int_0^T (f_z(t, y, y')(z'(t) - z'_{\lambda_k}(t)), \phi(t))dt \to 0.$$

Hence we deduce from (3.28)-(3.31) the desired convergence (3.19) and (3.20). Now let us take $k \to \infty$ in (3.8) with $\lambda = \lambda_k$ by using (3.18), (3.19) and (3.20). Then the element $z \in W(0, T)$ satisfies the equation (3.7) in the weak sense. If we take $f(t, \xi, \eta) \equiv f_y(t, y(t), y'(t))\xi + f_z(t, y(t), y'(t))\eta + B(w)(t) \in L^2(0, T; H)$ in Theorem 2.1, then this equation has a unique weak solution $z \in W(0, T)$. In fact, by (A4) and (A5) the term $f_y(t, y(t), y'(t))\xi$ and $f_z(t, y(t), y'(t))\eta$ satisfies

$$\begin{aligned} |f_y(t, y(t), y'(t))\xi| &\leq \|f_y(t, y(t), y'(t))\|_{\mathcal{L}(V,H)} \|\xi\| \\ &\leq \beta_1(t)(\|y\|_{L^\infty(0,T;V)} + \|y'\|_{L^\infty(0,T;H)} + 1)\|\xi\| \end{aligned}$$

and

$$|f_z(t, y(t), y'(t))\eta| \leq \beta_2(t)(\|y\|_{L^\infty(0,T;V)} + \|y'\|_{L^\infty(0,T;H)} + 1)|\eta|,$$

so that the assumption (A2) is satisfied. This means that z is a weak solution of (3.7). Hence by (3.18), (3.19) and (3.20) we see that $z_\lambda \to z = Dy(u)w$ weakly in $W(0, T)$ as $\lambda \to 0$.

It remains now to show the strong convergency of z_λ in $W(0, T)$. Since z and z_λ are weak solutions of equations (3.7) and (3.8) respectively, the difference $\chi_\lambda = z - z_\lambda$ satisfies the following equation

(3.32)
$$\begin{cases} \chi_\lambda'' + A(t)\chi_\lambda \int_0^t K(t, s)\chi_\lambda(s)ds \\ \quad = (f_z(t, y, y') - \Phi_\lambda(t))\chi_\lambda' + (f_y(t, y, y') - \Psi_\lambda(t))\chi_\lambda \\ \qquad + (f_y(t, y, y') - \Psi_\lambda(t))z_\lambda + (f_z(t, y, y') - \Phi_\lambda(t))z_\lambda' \\ \qquad + \Phi_\lambda(t)\chi_\lambda' \quad \text{in } (0, T), \\ \chi_\lambda(0) = 0, \quad \chi_\lambda'(0) = 0 \end{cases}$$

in the weak sense. Applying again the energy equality (2.6) to (3.32), we can deduce

$$\begin{aligned} & a(t; \chi_\lambda(t), \chi_\lambda(t)) + |\chi_\lambda'(t)|^2 \\ & = \int_0^t a'(s; \chi_\lambda(s), \chi_\lambda(s))ds + 2\int_0^t \int_0^s \partial_t k(s, \sigma; \chi_\lambda(\sigma), \chi_\lambda(s))d\sigma ds \\ & \quad - 2\int_0^t k(t, s; \chi_\lambda(s), \chi_\lambda(t))ds + 2\int_0^t k(s, s; \chi_\lambda(s), \chi_\lambda(s))ds \\ & \quad + 2\int_0^t (f_z(s, y, y') - \Phi_\lambda(s))\chi_\lambda'(s), \chi_\lambda'(s))ds \\ & \quad + 2\int_0^t ((f_y(s, y, y') - \Psi_\lambda(s))\chi_\lambda(s), \chi_\lambda'(s))ds \\ & \quad + 2\int_0^t ((f_y(s, y, y') - \Psi_\lambda(s))z_\lambda(s), \chi_\lambda'(s))ds \\ & \quad + 2\int_0^t (\Phi_\lambda(s)\chi_\lambda'(s), \chi_\lambda'(s))ds + 2\int_0^t ((f_z(s, y, y') - \Psi_\lambda(s))z_\lambda'(s), \chi_\lambda'(s))ds. \end{aligned}$$

To show the strong convergence of z_λ in $W(0, T)$, we set

(3.34) $\qquad \epsilon_1(\lambda) = \left\{ \int_0^T |(\Psi_\lambda(t) - f_y(t, y, y'))z_\lambda(t)| dt \right\}^{\frac{1}{2}} \|\chi_\lambda'\|_{L^2(0,T;H)},$

(3.35) $\qquad \epsilon_2(\lambda) = \left\{ \int_0^T |(\Phi_\lambda(t) - f_z(t, y, y'))z_\lambda'(t)|^2 d\sigma \right\}^{\frac{1}{2}} \|\chi_\lambda'\|_{L^2(0,T;H)}.$

Further, if we set

$$C_1 = \sup_{-1 \leq \lambda \leq 1, 0 \leq t \leq T} \{\|y_\lambda(t)\| + \|y(t)\| + |y_\lambda'(t)| + |y'(t)| + 1\},$$

then by the assumptions (A4) and (A5) and the definitions of Ψ_λ and Φ_λ we easily verify that

$$\|\Psi_\lambda(t)\|_{\mathcal{L}(V,H)} \leq C_1 \beta_1(t), \quad \|\Phi_\lambda(t)\|_{\mathcal{L}(H)} \leq C_1 \beta_2(t).$$

Hence we can deduce that

$$|(\Psi_\lambda(t) - f_y(t, y, y'))z(s)| \leq 2C_1\beta_1(t)\|z(t)\| \leq 2C_1\beta_1(t) \sup_{0 \leq t \leq T} \|z(t)\|,$$

$$|(\Phi_\lambda(t) - f_z(t, y, y'))z'(t)| \leq 2C_1\beta_2(t)|z'(t)| \leq 2C_1\beta_2(t) \sup_{0 \leq t \leq T} |z'(t)|$$

and

$$\lim_{\lambda \to 0} \|\Psi_\lambda(t) - f_y(t, y, y')\|_{\mathcal{L}(V,H)} \to 0 \quad \text{a.e. } t \in [0, T],$$

$$\lim_{\lambda \to 0} \|\Phi_\lambda(t) - f_z(t, y, y')\|_{\mathcal{L}(H)} \to 0 \quad \text{a.e. } t \in [0, T].$$

Therefore by the Lebesgue dominated convergence theorem

(3.36) $\qquad \begin{cases} \lim_{\lambda \to 0} \int_0^T |(\Psi_\lambda(t) - f_y(t, y, y'))z(t)|^2 dt = 0, \\ \lim_{\lambda \to 0} \int_0^T |(\Phi_\lambda(t) - f_z(t, y, y'))z'(t)|^2 dt = 0. \end{cases}$

Since $\{z_\lambda\}$ and $\{z_\lambda'\}$ are bounded in $L^2(0, T; V)$ and $L^2(0, T; H)$, respectively, we have by (3.34) and (3.35)

$$\lim_{\lambda \to 0} \epsilon_1(\lambda) = \lim_{\lambda \to 0} \epsilon_2(\lambda) = 0.$$

From (3.33), as in the derivation of (3.15), it follows by using Schwartz inequality that

(3.37)
$$\|\chi_\lambda(t)\|^2 + |\psi_\lambda'(t)|^2$$
$$\leq C\left(\epsilon_1(\lambda) + \epsilon_2(\lambda) + \int_0^t (1 + \beta_1(\sigma)^2 + \beta_2(\sigma)^2)(\|\chi_\lambda\|^2 + |\chi_\lambda'|^2) d\sigma \right)$$

for some $C > 0$. Hence by applying the Gronwall's inequality to (3.37) we reach

(3.38)
$$\|\chi_\lambda\|^2 + |\chi_\lambda'(t)|^2$$
$$\leq C(\epsilon_1(\lambda) + \epsilon_2(\lambda)) \exp\left\{ C \int_0^t (1 + \beta_1(\sigma)^2 + \beta_2(\sigma)^2) d\sigma \right\}.$$

Consequently, we obtain

(3.39) $\qquad z_\lambda(t) \to z(t) \quad \text{in } V \quad \text{and} \quad z_\lambda'(t) \to z'(t) \quad \text{in } H \quad \text{uniformly in } t \in [0, T].$

At the same time we see

(3.40) $z_\lambda' \to z'$ strongly in $L^2(0, T; H)$,

and hence

(3.41) $z_\lambda'' \to z''$ strongly in $L^2(0, T; V')$.

This shows the strong convergence $z_\lambda \to z$ in $W(0, T)$. This completes the proof.   □

In what follows we suppose the existence of an optimal control u for the cost (3.4). In order to give the form of necessary optimality conditions, we assume that the observer C is the product of distributive and terminal values observations. That is, the observation space M is a product of Hilbert spaces M_1 and M_2 and the observer C is given by

(3.42) $C = (C_1, C_2) \in \mathcal{L}(L^2(0, T; H), M_1) \times \mathcal{L}(H, M_2)$

and the cost is given by

(3.43) $J(v) = \|C_1 y(v) - z_d^1\|_{M_1}^2 + \|C_2 y(v; T) - z_d^2\|_{M_2}^2 + (Rv, v)_{\mathcal{U}}, \quad \forall v \in \mathcal{U}$,

where $z_d = (z_d^1, z_d^2) \in M_1 \times M_2$. Since $y \in C([0, T]; V) \cap C^1([0, T]; H)$ by Theorem 3.1, the above observation is meaningful. For the cost (3.43), the necessary condition on u is given by

(3.44)
$$\int_0^T (C_1^* \Lambda_{M_1}(C_1 y(u) - z_d^1)(t), z(t))dt$$
$$+ \ (C_2^* \Lambda_{M_2}(C_2 y(u; T) - z_d^2), z(T)) + (Ru, v - u)_{\mathcal{U}} \geq 0, \quad \forall v \in \mathcal{U}_{ad},$$

where $z = Dy(u)(v - u)$ and Λ_{M_i} is the canonical isomorphism from M_i onto M_i', $i = 1, 2$. In the above observation C in (3.42), we cannot construct an adjoint equation directly by the weak solution approach as in [6]. Because they must involve the differentiation term $\frac{d}{dt}(f_z^*(t, y(u; t), y'(u; t)))p$, where p is an adjoint state and $f_z^*(t, y, z)$ denotes the adjoint operator of $f_z(t, y, z)$. However, the desired regularity of the term to guarantee the existence of p as a weak solution of the adjoint system is missing. So we introduce a modified transposition suitable for the semilinear Volterra equations involving $f(t, y, y')$ term.

For each $g \in L^2(0, T; H)$, we have a unique weak solution $\psi = \psi(g) \in W(0, T)$ of the following equation

(3.45)
$$\begin{cases} \psi'' + A(t)\psi + \int_0^t K(t, s)\psi(s)ds \\ \quad = f_y(t, y(u; t), y'(u; t))\psi + f_z(t, y(u; t), y'(u; t))\psi' + g \quad \text{in } (0, T), \\ \psi(0) = 0, \quad \chi'(0) = 0, \end{cases}$$

because $f(t, \xi, \eta) = f_y(t, y(u; t), y'(u; t))\xi + f_z(t, y(u; t), y'(u; t))\eta + g(t)$ satisfies the assumptions (A1)-(A3) by (A4) and (A5).

Let us define the space

$$X \equiv \{\psi = \psi(g) \mid \psi \text{ satisfies (3.45) with } g \in L^2(0, T; H)\}.$$

It is seen in Theorem 2.1 that $X \subset W(0, T) \cap C([0, T]; V) \cap C^1([0, T]; H)$. We give an inner product $(\cdot, \cdot)_X$ on X by $(\psi_1, \psi_2)_X = (g_1, g_2)_{L^2(0, T; H)}$, where ψ_1, ψ_2 are the weak solutions of (3.45) for given $g = g_1, g_2 \in L^2(0, T; H)$, respectively. We see easily that

$(X, (\cdot, \cdot)_X)$ is a Hilbert space. Further, we can see that the mapping $\Theta : X \to L^2(0, T; H)$ defined by

(3.46)
$$\Theta\psi = \psi'' + A(\cdot)\psi + \int_0^\cdot K(\cdot, s)\psi(s)ds - f_y(\cdot, y(u), y'(u))\psi$$
$$- f_z(\cdot, y(u), y'(u))\psi'$$

is an isomorphism. Hence for each continuous linear functional $L : X \to \mathbf{R}$, there exists uniquely a $p = p_L \in L^2(0, T; H)$ such that

(3.47)
$$\int_0^T (p(t), \Theta\psi(t))dt = L(\psi), \quad \forall\psi \in X.$$

For $g \in L^1(0, T; V')$, $p_0 \in H$ and $p_1 \in V'$, let us define the functional $L = L(g, p_0, p_1)$ by

(3.48)
$$L(\psi) = \int_0^T \langle g(t), \psi(t)\rangle dt + \langle p_1, \psi(T)\rangle - (p_0, \psi'(T)).$$

Then this L is linear on X. Next we shall show the boundedness of L. It is easily checked from the fact $\psi \in X \subset C([0, T]; V) \cap C^1([0, T]; H)$ that

(3.49) $|L(\psi)| \leq (\|g\|_{L^1(0,T;V')} + \|p_1\|_{V'} + |p_0|)(\|\psi\|_{C([0,T];V)} + \|\psi(T)\| + |\psi'(T)|).$

Since $\psi \in X$, by the definition of X there exists a $g(\psi) \in L^2(0, T; H)$ such that (3.45) holds with $g = g(\psi)$. If we take $f(t, \psi, \phi) = f_y(t, y(u;t), y'(u;t))\psi + f_z(t, y(u;t), y'(u;t))\phi + g(\psi)(t)$, then by (A4) and (A5) this f satisfies (A2) and (A3) in which $\beta(t)$ and $\gamma(t)$ are given by $\beta(t) = (\beta_1(t)+\beta_2(t))(\|y(u)\|_{L^\infty(0,T;V)}+|y'(u)|_{L^\infty(0,T;H)}+1)$ and $\gamma(t) = |g(\psi;t)|$, respectively. Hence by the the energy equality in Theorem 2.1 we can verify that ψ satisfies

(3.50)
$$\|\psi(t)\|^2 + |\psi'(t)|^2 \leq c\|g(\psi)\|^2_{L^2(0,T;H)},$$

where c is a constant depending only on the above β. Since $\psi \in C([0, T]; V) \cap C^1([0, T]; H)$, we have from (3.50)

(3.51)
$$\|\psi\|_{C([0,T];V)} + \|\psi'\|_{C([0,T];H)} \leq K_1\|g(\psi)\|_{L^2(0,T;H)},$$

where $K_1 > 0$ is some constant independent of ψ. Consequently, it follows by (3.49) and (3.51) that

$$|L(\psi)| \leq K_2\|g(\psi)\|_{L^2(0,T;H)} = K_2\|\psi\|_X,$$

which proves that L is a continuous linear functional on X, where K_2 is a positive constant depending on (g, p_0, p_1). Therefore we have the following proposition.

Proposition 3.1. *For $g \in L^1(0, T; V')$, $p_0 \in H$ and $p_1 \in V'$, there is a unique solution $p \in L^2(0, T; H)$ such that*

(3.52)
$$\int_0^T (p(t), \Theta\psi(t))dt = \int_0^T \langle g(t), \psi(t)\rangle dt + \langle p_1, \psi(T)\rangle - (p_0, \psi'(T)), \quad \forall\psi \in X,$$

where $\Theta\psi$ is given in (3.46).

We are now in the position to apply (3.52) to the necessary condition of $C = (C_1, C_2)$ observation.

Now we will formulate the adjoint system to describe the optimality condition by applying Proposition 3.1. Since $C_1^* \Lambda_{M_1}(C_1 y(u) - z_d^1) \in L^2(0, T; H) \subset L^1(0, T; V')$ and $C_2^* \Lambda_{M_2}(C_2 y(u; T) - z_d^2) \in H$, there exists a $p(u) \in L^2(0, T; H)$ satisfying

(3.53)
$$\int_0^T (p(t), \Theta \psi(t)) dt$$
$$= \int_0^T (C_1^* \Lambda_{M_1}(C_1 y(u) - z_d^1)(t), \psi(t)) dt$$
$$+ (C_2^* \Lambda_{M_2}(C_2 y(u; T) - z_d^2), \psi'(T)), \quad \forall \psi \in X,$$

where $\Theta \psi$ is given by (3.46). We note that $\psi \in X$ is characterized by $\Theta \psi \in L^2(0, T; H)$ and $\psi(0) = \psi'(0) = 0$ from the isomorphism of X onto $L^2(0, T; H)$. Especially, the Gâteaux derivative $z = Dy(u)(v - u)$ satisfies

$$\Theta z = B(v - u) \in L^2(0, T; H), \quad z(0) = 0, \quad z'(0) = 0,$$

so that this z belongs to X. Hence, if we take $\psi = z = Dy(u)(v - u)$ in (3.53), then we have

$$\int_0^T (C_1^* \Lambda_{M_1}(C_1 y(u) - z_d^1)(t), z(t)) dt + (C_2^* \Lambda_{M_2}(C_2 y(u; T) - z_d^2), z(T))$$
$$= \int_0^T (p(u; t), \Theta z(t)) dt$$
$$= \int_0^T (p(u; t), B(v - u)(t)) \, dt = (\Lambda_{\mathcal{U}}^{-1} B^* p(u), v - u)_{\mathcal{U}}.$$

Thus we conclude that the optimality condition (3.44) is equivalent to

$$(\Lambda_{\mathcal{U}}^{-1} B^* p(u) + Ru, v - u)_{\mathcal{U}} \geq 0, \quad \forall v \in \mathcal{U}_{ad}.$$

Therefore, we prove the following theorem.

Theorem 3.2. *Assume that the observer $C = (C_1, C_2)$ is given by (3.42) and the cost J is given by (3.43) Then the optimal control u is characterized by the following system of equations and inequality:*

$$\begin{cases} y''(u) + A(t)y(u) + \int_0^t K(t, s)y(s)ds = f(t, y(u), y'(u)) + Bu \quad \text{in } (0, T), \\ y(u; 0) = y_0 \in V, \quad y'(u; 0) = y_1 \in H; \end{cases}$$

$$\begin{cases} \int_0^T \big(p(u; t), \psi''(t) + A(t)\psi(t) + \int_0^t K(t, s)\psi(s)ds \\ \quad -f_y(t, y(u), y'(u))\psi(t) - f_z(t, y(u), y'(u))\psi'(t)\big) dt \\ = \int_0^T (C_1^* \Lambda_{M_1}(C_1 y(u) - z_d^1)(t), \psi(t)) dt + (C_2^* \Lambda_{M_2}(C_2 y(u; T) - z_d^2)(t), \psi(T)), \\ \forall \psi \quad \text{such that} \\ \psi'' + A(t)\psi + \int_0^t K(t, s)\psi(s)ds - f_y(t, y(u), y'(u))\psi(t) \\ \quad -f_z(t, y(u), y'(u))\psi'(t) \in L^2(0, T; H), \\ \psi(0) = 0, \quad \psi'(0) = 0; \end{cases}$$

$$(\Lambda_{\mathcal{U}}^{-1} B^* p(u) + Ru, v - u)_{\mathcal{U}} \geq 0, \quad \forall v \in \mathcal{U}_{ad}.$$

Next, we consider the special form of nonlinear term f in (3.2) independent on y', i.e., $f(t, y, y') = f(t, y)$. For the nonlinear function $f(t, y)$ we suppose the following assumptions:

(B1) $t \to f(t, y)$ is strongly measurable in H for all $y \in V$;

(B2) there exists a $\beta \in L^2(0, T; \mathbf{R}^+)$ such that

$$|f(t, \xi_1) - f(t, \xi_2)| \leq \beta(t)\|\xi_1 - \xi_2\| \quad \text{a.e. } t \in [0, T] \quad \text{for all } \xi_i \in V, \ i = 1, 2;$$

(B3) there exists a $\gamma \in L^2(0, T; \mathbf{R}^+)$ such that $|f(t, 0)| \leq \gamma(t)$ a.e. $t \in [0, T]$;

(b4) for each $t \in [0, T]$, $f(t, y) \in C^1(V, H)$, and for each $t \in [0, T]$, $f_y(t, y) \in C(V, \mathcal{L}(V, H))$ and there is $\beta_1 \in L^2(0, T; \mathbf{R}^+)$ such that

$$\|f_y(t, y)\|_{\mathcal{L}(V, H)} \leq \beta_1(t)(\|y\| + 1) \quad \text{a.e. } t \in [0, T].$$

For this special f, we can establish the necessary optimality conditions without using the transposition method.

Theorem 3.3. *Assume that the observer $C = (C_1, C_2)$ is given by (3.42) and the cost J is given by (3.43) and the nonlinear term $f(t, y)$ satisfies (B1)-(B4). Then the optimal control u is characterized by the following system of equations and inequality:*

$$\begin{cases} y''(u) + A(t)y(u) + \int_0^t K(t, s)y(s)ds = f(t, y(u)) + Bu & \text{in } (0, T), \\ y(u; 0) = y_0 \in V, \quad y'(u; 0) = y_1 \in H; \end{cases}$$

$$\begin{cases} p''(u) + A(t)p(u) + \int_t^T K^*(s, t)p(u; s)ds \\ \quad = f_y^*(t, y(u))p(u) + C_1^*\Lambda_{M_1}(C_1y(u) - z_d^1) & \text{in } (0, T), \\ p(u; T) = 0, \quad p'(u; T) = C_2^*\Lambda_{M_2}(C_2y(u; T) - z_d^2); \end{cases}$$

$$(\Lambda_{\mathcal{U}}^{-1}B^*p(u) + Ru, v - u)_{\mathcal{U}} \geq 0, \quad \forall v \in \mathcal{U}_{ad}.$$

Proof. Multiplying both sides of the adjoint equation for $p(u)$ by z and integrating it by parts on $[0, T]$, we have that

$$\int_0^T (C_1^*\Lambda_{M_1}(C_1y(u) - z_d^1)(t), z(t))dt + (C_2^*\Lambda_{M_2}(C_2y(u; T) - z_d^2), z(T))$$

$$= \int_0^T \left\langle p''(u; t) + A(t)p(u; t) + \int_t^T K^*(s, t)p(u; s)ds - f_y^*(t, y(u))p(u), z(t) \right\rangle dt$$

$$= \int_0^T \left\langle p(u; t), \left(\frac{d^2}{dt^2} + A(t)\right)z(t) + \int_0^t K(t, s)z(s)ds - f_y(t, y(u))z(t) \right\rangle dt$$

$$= \int_0^T (p(u; t), B(v - u)(t))dt = (\Lambda_{\mathcal{U}}^{-1}B^*p(u), v - u)_{\mathcal{U}}.$$

Hence, the necessary condition follows. $\qquad\square$

4. An Application to the Semilinear wave Equation with a Fading Memory

We give an application to the semilinear wave problem with a fading memory. Let Ω be an open subset in $\mathbf{R}^3$ with a smooth boundary $\Gamma = \partial\Omega$. Let $Q = (0, T) \times \Omega$ and

$\Sigma = (0, T) \times \Gamma$. The semilinear wave problem with a fading memory is described by the following Dirichlet problem

(4.1)
$$\begin{cases} \partial_t^2 y - \alpha \Delta y + \int_0^t b(t-s) \Delta y(s, x) ds = f(t, x, y) + v & \text{in } Q, \\ y = 0 & \text{on } \Sigma, \\ y(0, x) = y_0(x), \quad \partial_t y(0, x) = y_1(x) & \text{in } \Omega, \end{cases}$$

where $\alpha > 0$. The scalar kernel function $b(\cdot)$ denotes the fading rate of a memory effect in isotrophic material such that $b(\cdot)$, and $b'(\cdot) \in L^\infty(0, T)$. And v is the control variables of forcing functions in $L^2(Q)$. We take $V = H_0^1(\Omega)$ and $H = L^2(\Omega)$ and introduce two bilinear forms

$$a(\phi, \varphi) = \int_\Omega \alpha \nabla \phi \cdot \nabla \varphi dx, \quad \forall \phi, \varphi \in H_0^1(\Omega)$$

and

$$k(\cdot; \phi, \varphi) = \int_\Omega b(\cdot) \nabla \phi \cdot \nabla \varphi dx, \quad \forall \phi, \varphi \in H_0^1(\Omega).$$

And we assume the function $f : [0, T] \times \Omega \times \mathbf{R} \to \mathbf{R}$ satisfying the following conditions:

(a) $f(\cdot, x, y)$ is measurable on $[0, T]$ for all $x \in \Omega$ and $y \in \mathbf{R}$;

(b) there is a $\beta(t, x) \in L^2(0, T; L^\infty(\Omega))$ such that for each $t \in [0, T]$

$$|f(t, x, y_1) - f(t, x, y_2)| \leq \beta(t, x)|y_1 - y_2| \quad \text{a.e. } x \in \Omega, \quad \forall y_1, y_2 \in \mathbf{R};$$

(c) there is a $\gamma(t, x) \in L^2(0, T; L^\infty(\Omega))$ such that for each $t \in [0, T]$,

$$|f(t, x, 0)| \leq \gamma(t, x) \quad \text{a.e. } x \in \Omega;$$

(d) $f(t, x, y)$ is once continuously differentiable in $y \in \mathbf{R}$ for all $x \in \Omega$, and there is a $\beta_1(t, x) \in L^2(0, T; L^\infty(\Omega))$ such that for all $t \in [0, T]$ and $y \in \mathbf{R}$, the derivative $\partial_y f(t, x, y)$ satisfies

$$|\partial_y f(t, x, y)| \leq \beta_1(t, x) \quad \text{a.e. } x \in \Omega.$$

Under the above assumptions on f, we can verify that the function

$$f(t, y) = f(t, x, y(x)) : [0, T] \times V \to H$$

is Fréchet differentiable in y for each $t \in [0, T]$ and satisfy the assumption (B4). By Theorem 2.1, for $y_0 \in V$ and $y_1 \in H$, there exists a unique y satisfying (4.1) in the weak sense and $y \in C([0, T]; V) \cap C^1([0, T]; H)$. We consider optimal control problem with the following cost subject to (4.1)

(4.2) $J(v) = \int_Q (y(v) - z_d)^2 dx dt + \int_\Omega (y(v; T) - z_d(T))^2 dx + \int_Q v^2 dx dt, \quad \forall v \in L^2(Q).$

Let u be an optimal control for the cost (4.2). Since $f(t, y)$ satisfies (B4), it can be verified that the map $v \to y(v)$ of $L^2(Q) \to W(0, T)$ is Gâteaux differentiable at $v = u$ and such the Gâteaux derivative of $y(v)$ at $v = u$ in the direction $v - u$, say

$z = Dy(u)(v - u)$ is a unique solution satisfying the following equation

$$(4.3) \quad \begin{cases} \partial_t^2 z - \alpha \Delta z + \int_0^t b(t - s)\Delta z(s, x)ds = \partial_y f(t, x, y(u))z + (v - u) & \text{in } Q, \\ z = 0 \quad \text{on } \Sigma, \\ z(0, x) = 0, \quad \partial_t z(0, x) = 0 \quad \text{in } \Omega. \end{cases}$$

Then the optimality condition is given by

$$(4.4) \quad \begin{aligned} & \int_Q (y(u) - z_d)z dx dt + \int_\Omega (y(u; T) - z_d(T))z(T)dx \\ & + \int_Q u(v - u)dx dt \geq 0, \quad \text{for all } v \in L^2(Q), \end{aligned}$$

where u is a optimal control, $z_d \in L^2(Q)$ and $z_d(T) \in L^2(\Omega)$ are desired values, and z is the weak solution of (4.3) and $z(T)$ is the terminal value of it. According to the optimality condition (4.4), we can apply Theorem 3.3 to formulate the adjoint system in order to obtain the optimality condition. We define $p(u)$ as a unique solution in of the adjoint equation

$$(4.5) \quad \begin{cases} \partial_t^2 p(u) - \alpha \Delta p(u) + \int_t^T b(t - s)\Delta p(u; s)ds \\ \quad = \partial_y f(t, x, y(u))p(u) + y(u) - z_d \quad \text{in } Q, \\ p(u) = 0 \quad \text{on } \Sigma, \\ p(u; T, x) = 0, \quad \partial_t p(u; T, x) = y(u; T) - z_d(T) \quad \text{in } \Omega. \end{cases}$$

Then by Theorem 3.3, we have the following optimality condition

$$(4.6) \quad \int_Q (p(u) + u)(v - u)dx dt \geq 0, \quad \forall v \in L^2(Q).$$

REFERENCES

1. N. U. Ahmed and K. L. Teo, *Optimal Control of Distributed Parameter Systems*, North Holland, 1981.
2. M. M. Cavalcanti and H. P. Oquendo, *Frictional versus viscoelastic damping in a semilinear wave equation*, SIAM J. Control Optim, **42(4)** (2003), 1310–1324.
3. G. Da Prato and M. Iannelli (Editors), *Volterra Integrodifferential Equations in Banach Spaces and Applications*, Pitman Research Notes in Mathematics Series 190, Longman Scientific and Technical, 1989.
4. R. Dautary and J. L. Lions, *Mathematical Analysis and Numerical Methods for Science and Technology, Vol. 5, Evolution Problems I*, Springer-Verlag, 1992.
5. C. Giorgi, J. E. M. Rivera and V. Pata, *Global attracters for a semilinear hyperbolic equation in viscoelasticiy*, J. Math. Anal. Appl. 260 (2001), 83–99.
6. J. Ha and S. Nakagiri, *Optimal control problems for nonlinear hyperbolic distributed parameter systems with damping terms*, Funcialaj Ekvacioj **47** (2004), 1–23.
7. J. Hwang and S. Nakagiri, *Optimal control problems for second order Volterra Integro-differential equations in Hilbert space*, Mem. Grad. School Sci. Technol., Kobe Univ. **22-A** (2004), 77–90.
8. J. Hwang and S. Nakagiri, *On semi-linear second order Volterra integro-differential equations in Hilbert space*, submitted in Taiwanese J. Math.

9. J. L. Lions, *Optimal Control of Systems Governed by Partial Differential Equations*, Springer-Verlag Berlin Heidelberg New York, 1971.

10. J. L. Lions and E. Magenes, *Non-Homogeneous Boundary Value Problems and Applications I, II*, Springer-Verlag, Berlin-Heidelberg-New York, 1972.

11. S. Omatu and J. H. Seinfeld, *Distributed parameter systems, Theory and applications*, Oxford Science Publications, 1989.

12. R. Renardy, W. J. Hrusa, and Nohel, *Mathematical Problems in Viscoelasticity*, Longman Scientific and Technical, Harlow/New York, 1987.

13. J. E. M. Rivera, M. G. Naso and F. M. Vegni, *Asymptotic behaviour of the energy for a class of weakly dissipative second-order system with memory*, J. Math. Anal. Appl. **286** (2003), 692–704.

14. T. I. Seidman and S. S. Antman, *Optimal control of a nonlinearly viscoelatic rod, operators, in "Control of nonlinear distributed parameter systems"*, Lecture Notes in Pure and Applied Mathematics, Vol. 218, pp. 273–283, Marcel Dekker, New York, 2001.

15. Q. Tiehu and N. Guoxi, *Three-dimensional travelling waves for nonlinear viscoelastic materials with memory*, J. Math. Anal. Appl. **284** (2003), 76–88.

Nonlinear Functional Analysis and Applications, Volume 1, 193–200

SOME PROPERTIES OF A GENERALIZED FRACTIONAL BROWNIAN MOTION

Zhengyan Lin[1,*] and Jing Zheng[1,2]

ABSTRACT. In the paper, we give a new generalization of fraction Browian motion (gfBm). We study the existence of the local nondeterminism and the joint continuity of the local time of gfBm, and we get upper and lower bounds of Hausdroff dimensions of the level sets of a gfBm.

1. Introduction

Given a constant $H \in (0,1)$, a fractional Brownian motion in R^+ with index H is a real valued, centered Gaussian process $W_H = \{W_H(t), t \in R\}$, with the covariance function

$$E[W_H(s)W_H(t)] = \frac{1}{2}(|s|^{2H} + |t|^{2H} - |s-t|^{2H}).$$

Fractional Brownian motion (fBm) was introduced by Mandelbrot and Van Ness (1968) as a moving average Gaussian process. From the covariance function, it is easy to verify that W_H is a self similar process and it has stationary increments. Pitt (1978) discovered the strong local nondeterminism of the fBm. Xiao (1997) proved the Hölder conditions for the local times and the Hausdorff measure of the level sets of the fBm. Since H is independent of the time parameter t, the regularity of the fBm is the same all along its paths. This property is undesirable when we model some phenomena that do not admit a constant Hölder exponent; for instance, the use of the fBm for synthesizing artificial mountains does not allow to take into account erosion phenomena. To relax this restriction, as a generalization of the fBm, Peltier and Lévy (1995) and Benassi et al.(1997) independently

Received October 26, 2007.   * Corresponding author.

2000 *Mathematics Subject Classification.* 60F15, 60G15, 60G17.

Key words and phrases. Hausdorff dimension, strong local nondeterminism, lever set, local time, generalized fractional Brownian motion.

This project is supported by NSFC (10571159) and SRFDP(2002335090).

[1] Department of Mathematics, Zhejiang University, Hangzhou, 310027, People's Republic of China (*E-mail*: zlin@zju.edu.cn)

[2] Department of Mathematics, Ningbo University, Ningbo 315211, People's Republic of China (*E-mail*: tongchangqing@nbu.edu.cn)

introduced the following definition: Let $H_t = H(t) : [0, \infty) \to (0, 1)$ be a Hölder function of exponent $\beta > 0$, i.e. for any $t_1, t_2 \in [0, \infty)$ such that $|t_1 - t_2| < 1$, there exists a constant $C > 0$ such that

$$|H(t_1) - H(t_2)| \leq C|t_1 - t_2|^\beta.$$

Then

$$(1.1) \qquad \tilde{W}(t) := W_{H_t}(t) = \frac{1}{\Gamma(H_t + 1/2)} \int_{-\infty}^{t} [(t - u)_+^{H_t - 1/2} - (-u)_+^{H_t - 1/2}] dW(u)$$

is called a *multifractional Brownian motion* (mBm), where $W(u)$ is a Brownian motion. Peltier and Lévy (1995) showed that a mBm has continuous sample paths with probability one and studied its local Hölder properties. Lin (2002) studied large increment behavior of a mBm, and had the following result:

Let $0 \leq a_T \leq T$, $a_T \to \infty$ as $T \to \infty$ and $D^2(s, t) = E(\tilde{W}(t) - \tilde{W}(s))^2$ for $0 \leq s < t$, under the condition:

$$(1.2) \qquad |H_t - H_s| = o((1 - s/t)^{H_t} (\log t)^{-1}) \quad \text{as } 0 < t - s \to 0,$$

with $\beta(t, h) = \{2D^2(t, t + h)(\log 1/h + \log \log 1/h)\}^{-1/2}$,

$$\limsup_{h \to \infty} \sup_{0 \leq t \leq T - a_T} \sup_{0 \leq s \leq a_T} \beta(t, T)|\tilde{W}(t + s) - \tilde{W}(t)| = 1 \quad a.s.$$

In this paper, we consider a generalization of fBm. Suppose that $H(t)$ is a continuous non-decreasing function with $a < H(t) \leq 1 - a$ for some $0 < a < 1/2$ and any $t \geq 0$. We call $\tilde{W}$ with such $H(t)$ as generalized fractional Brownian motion (gfBm) and study some properties of a gfBm. We investigate the existence of the local non-determinism of a gfBm in section 2. In section 3, we study the existence and the joint continuity of the local time of a gfBm. In section 4, the upper and lower bounds of Hausdroff dimensions of level sets of a gfBm are given under the condition (1.2).

In this paper, C always stands for a positive constant, whose value is irrelevant. $\dim_H A$ is denoted the Hausdorff dimension of set A.

2. The Local Nondeterminism of a gfBm

The concept of local nondeterminism was introduced by Berman (1973). A process is locally nondeterministic if a future observation is "relatively unpredictable" on the basis of a finite set of observations from the immediate past. For a Gaussian process, the local nondeterminism is closely related to the existence of a continuous local time. As a property of Gaussian process, it is of independent interest. Pitt (1978) discovered the strong local nondeterminism of a fractional Brownian motion: there exists a constant $0 < K_1 < \infty$, depending on H only, such that for all $t \in R$ and $0 \leq r \leq |t|$,

$$Var(W_H(t)|W_H(s) : |s - t| \geq r) = K_1 r^{2H}.$$

In his proof, the self-similarity and the stationarity play crucial roles. It is obvious that a gfBm has neither the self-similarity nor the stationarity. How about the local nondeterminism for a gfBm?

Definition 2.1. Let J be an open interval on the t-axis. we assume that there exists $d > 0$ such that

(2.1)
$$E(X(t) - X(s))^2 > 0,$$

where $s, t \in J$ and $0 < |t - s| \le d$,

(2.2)
$$EX^2(t) > 0 \quad \text{for all } t \in J.$$

For $m \ge 2$, let $t_1, \cdots, t_m$ be arbitrary points in J with $t_1 < \cdots < t_m$. Let

$$V_m = \frac{Var\{X(t_m) - X(t_{m-1})|X(t_1), \cdots, X(t_{m-1})\}}{Var\{X(t_m) - X(t_{m-1})\}}.$$

The process is called locally nondeterministic on J if for every integer $m \ge 2$

(2.3)
$$\lim_{c \downarrow 0} \inf_{t_m - t_1 \le c} V_m > 0.$$

For a gfBm, conditions (2.1) and (2.2) are obvious, we prove (2.3). Let

$$F(t, u) = \frac{1}{\Gamma(H_t + 1/2)} \{(t - u)_+^{H_t - 1/2} - (-u)_+^{H_t - 1/2}\},$$

then

(2.4)
$$\tilde{W}(t) = \int_{-\infty}^{t} F(t, u) dW(u)$$

and

$$Var\{\tilde{W}(t)\} = \int_{-\infty}^{t} F^2(t, u) du.$$

It is apparent from the representation (2.4) that

$$\{W(u), u \le s\} \supset \{\tilde{W}(u), u \le s\}.$$

Therefore, for any time set A,

(2.5)
$$Var\{\tilde{W}(t) - \tilde{W}(s)| \tilde{W}(u), u \in A, u \le s\}$$
$$\ge Var\{\tilde{W}(t) - \tilde{W}(s)|W(u), u \le s\}.$$

By the independence of increments of a Brownian motion, we have

(2.6)
$$Var\{\tilde{W}(t) - \tilde{W}(s)|W(u), u \le s\}$$
$$= Var\left(\int_s^t F(t, u) dW(u)\right) = \int_s^t F^2(t, u) du, \quad 0 < s < t.$$

Lemma 2.1. For $0 \le s \le t < T$ and $t - s \to 0$,

$$Var\{\tilde{W}(t) - \tilde{W}(s)\} = O\big((t - s)^{2H_t} + (H_t - H_s)^2 t^{2H_t} \log^2 t\big).$$

Proof. The proof is similar to Lin (2002). $\qquad \square$

Now, we show the following theorem.

Theorem 2.1. *Let $\widetilde{W}(t)$ be a gfBm, Suppose that*

(2.7) $|H_t - H_s| = O((1 - s/t)^{H_t}(\log t)^{-1}) \quad as\ 0 < t - s \to .$

Then $\widetilde{W}(t)$ is locally nondeterministic on open interval $(0, T)$.

Proof. By (2.5) and (2.6), V_m is at least

(2.8) $$\frac{\int_s^t F^2(t, u)du}{Var\{\widetilde{W}(t) - \widetilde{W}(s)\}},$$

where $t = t_m$ and $s = t_{m-1}$. Moreover

(2.9) $$\int_s^t F^2(t, u)du = \frac{1}{\Gamma(H_t + 1/2)^2 H_t}(t - s)^{2H_t}.$$

By Lemma 2.1 and (2.9), (2.8) is at least

$$\frac{(t - s)^{2H_t}/(\Gamma(H_t + 1/2)^2 H_t)}{((t - s)^{2H_t} + (H_t - H_s)^2 t^{2H_t} \log^2 t)} > 0$$

as $0 \le s \le t < T$ and $t - s \to 0$. The proof is complete.   □

3. The Jointly Continuity of the Local Time of a gfBm

Berman (1973) gave the sufficient conditions for joint continuity of local time on a Gaussian process (See also German (1980)).

Lemma 3.1. *Let $X(t)$, $0 \le t \le T$, be a Gaussian process with mean 0 and satisfy the following three conditions:*
  (a) *$X(t) = 0$ almost surely;*
  (b) *$X(t)$ is locally nondeterministic on $(0, T)$;*
  (c) *There exists $b(t)$ such that $b^2(t) \le E(X(s + t) - X(s))^2$ for all s and*

(3.1) $$\int_0^\epsilon \frac{dt}{\{b(t)\}^{1+\delta}} < \infty \quad for\ some\ \epsilon, \delta > 0.$$

Then the local time of $X(t)$ exists and is joint continuous in the sense that $(x, t) \to L(x, t)$ is continuous on $R \times [0, T]$.

We just verify (3.1). First we quote an well known inequality (cf., e.g., Lin et al. (1999)).

Lemma 3.2. *If X and Y are independent, $EX = 0$, $E|X|^p < \infty$, $E|Y|^p < \infty$, $1 \le p \le \infty$, then*

(3.2) $$E|Y|^p \le E|X + Y|^p.$$

Now, for $0 \le s < t$, let

$$X = W_{H_t}(s) - W_{H_s}(s), \quad Y = \frac{1}{\Gamma(H_t + 1/2)} \int_s^t (t - u)^{H_t - 1/2} dW(u).$$

Then

$$\widetilde{W}(t) - \widetilde{W}(s) = X + Y.$$

By (3.2), we have

$$E|\tilde{W}(t) - \tilde{W}(s)|^2 \ge E|Y|^2$$

$$= \frac{1}{\Gamma^2(H_t + 1/2)} \int_s^t (t-u)^{2H_t-1} du$$

$$= \frac{1}{\Gamma^2(H_t + 1/2)(2H_t)}(t-s)^{2H_t}.$$

Note that $\frac{1}{\Gamma^2(H_t+1/2)(2H_t)} \ge C^2 > 0$ for any $t \ge 0$. Let $b(t) = Ct^{H_t}$, we have (3.1) for some $\delta > 0$ small enough.

Now, Lemma 3.1 implies the following theorem.

Theorem 3.1. *Let $\tilde{W}(t)$ be a gfBm and satisfy the condition (2.7). Then it has a joint continuous local time.*

4. The Hausdorff Dimensions of the Level Sets of a gfBm

The set $E(x, T) = \{t \in [0, T], \tilde{W}(t) = x\}$ is called the level set of $\tilde{W}(t)$ in x, where x is the interior of range of $\tilde{W}(t)$. At first, we show the upper bound of the Hausdorff dimensions of the level sets. The following lemma is similar to Theorem 3.1 of Lin (2002).

Lemma 4.1. *Suppose that the gfBm $\tilde{W}(\cdot)$ satisfies condition (1.2). Let $D^2(s, t) = E(\tilde{W}(t) - \tilde{W}(s))^2$ for $0 \le s < t$. Then, with $\beta(t, h) = \{2D^2(t, t+h)(\log 1/h + \log\log 1/h)\}^{-1/2}$,*

$$(4.1) \qquad \limsup_{h \to 0} \sup_{0 \le t \le 1-h} \sup_{0 \le s \le h} \beta(t, h)|\tilde{W}(t+s) - \tilde{W}(t)| = 1 \quad a.s.$$

Theorem 4.1. *Suppose that the gfBm $\tilde{W}(\cdot)$ satisfies condition of Lemma 4.1, then for any $T \in R^+$ and almost all $x \in R$, the Hausdorff dimension*

$$(4.2) \qquad \dim_H E(x, T) \le 1 - H(0).$$

Proof. By Lemma 2.1 and the condition of Lemma 4.1, $D^2(t+s, t) = O(s^{2H_{t+s}})$, so the order of Hölder exponent of $D(\cdot, \cdot)$ is at least $\min_{t\in[0,T]} H_t = H(0)$. For any $\epsilon > 0$, it follows from (4.1) that for small $s > 0$,

$$|\tilde{W}(t+s) - \tilde{W}(t)| \le CD^{1+\epsilon}(t+s, t) \quad a.s.$$

By an argument similar to the proof of Lemma 8.2.2 of Adler (1981), we have (4.2). □

Next we consider the lower bound. First, we show the following inequality.

Lemma 4.2. *Let $\tilde{W}(t)$ be a gfBm, $0 < s < t$, then*

$$(4.3) \qquad \int_R \int_R E\exp\{i[u_1\tilde{W}(t) + u_2\tilde{W}(s)]\}du_1 du_2 \le C[(t-s)^{-H_t}s^{-H_s}].$$

Proof. Since $\tilde{W}(t)$ is a Gaussian process, the left of (4.3) equals

(4.4) $$\int_R \int_R \exp\left\{-\frac{1}{2}E[u_1 \tilde{W}(t) + u_2 \tilde{W}(s)]^2\right\}du_1 du_2.$$

Let

$$v_1 = \left[E \tilde{W}^2(t)\right]^{1/2} u_1 + \left(E \tilde{W}(s) \tilde{W}(t) / \left[E \tilde{W}^2(t)\right]^{1/2}\right) u_2,$$

$$v_2 = \left\{\frac{\Delta}{E \tilde{W}^2(t)}\right\}^{1/2} u_2,$$

where

$$\Delta = E \tilde{W}^2(t)E \tilde{W}^2(s) - [E \tilde{W}(t) \tilde{W}(s)]^2.$$

Then (4.4) equals

(4.5) $$\Delta^{-1/2}\int_R \int_R \exp\{\frac{1}{2}(v_1^2 + v_2^2)\}dv_1 dv_2 = (2\pi)^2 \Delta^{-1/2}.$$

Next, we calculate Δ. Let $W_{H_t}(s) = \int_{-\infty}^s F(t,u)dW(u)$, then

(4.6) $$[E \tilde{W}(t) \tilde{W}(s)]^2 = [EW_{H_t}(s)W_{H_s}(s)]^2 \leq EW_{H_t}^2(s)EW_{H_s}^2(s).$$

Let $W_{H_t}(s,t) = \frac{1}{\Gamma(H_t+1/2)}\int_s^t (t-u)^{H_t-1/2}dW(u)$, then

(4.7) $$E \tilde{W}^2(t)E \tilde{W}^2(s) = EW_{H_t}^2(s)EW_{H_s}^2(s) + EW_{H_t}^2(s,t)EW_{H_s}^2(s).$$

Moreover

(4.8)
$$EW_{H_s}^2(s) = \frac{1}{\Gamma^2(H_s+1/2)}\left\{\int_{-\infty}^0 [(s-u)^{H_s-1/2} - (-u)^{H_s-1/2}]^2 du\right.$$
$$\left. + \int_0^s (s-u)^{2H_s-1}du\right.$$
$$\geq Cs^{2H_s}.$$

Similarly

(4.9) $$EW_{H_t}^2(s,t) \geq C(t-s)^{2H_t}.$$

Combining (4.6)-(4.9), we have

$$\Delta \geq EW_{H_t}^2(s,t)EW_{H_s}^2(s) \geq C(t-s)^{2H_t}s^{2H_s}.$$

(4.3) follows from (4.4) and (4.5).   □

Lemma 4.3. *Let $\tilde{W}(t)$ be a gfBm and satisfy condition* (2.7), *$L(x,T)$, $T \in R$, be the local time of $\tilde{W}(t)$, then the zero set of $L(x,T)$*

$$\{x : L(x,T) = 0\}$$

is non-dense everywhere.

Proof. By Theorem 3.1, $\tilde{W}(t)$ has jointly continuous local time, hence from Adler (1981), our conclusion is obvious.   □

Now, we can show the lower bound of the Hausdorff dimension.

Theorem 4.2. *Let $\widetilde{W}(t)$ be a gfBm and satisfy condition (2.7), then for any $T \in R$ and almost every x,*

$$\dim_H E(x,T) \geq 1 - H_T.$$

Proof. By Lemma 4.3, for any $T \in R$, we have $L(x,T) > 0$ for almost every x. So we can define random measure u:

$$u(B) = \frac{L(x,[0,T] \cap B)}{L(x,T)}.$$

Then u is a probability measure on $E(x,T)$. By the energy integration formulation, we will just show for $\beta < 1 - H_T$, the energy integration

$$(4.10) \qquad I_\beta(u) = \int_{E^2(x,T)} |s-t|^{-\beta} du(s) du(t) < \infty$$

By an argument similar to the proof of Proposition 3.1 of Pitt (1978) (see also Theorem 8.7.4 of Adler (1981)), we have

$$(4.11) \qquad \begin{aligned} EI_\beta(u) = \frac{1}{L^2(x,T)(2\pi)^2} \int_0^T \int_0^T \int_R \int_R \exp\{-i(xu_1 + xu_2)\} E \exp\{iu_1 \widetilde{W}(s) \\ + iu_2 \widetilde{W}(t)\} |t-s|^{-\beta} du_1 du_2 ds dt. \end{aligned}$$

Therefore in order to prove (4.10), it suffices to show that the right side of (4.11) is finite. By Lemma 4.2, the right side of (4.11) is no more than

$$\frac{1}{L^2(x,T)(2\pi)^2} \int_0^T \int_0^T \int_R \int_R |E \exp\{iu_1 \widetilde{W}(s) + iu_2 \widetilde{W}(t)\}| |t-s|^{-\beta} du_1 du_2 ds dt$$

$$\leq \frac{C}{L^2(x,T)(2\pi)^2} \int_0^T \left\{ \int_0^t |t-s|^{-H_t} s^{-H_s} ds + \int_t^T |t-s|^{-H_s} t^{-H_t} ds \right\} |t-s|^{-\beta} dt$$

$$\leq \frac{C}{L^2(x,T)(2\pi)^2} \int_0^T \left\{ \int_0^T |t-s|^{-H_t} s^{-H_s} ds + \int_0^T |t-s|^{-H_s} t^{-H_t} ds \right\} |t-s|^{-\beta} dt$$

$$= \frac{2C}{L^2(x,T)(2\pi)^2} \int_0^T \int_0^T |t-s|^{-H_t} s^{-H_s} |t-s|^{-\beta} ds dt$$

$$\leq \frac{C'}{L^2(x,T)(2\pi)^2} \int_0^T |u|^{-H_t - \beta} du < \infty$$

for almost every x. Hence

$$\dim_H E(x,T) \geq 1 - H_T \quad a.s.$$

This completes the proof of Theorem 4.1. $\square$

The lower bound of the Hausdorff dimension is not sharp, but for a fBm, we have the following corellary by combining Theorem 4.1 and Theorem 4.2.

Corollary. *Let $W_H(t)$ be a fractional Browian motion, then for almost every x in the range of $W_H(t)$,*

$$\dim_H E(x,T) = 1 - H.$$

REFERENCES

1. R. J. Adler, *The Geometry of Random Fields,* New York, Wiley, 1981.

2. A. Benassi, S. Jaffard and D. Roux, *Gaussian processes and pseudo-differential elliptic operators,* Rev. Mat. Iberoamericana **12** (1997), 19–81.

3. S. M. DBerman, *Local nondeterminism and local times of Gaussian processes,* Indiana Univ. Math. J. **23** (1973), 69–94.

4. K. J. Facloner, *Fractal Geometry-Mathematical Fundations and Applications,* New York, Wiley and Sons, 1990.

5. D. Geman and J. Horowitz, *Occupation densities,* Ann. Probab. **8** (1980), 1–67.

6. D. Khoshnevisan and Y. M. Xiao, *Lever set of additive Lévy processes,* Ann. Probab. **27** (1980), 62–100.

7. Z. Y. Lin, *How big are the increments of a multifractional Brownian motion?* Science in China (A) **45** (2002), 1292–1300.

8. Z. Y. Lin and R. M. Zhang, *The Hausdorff dimension of level set for a fractional Brownian sheet,* Stoch. Anal. Appl. **22** (2004), 1511–1523.

9. Z. Y. Lin, C. R. Lu and Z. G. Su, *Limit Theorem of Probability,* Higher Education Press, 1999.

10. R. F. Peltier and V. J. Lévy, *Multifractional Bromnian motion: definition and preliminary results,* Rapport de Recherche de l'INRIA, No. 2645, 1995.

11. L. D. Pitt, *Local times for Gaussian vector fields,* Indiana Univ. Math. J. **27** (1978), 309–330.

12. M. Talagrand, *Hausdorff measure of the trajectories of Multiparameter fractional Brownian motion,* Ann. Probab. **23** (1995), 767–775.

13. S. J. Taylor and J. G. Wendel, *The exact Hausdorff measure of the level set of a stable process,* Z. Wahrsch. Verw. Gebiete **6** (1966), 170–180.

14. Y. M. Xiao, *Hölder conditions for the local times and the Hausdorff measure of the level sets of Gaussian random fields,* Probab. Theory Relat. Fields **109** (1997), 129–157.

15. Y. M. Xiaoand T. S. Zhang, *Local times of fractional Brownian sheets,* Probab. Theory Relat. Fields **124** (2002), 204–226.

Nonlinear Functional Analysis and Applications, Volume 1, 201–209

SOME COMMON FIXED POINT THEOREMS FOR WEAKLY COMMUTING MAPPINGS IN MENGER PROBABILISYIC METRIC SPACES

REZA SAADATI[1]

ABSTRACT. In this paper we prove two common fixed point theorems in Menger probabilistic metric space.

1. Introduction and Preliminaries

K. Menger introduced the notion of a probabilistic metric space in 1942 and since then the theory of probabilistic metric spaces has developed in many directions [11]. The idea of K. Menger was to use distribution functions instead of nonnegative real numbers as values of the metric. The notion of a probabilistic metric space corresponds to situations when we do not know exactly the distance between two points, but we know probabilities of possible values of this distance. A probabilistic generalization of metric spaces appears to be interest in the investigation of physical quantities and physiological thresholds. It is also of fundamental importance in probabilistic functional analysis. Probabilistic normed spaces were introduced by Šerstnev in 1962 [12] by means of a definition that was closely modelled on the theory of (classical) normed spaces, and used to study the problem of best approximation in statistics. In the sequel, we shall adopt the usual terminology, notation and conventions of the theory of probabilistic normed spaces, as in [1, 2-4, 9, 11]. Throughout this paper, the space of all probability distribution functions (briefly, d.f.'s) is denoted by $\Delta^+ = \{F : \mathbf{R} \cup \{-\infty, +\infty\} \longrightarrow [0,1] : F$ is left-continuous and non-decreasing on $\mathbf{R}$, $F(0) = 0$ and $F(+\infty) = 1\}$ and the subset $D^+ \subseteq \Delta^+$ is the set $D^+ = \{F \in \Delta^+ : l^- F(+\infty) = 1\}$. Here $l^- f(x)$ denotes the left limit of the function f at the point x, $l^- f(x) = \lim_{t \to x^-} f(t)$. The space Δ^+ is partially ordered by the usual point-wise ordering of functions, i.e., $F \leq G$ if and only if $F(t) \leq G(t)$ for all t in $\mathbf{R}$.

Received October 30, 2007.

2000 *Mathematics Subject Classification.* 54E40, 54E35, 54H25.

Key words and phrases. Probabilistic contractive mapping, complete Menger probabilistic metric space, fixed point theorem.

[1] Faculty of Sciences, University of Shomal, Amol, P.O.Box 731, Iran (*E-mail:* `rsaadati@eml.cc`)

The maximal element for Δ^+ in this order is the d.f. given by

$$\varepsilon_0(t) = \begin{cases} 0, & \text{if } t \leq 0, \\ 1, & \text{if } t > 0. \end{cases}$$

Definition 1.1. ([11]) A mapping $T : [0,1] \times [0,1] \longrightarrow [0,1]$ is a *continuous t-norm* if T satisfies the following conditions:

(1) T is commutative and associative;
(2) T is continuous;
(3) $T(a,1) = a$ for all $a \in [0,1]$;
(4) $T(a,b) \leq T(c,d)$ whenever $a \leq c$ and $c \leq d$ for all $a,b,c,d \in [0,1]$.

Two typical examples of continuous t-norm are $T(a,b) = ab$ and $T(a,b) = \min(a,b)$. Now t-norms are recursively defined by $T^1 = T$ and

$$T^n(x_1, \cdots, x_{n+1}) = T(T^{n-1}(x_1, \cdots, x_n), x_{n+1})$$

for $n \geq 2$ and $x_i \in [0,1]$, $i \in \{1, 2, \cdots, n+1\}$.

Definition 1.2. A *Menger Probabilistic Metric space* (briefly, Menger PM-space) is a triple $(X, \mathcal{F}, T)$, where X is a nonempty set, T is a continuous t-norm, and $\mathcal{F}$ is a mapping from $X \times X$ into D^+ such that, if $F_{x,y}$ denotes the value of $\mathcal{F}$ at the pair (x,y), the following conditions hold: for all x, y, z in X,

(PM1) $F_{x,y}(t) = \varepsilon_0(t)$ for all $t > 0$ if and only if $x = y$;
(PM2) $F_{x,y}(t) = F_{y,x}(t)$;
(PM3) $F_{x,z}(t+s) \geq T(F_{x,y}(t), F_{y,z}(s))$ for all $x, y, z \in X$ and $t, s \geq 0$.

Definition 1.3. Let $(X, \mathcal{F}, T)$ be a Menger PM-space.

(1) A sequence $\{x_n\}_n$ in X is said to be *convergent* to x in X if, for every $\epsilon > 0$ and $\lambda > 0$, there exists positive integer N such that $F_{x_n,x}(\epsilon) > 1 - \lambda$ whenever $n \geq N$.

(2) A sequence $\{x_n\}_n$ in X is called *Cauchy sequence* if, for every $\epsilon > 0$ and $\lambda > 0$, there exists positive integer N such that $F_{x_n,x_m}(\epsilon) > 1 - \lambda$ whenever $n, m \geq N$.

(3) A Menger PM-space $(X, \mathcal{F}, T)$ is said to be *complete* if and only if every Cauchy sequence in X is convergent to a point in X.

Definition 1.4. Let $(X, \mathcal{F}, T)$ be a Menger PM space. For each p in X and $\lambda > 0$, the *strong λ-neighborhood* of p is the set

$$N_p(\lambda) = \{q \in X : F_{p,q}(\lambda) > 1 - \lambda\}$$

and the strong neighborhood system for X is the union $\bigcup_{p \in V} \mathcal{N}_p$, where $\mathcal{N}_p = \{N_p(\lambda) : \lambda > 0\}$.

The strong neighborhood system for X determines a Hausdorff topology for X.

Lemma 1.5. *Let $(X, \mathcal{F}, T)$ be a Menger PM space and define $E_{\lambda,F} : X^2 \longrightarrow \mathbf{R}^+ \cup \{0\}$ by*

$$E_{\lambda,F}(x,y) = \inf\{t > 0 : F_{x,y}(t) > 1 - \lambda\}$$

for each $\lambda \in]0,1[$ and $x, y \in X$. Then we have

(a) *For any* $\mu \in]0, 1[$, *there exists* $\lambda \in]0, 1[$ *such that*

$$E_{\mu,F}(x_1, x_n) \leq E_{\lambda,F}(x_1, x_2) + \cdots + E_{\lambda,F}(x_{n-1}, x_n)$$

for any $x_1, ..., x_n \in X$;

(b) *A sequence* $\{x_n\}$ *is convergent with respect to Menger PM* $\mathcal{F}$ *if and only if* $E_{\lambda,F}(x_n, x) \rightarrow 0$. *Also, sequence* $\{x_n\}$ *is a Cauchy sequence with respect to Menger PM* $\mathcal{F}$ *if and only if it is a Cauchy sequence with* $E_{\lambda,F}$.

Proof. (a) For every $\mu \in]0, 1[$, we can find a $\lambda \in]0, 1[$ such that $T^{n-1}(1-\lambda, \cdots, 1-\lambda) > 1 - \mu$. By the triangular inequality, we have

$$F_{x,x_n}(E_{\lambda,F}(x_1, x_2) + \cdots + E_{\lambda,F}(x_{n-1}, x_n) + n\delta)$$
$$\geq T^{n-1}(F_{x_1,x_2}(E_{\lambda,F}(x_1, x_2) + \delta), \cdots, F_{x_{n-1},x_n}(E_{\lambda,M}(x_{n-1}, x_n) + \delta))$$
$$\geq T^{n-1}(1 - \lambda, \cdots, 1 - \lambda) > 1 - \mu$$

for every $\delta > 0$, which implies that

$$E_{\mu,F}(x_1, x_n) \leq E_{\lambda,F}(x_1, x_2) + E_{\lambda,F}(x_2, x_3) + \cdots + E_{\lambda,F}(x_{n-1}, x_n) + n\delta.$$

Since $\delta > 0$ is arbitrary, we have

$$E_{\mu,F}(x_1, x_n) \leq E_{\lambda,F}(x_1, x_2) + E_{\lambda,F}(x_2, x_3) + \cdots + E_{\lambda,F}(x_{n-1}, x_n).$$

(b) We have $F_{x_n,x}(\eta) > 1 - \lambda \Longleftrightarrow E_{\lambda,F}(x_n, x) < \eta$ for every $\eta > 0$. This completes the proof. $\qquad\square$

Lemma 1.6. *If a Menger PM space* $(X, \mathcal{F}, T)$ *satisfies the following condition:*

$$F_{x,y}(t) = C$$

for all $t > 0$. *Then we have* $C = \varepsilon_0(t)$ *and* $x = y$.

Proof. Let $F_{x,y}(t) = C$ for all $t > 0$. Then, since $F \in D^+$, we have $C = \varepsilon_0(t)$ and, by (PM1), we conclude that $x = y$. $\qquad\square$

Lemma 1.7. *([6, 7]) Let the function* $\phi(t)$ *satisfy the following condition:*
(Φ) $\quad \phi(t) : [0, \infty) \longrightarrow [0, \infty)$ *is nondecreasing and* $\sum_{n=1}^{\infty} \phi^n(t) < \infty$ *for all* $t > 0$, *where* $\phi^n(t)$ *denotes the* n-*th iterative function of* $\phi(t)$. *Then* $\phi(t) < t$ *for all* $t > 0$.

Lemma 1.8. *Suppose that the function* $\phi(t) : [0, \infty) \longrightarrow [0, \infty)$ *is onto and strictly increasing and let* $(X, \mathcal{F}, T)$ *be a Menger PM space. Then*

$$\inf\{\phi^n(t) > 0 : F_{x,y}(t) > 1 - \lambda\} \leq \phi^n(\inf\{t > 0 : F_{x,y}(t) > 1 - \lambda\})$$

for every $x, y \in X$, $\lambda \in (0, 1)$ *and* $n \in \{1, 2, \cdots\}$.

Proof. Fix $t \in [0, \infty)$ with $F_{x,y}(t) > 1 - \lambda$. Then $\phi^n(t) > 0$. Also

$$\phi^n(t) \geq \inf\{\phi^n(s) > 0 : F_{x,y}(s) > 1 - \lambda\}.$$

Now, since ϕ^n is onto and strictly increasing, we have

$$t \geq (\phi^n)^{-1}(\inf\{\phi^n(s) > 0 : F_{x,y}(s) > 1 - \lambda\}).$$

Thus we have

$$\inf\{t > 0 : F_{x,y}(t) > 1 - \lambda\} \geq (\phi^n)^{-1}(\inf\{\phi^n(s) > 0 : F_{x,y}(s) > 1 - \lambda\})$$

and so

$$\inf\{\phi^n(t) > 0 : F_{x,y}(t) > 1 - \lambda\} \leq \phi^n(\inf\{t > 0 : F_{x,y}(t) > 1 - \lambda\}).$$

This completes the proof.   □

Remark 1.9. Suppose that, for every $\mu \in (0,1)$, there exists $\lambda \in (0,1)$ (which does not depend on n) with

(1.1) $$T^{n-1}(1 - \lambda, \cdots, 1 - \lambda) > 1 - \mu$$

for each $n \in \{1, 2, \cdots\}$. In this case, the λ in Lemma 1.5 (a) does not depend on n and so, if (1.1) is assumed, then $E_F(u_0, u_1) < \infty$ is not needed in Theorem 2.4.

Lemma 1.10. *Let $(X, \mathcal{F}, T)$ be a complete Menger PM space. Suppose*

$$F_{x_n, x_{n+1}}(t) \geq F_{x_0, x_1}(k^n t)$$

for some $k > 1$ and $n \in \mathbb{N}$. Also, assume that

$$E_F(x_0, x_1) = \sup\{E_{\gamma, F}(x_0, x_1) : \gamma \in (0,1)\} < \infty.$$

Then $\{x_n\}$ is a Cauchy sequence.

Proof. For every $\lambda \in (0,1)$, we have

$$
\begin{aligned}
E_{\lambda, F}(x_{n+1}, x_n) &= \inf\{t > 0 \ : \ F_{x_{n+1}, x_n}(t) > 1 - \lambda\} \\
&\leq \inf\{t > 0 \ : \ F_{x_0, x_1}(k^n t) > 1 - \lambda\} \\
&= \inf\left\{\frac{t}{k^n} \ : \ F_{x_0, x_1}(t) > 1 - \lambda\right\} \\
&= \frac{1}{k^n} \inf\{t > 0 \ : \ F_{x_0, x_1}(t) > 1 - \lambda\} \\
&= \frac{1}{k^n} E_{\lambda, F}(x_0, x_1).
\end{aligned}
$$

From Lemma 1.5, for every $\mu \in (0,1)$, there exists $\lambda \in (0,1)$ such that

$$
\begin{aligned}
E_{\mu, F}(x_n, x_m) &\leq E_{\lambda, F}(x_n, x_{n+1}) + E_{\lambda, F}(x_{n+1}, x_{n+2}) + \cdots + E_{\lambda, F}(x_{m-1}, x_m) \\
&\leq \frac{1}{k^n} E_{\lambda, F}(x_0, x_1) + \frac{1}{k^{n+1}} E_{\lambda, F}(x_0, x_1) + \cdots + \frac{1}{k^{m-1}} E_{\lambda, F}(x_0, x_1) \\
&\leq E_F(x_0, x_1) \sum_{j=n}^{m-1} \frac{1}{k^j} \longrightarrow 0.
\end{aligned}
$$

Hence the sequence $\{x_n\}$ is a Cauchy sequence. This completes the proof.   □

2. Main Results

First, we give two definitions for our main results.

Definition 2.1. Let f and g be maps from a Menger PM space $(X, \mathcal{F}, T)$ into itself. The mappings f and g are said to be *weakly commuting* if

$$F_{fgx,gfx}(t) \geq F_{fx,gx}(t)$$

for all $x \in X$ and $t > 0$.

For the remainder of the paper, let Φ be the set of all onto and strictly increasing functions

$$\phi : [0, \infty) \longrightarrow [0, \infty)$$

which satisfy $\lim_{n \to \infty} \phi^n(t) = 0$ for all $t > 0$.

Definition 2.2. Let f and g be maps from a Menger PM space $(X, \mathcal{F}, T)$ into itself. The mappings f and g are said to be ϕ-*weakly commuting* if there exists $\phi \in \Phi$ such that

$$F_{fgx,gfx}(\phi(t)) \geq F_{fx,gx}(t)$$

for all $x \in X$ and $t > 0$.

Remark 2.3. (1) First notice that if $\phi \in \Phi$, then $\phi(t) < t$ for all $t > 0$. To see this, suppose that there exists $t_0 > 0$ with $t_0 \leq \phi(t_0)$. Then, since ϕ is nondecreasing we have $t_0 \leq \phi^n(t_0)$ for each $n \in \{1, 2, \cdots\}$, which is a contradiction. Note also that $\phi(0) = 0$.
(2) ϕ-weak commutativity implies weak commutativity in a Menger PM space.

Theorem 2.4. *Let $(X, \mathcal{F}, T)$ be a complete Menger PM space and let f, g be weakly commuting self-mappings of X satisfying the following conditions:*
(a) $f(X) \subseteq g(X)$;
(b) f or g is continuous;
(c) $F_{fx,fy}(\phi(t)) \geq F_{gx,gy},(t)$ *where* $\phi \in \Phi$;
(d) Assume that there exists $x_0 \in X$ with

$$E_F(gx_0, fx_0) = \sup\{E_{\gamma,F}(gx_0, fx_0) : \gamma \in (0, 1)\} < \infty.$$

Then f and g have a unique common fixed point.

Proof. Choose $x_0 \in X$ with $E_F(gx_0, fx_0) < \infty$. Choose $x_1 \in X$ with $fx_0 = gx_1$. In general, choose x_{n+1} such that $fx_n = gx_{n+1}$. Now

$$F_{fx_n,fx_{n+1}}(\phi^{n+1}(t)) \geq F_{gx_n,gx_{n+1}}(\phi^n(t)) = F_{fx_{n-1},fx_n}(\phi^n(t))$$
$$\geq \cdots \geq F_{gx_0,gx_1}(t).$$

Note (see Lemma 1.8) that, for each $\lambda \in (0, 1)$,

$$E_{\lambda,F}(fx_n, fx_{n+1}) = \inf\{\phi^{n+1}(t) > 0 : F_{fx_n,fx_{n+1}}(\phi^{n+1}(t)) > 1 - \lambda\}$$
$$\leq \inf\{\phi^{n+1}(t) > 0 : F_{gx_0,fx_0}(t) > 1 - \lambda\}$$
$$\leq \phi^{n+1}(\inf\{t > 0 : F_{gx_0,fx_0}(t) > 1 - \lambda\})$$
$$= \phi^{n+1}(E_{\lambda,F}(gx_0, fx_0))$$
$$\leq \phi^{n+1}(E_F(gx_0, fx_0)).$$

Thus $E_{\lambda,F}(fx_n, fx_{n+1}) \leq \phi^{n+1}(E_F(gx_0, fx_0))$ for all $\lambda \in (0, 1)$ and so

$$E_F(fx_n, fx_{n+1}) \leq \phi^{n+1}(E_F(gx_0, fx_0)).$$

Let $\epsilon > 0$. Choose $n \in \{1, 2, \cdots\}$ so that $E_F(fx_n, fx_{n+1}) < \epsilon - \phi(\epsilon)$. For all $\lambda \in (0, 1)$ there exists $\mu \in (0, 1)$ such that

$$
\begin{aligned}
E_{\lambda, F}(fx_n, fx_{n+2}) &\leq E_{\mu, F}(fx_n, fx_{n+1}) + E_{\mu, F}(fx_{n+1}, fx_{n+2}) \\
&\leq E_{\mu, F}(fx_n, fx_{n+1}) + \phi(E_{\mu, F}(fx_n, fx_{n+1})) \\
&\leq E_F(fx_n, fx_{n+1}) + \phi(E_F(fx_n, fx_{n+1})) \\
&\leq \epsilon - \phi(\epsilon) + \phi(\epsilon - \phi(\epsilon)) \\
&\leq \epsilon
\end{aligned}
$$

and so

$$E_F(fx_n, fx_{n+2}) \leq \epsilon.$$

For all $\lambda \in (0, 1)$, there exists $\mu \in (0, 1)$ such that

$$
\begin{aligned}
E_{\lambda, F}(fx_n, x_{n+3}) &\leq E_{\mu, F}(fx_n, fx_{n+1}) + E_{\mu, F}(fx_{n+1}, fx_{n+3}) \\
&\leq E_{\mu, F}(fx_n, fx_{n+1}) + \phi(E_{\mu, F}(fx_n, fx_{n+2})) \\
&\leq E_F(fx_n, fx_{n+1}) + \phi(E_F(fx_n, fx_{n+2})) \\
&\leq \epsilon - \phi(\epsilon) + \phi(\epsilon) = \epsilon
\end{aligned}
$$

and so

$$E_F(fx_n, fx_{n+3}) \leq \epsilon.$$

By induction,

$$E_F(fx_n, fx_{n+k}) \leq \epsilon$$

for $k \in \{1, 2, \cdots\}$. Thus $\{fx_n\}$ is a Cauchy sequence and, by the completeness of X, $\{fx_n\}$ converges to a point $z \in X$. Also, $\{gx_n\}$ converges to the point z. Let us now suppose that the mapping f is continuous. Then $\lim_{n \to \infty} ffx_n = fz$ and $\lim_{n \to \infty} fgx_n = fz$. Furthermore, since f and g are weakly commuting, we have

$$F_{fgx_n, gfx_n}(t) \geq F_{fx_n, gx_n}(t).$$

Letting $n \to \infty$ in the above inequality, we get $\lim_{n \to \infty} gfx_n = fz$ by continuity of $\mathcal{F}$.

We now prove that $z = fz$. Suppose that $z \neq fz$. By (c), for any $t > 0$, we have

$$F_{fx_n, ffx_n}(\phi^{k+1}(t)) \geq F_{gx_n, gfx_n}(\phi^k(t))$$

for $k \in \{1, 2, \cdots\}$. Letting $n \to \infty$ in the above inequality, we get

$$F_{z, fz}(\phi^{k+1}(t)) \geq F_{z, fz}\phi^k(t)).$$

Also, we have

$$F_{z, fz}(\phi^k(t)) \geq F_{z, fz}(\phi^{k-1}(t))$$

and

$$F_{z, fz}(\phi(t)) \geq F_{z, fz}(t).$$

Therefore, we have

$$F_{z, fz}(\phi^{k+1}(t)) \geq F_{z, fz}(t).$$

On the other hand, we have (see Remark 2.3)

$$F_{z, fz}(\phi^{k+1}(t)) \leq F_{z, fz}(t).$$

Then $F_{z,fz}(t) = C$ and, by Lemma 1.6, $z = fz$. Since $f(X) \subseteq g(X)$, we can find z_1 in X such that $z = fz = gz_1$. Now,

$$F_{ffx_n,fz_1}(t) \geq F_{gfx_n,gz_1}(\phi^{-1}(t)).$$

Taking the limit as $n \to \infty$, we get

$$F_{fz,fz_1}(t) \geq F_{fz,gz_1}(\phi^{-1}(t)) = \varepsilon_0(t),$$

which implies that $fz = fz_1$, i.e., $z = fz = fz_1 = gz_1$. Also, for any $t > 0$, since f and g are weakly commuting,

$$F_{fz,gz}(t) = F_{fgz_1,gfz_1}(t) \geq F_{fz_1,gz_1}(t) = \varepsilon_0(t),$$

which again implies that $fz = gz$. Thus z is a common fixed point of f and g.

Now, to prove the uniqueness of the point z, suppose that $z' \neq z$ is another common fixed point of f and g. Then, for any $t > 0$, we have

$$F_{z,z'}(\phi^{n+1}(t)) = F_{fz,fz'}(\phi^{n+1}(t)) \geq F_{gz,gz'}(\phi^n(t)) = F_{z,z'}(\phi^n(t)).$$

Also, we have

$$F_{z,z'}(\phi^n(t)) \geq F_{z,z'}(\phi^{n-1}(t))$$

and

$$F_{z,z'}(\phi(t)) \geq F_{z,z'}(t).$$

Therefore, we have

$$F_{z,z'}(\phi^{n+1}(t)) \geq F_{z,z'}(t).$$

On the other hand, we have

$$F_{z,z'}(t) \leq F_{z,z'}(\phi^{n+1}(t)).$$

Then $F_{z,z'}(t) = C$ and by Lemma 1.6, $z = z'$, which is contradiction. Therefore z is the unique common fixed point of f and g. This completes the proof. $\qquad\square$

Corollary 2.5. (Boyd-Wong Probabilistic Version) *Let* $(X, \mathcal{F}, T)$ *be a complete Menger PM space and let* f *be weakly commuting self-mapping of* X *satisfying the following conditions:*

(a) f *is continuous;*

(b) $F_{fx,fy}(\phi(t)) \geq F_{x,y}(t)$, *where* $\phi \in \Phi$. *Also, assume that, there exists* $x_0 \in X$ *with*

$$E_F(x_0, fx_0) = \sup\{E_{\gamma,F}(x_0, fx_0) : \gamma \in (0,1)\} < \infty.$$

Then f *has a unique common fixed point.*

Theorem 2.6. *Let* $(X, \mathcal{F}, T)$ *be a complete Menger PM space and let* S, T *be two self-mappings of* X *satisfying the following conditions:*

(a) $F_{Sx,TSy}(t) \geq F_{x,Sy}(kt)$ *for some* $k > 1$;

(b) S *or* T *is continuous;*

(c) *Assume that there exists* $x_0 \in X$ *with*

$$E_F(x_0, Tx_0) = \sup\{E_{\gamma,F}(x_0, Tx_0) : \gamma \in (0,1)\} < \infty.$$

Then S *and* T *have a unique common fixed point.*

Proof. Let $x_0 \in X$ be such that $E_\mathcal{M}(x_0, Tx_0) < \infty$. Define

$$x_{2n} = Sx_{2n-1}, \quad \forall n \geq 1,$$
$$x_{2n+1} = Tx_{2n}, \quad \forall n \geq 0.$$

Now, for an even integer $n = 2m$, we have

$$\begin{aligned}
F_{x_{2m}, x_{2m+1}}(t) &= F_{Sx_{2m-1}, Tx_{2m}}(t) \\
&= F_{Sx_{2m-1}, TSx_{2m-1}}(t) \\
&\geq F_{x_{2m-1}, Sx_{2m-1}}(kt) \\
&= F_{x_{2m-1}, x_{2m}}(kt) = F_{x_{2m}, x_{2m-1}}(kt) \\
&= F_{Sx_{2m-1}, TSx_{2m-3}}(kt) \\
&\geq F_{x_{2m-1}, Sx_{2m-3}}(k^2 t) \\
&= F_{x_{2m-1}, x_{2m-2}}(k^2 t) \\
&\quad \cdots \\
&\geq F_{x_0, x_1}(k^n t).
\end{aligned}$$

Thus, by Lemma 1.10, $\{x_n\}$ is a Cauchy sequence and, by the completeness of X, $\{x_n\}$ converges to a point $x \in X$. Then

$$\lim_{n \to \infty} x_n = \lim_{n \to \infty} x_{2n} = \lim_{n \to \infty} Sx_{2n-1} = x$$

and

$$\lim_{n \to \infty} x_{2n+1} = \lim_{n \to \infty} Tx_{2n} = x.$$

Let us suppose that the mapping S is continuous. Then

$$\lim_{n \to \infty} Sx_{2n-1} = S \lim_{n \to \infty} x_{2n-1} \Longrightarrow Sx = x.$$

Also, we have

$$F_{x, Tx}(t) = F_{Sx, TSx}(t) \geq F_{x, Sx}(kt) = F_{x, x}(kt) = \varepsilon_0(t)$$

anf so $Tx = x$.

Now, to prove the uniqueness of the point x, suppose that $y \neq x$ is another common fixed point of S and T. Then, for all $t > 0$, we have

$$\begin{aligned}
F_{x, y}(t) &= F_{Sx, TSy}(t) \\
&\geq F_{x, Sy}(kt) \\
&= F_{x, y}(kt) \\
&\quad \cdots \\
&\geq F_{x, y}(k^n t).
\end{aligned}$$

On the other hand, since $\mathcal{F}$ is non-decreasing, we have

$$F_{x, y}(t) \leq F_{x, y}(k^n t).$$

Thus $F_{x, y}(t) = C$ for all $t > 0$. Then, from Lemma 1.6, we have $x = y$. Therefore x is the unique common fixed point of S and T. This completes the proof. $\square$

REFERENCES

1. C. Alsina, B. Schweizer and A. Sklar, *On the definition of a probabilistic normed space,* Aequationes Math. **46** (1993), 91–98.

2. C. Alsina, B. Schweizer and A. Sklar, *Continuity properties of probabilistic norms,* J. Math. Anal. Appl. **208** (1997), 446–452.

3. A. Bharucha-Reid, *Fixed point theorems in probabilistic analysis,* Bull. Amer. Math. Soc. **82** (1976), 641–657.

4. S. S. Chang, Y. J. Cho and S. M. Kang, *Nonlinear Operator Theory in Probabilistic Metric Spaces,* Nova Science Publishers, Inc., New York, 2001.

5. J. X. Fang, *On fixed point theorems in fuzzy metric spaces,* Fuzzy Sets and Systems **46** (1992), 107–113.

6. O. Hadžić and E. Pap, *Fixed Point Theory in PM Spaces,* Kluwer Academic Publishers, Dordrecht, 2001.

7. O. Hadžić and E. Pap, *New classes of probabilistic contractions and applications to random operators,* in: Y. J. Cho, J. K. Kim, S. M. Kong (Eds.), Fixed Point Theory and Application, vol. 4, Nova Science Publishers, Hauppauge, NewYork, 2003, pp. 97–119.

8. G. Jungck, *Commuting maps and fixed points,* Amer. Math. Monthly **83** (1976) 261–263.

9. M. A. Khamsi and V. Y. Kreinovich, *Fixed point theorems for dissipative mappings in complete probabilistic metric spaces,* Math. Jap. **44** (1996), 513–520.

10. A. Razani and M. Shirdaryazdi, *Some results on fixed points in the fuzzy metric space,* J. Appl. Math. and Computing **20** (2006), 401–408.

11. B. Schweizer and A. Sklar, *Probabilistic Metric Spaces,* Elsevier North Holand, New York, 1983.

12. A. N. Šerstnev, *On the motion of a random normed space,* Dokl. Akad. Nauk SSSR **149** (1963), 280–283 (English translation in Soviet Math. Dokl. **4** (1963), 388–390.

13. B. Singh and S. Jain, *A fixed point theorem in Menger space through weak compatibility,* J. Math. Anal. Appl. **301** (2005), 439–448.

Nonlinear Functional Analysis and Applications, Volume 1, 211–218
© 2012 Nova Science Publishers, Inc.

GLOBAL CONVERGENCE THEOREMS FOR ACCRETIVE
AND PSEUDO-CONTRACTIVE MAPPINGS

Haiyun Zhou[1,*] and Jinti Guo[2]

ABSTRACT. Let E be a real uniformly smooth Banach space and $A : D(A) \subseteq E \to 2^E$ be a m-accretive mapping which satisfies a linear growth condition of the form $\|u\| \leq C(1 + \|x\|)$ for some constant $C > 0$ and for all $x \in E$, $u \in Ax$, $z \in \overline{D(A)}$ be an arbitrary element. Suppose $A^{-1}0 \neq \emptyset$. The sequence $\{x_n\} \subset D(A)$ is generated from arbitrary $x_0 \in D(A)$ by

$$x_{n+1} \in x_n - \lambda_n(u_n + \theta_n(x_n - z)), \quad \forall u_n \in Ax_n, \, n \geq 0,$$

where $\{\lambda_n\}$ and $\{\theta_n\}$ are acceptably paired. Then $\{x_n\}$ converges strongly to a point $x^* \in A^{-1}(0)$. As its application, a strong convergence theorem of fixed points for continuous pseudo-contractions is deduced.

1. Introduction

Let E be a real normed linear space with dual E^*. We denote by J the normalized duality mapping from E to 2^{E^*} defined by

$$J(x) = \{x^* \in E^* : \langle x, x^* \rangle = \|x\|^2 = \|x^*\|^2\}, \quad \forall x in E,$$

where $\langle \cdot, \cdot \rangle$ denotes the generalized duality pairing. It is well known that, if E^* is strictly convex, then J is single-valued. In the sequel, we shall denote the single-valued normalized duality mapping by j. A mapping $A : E \to 2^E$ is called *accretive* if, for all $x, y \in D(A)$, there exists $j(x - y) \in J(x - y)$ such that

$$\langle u - v, j(x - y) \rangle \geq 0 \quad \forall u \in Ax, \, v \in Ay.$$

Received December 19, 2007.   * Corresponding author.

2000 *Mathematics Subject Classification.* Primary 47H17; Secondary 47H05, 47H10.

Key words and phrases. Accretive mapping; Pseudo-contraction; Regularization iteration algorithm; Global convergence theorem.

This paper was supported by the National Science Foundation of China (Grant No.10771050)

[1] Institute of Nonlinear Analysis, North China Electric Power University, Baoding 071003, People's Republic of China (*E-mail:* witmann66@yahoo.com.cn)

[2] College of Mathematics and Statistics, Hebei University of Economics and Business, Shijiazhuang 071051, People's Republic of China

The mapping A is called *strongly accretive* if, for all $x, y \in D(A)$, there exist $j(x-y) \in J(x-y)$ and $k > 0$ such that

$$\langle u-v, j(x-y)\rangle \geq k\|x-y\|^2, \quad \forall u \in Ax, v \in Ay.$$

If E is a Hilbert space, accretive operator is also called *monotone*. An operator A is called *m-accretive* if it is accretive and $R(I+rA) = E$, range of $(I+rA)$, is E for all $r > 0$. The class closely related to the class of accretive mappings is the class of pseudo-contractions. An operator T with domain $D(T)$ in E and the range $R(T)$ in E is called *pesudo-contractive* if, for each $x, y \in D(T)$, there exist $j(x-y) \in J(x-y)$ such that

$$\langle u-v, j(x-y)\rangle \leq \|x-y\|^2, \quad \forall u \in Ax, v \in Ay.$$

and it is called *strongly pseudo-contractive* if, for each $x, y \in D(T)$, there exists $j(x-y) \in J(x-y)$ and constant $k \in (0,1)$ such that

$$\langle u-v, j(x-y)\rangle \leq k\|x-y\|^2, \quad \forall u \in Ax, v \in Ay.$$

Observe that A is accretive if and only if $I - A$ is pseudocontractive and thus a zero of A is a fixed point of $T := I - A$.

Interest of accretive mappings stems mainly from their firm connection with equations of evolution (for example, heat, wave, or Schrödinger equations). Since 1967, the theory of the accretive operator has been well developed and the studies on the existence and construction for zeros of the accretive operators have been attracting many excellent mathematicians. In 1974, Bruck [1] adopted a regularization iteration algorithm to construct zeros of maximal monotone mapping in a Hilbert space. To be specific, Bruck [1] proved the following convergence theorem.

Theorem B. ([1]) *Let A be a maximal monotone operator on a Hilbert space H with $0 \in R(A)$. Suppose $\{\lambda_n\}$ and $\{\theta_n\}$ are acceptably paired, $z \in H$ and the sequence $\{x_n\} \subset D(A)$ is generated from arbitrary $x_0 \in D(A)$ by*

$$x_{n+1} \in x_n - \lambda_n(u_n + \theta_n(x_n - z)), \quad \forall u_n \in Ax_n, n \geq 0,$$

If $\{x_n\}$ and $\{u_n\}$ are bounded, then $\{x_n\}$ converges strongly to x^*, the point of $A^{-1}(0)$ closest to z.

By applying Theorem B, Bruck [1] also obtained a local convergence theorem. In 1979, Nevanlinna [3] gave a global version of this theorem, still in Hilbert spaces, by assuming that A either satisfies a linear growth condition of the form $\|u\| \leq C(1+\|x\|)$ for some constant $C > 0$ and for all $x \in E$, $u \in Ax$ or is continuous. In 1980, Reich [5] developed the main results of Bruck [1] into uniformly smooth Banach spaces more general than Hilbert spaces. However, we remark that the main results of Reich [5] were proved under the more strict assumptions on the iterative parameters. Recently, Chidume and Zegeye [2] extended the main results of Nevanlinna [3] from Hilbert spaces to q-uniformly smooth Banach spaces.

The purpose of this paper is to extend the main results of Chidume and Zegeye [2] to uniformly smooth Banach spaces.

2. Preliminaries

Let E be a real normed linear space of dimension, $\dim E \geq 2$. The modulus of smoothness of E is defined by

$$\rho_E(\tau) = \sup\left\{ \frac{\|x+y\| + \|x-y\|}{2} - 1 : \|x\| = 1, \|y\| = \tau \right\}, \quad \tau > 0.$$

If $\rho_E(\tau) > 0$ for all $\tau > 0$, then E is said to be *smooth*. If there exist constant $c > 0$ and a real number $1 < q < \infty$ such that $\rho_E(\tau) \leq c\tau^q$, then E is said to be *q-uniformly smooth*. A Banach space E is called *uniformly smooth* if $\lim_{\tau \to 0} \rho_E(\tau)/\tau = 0$.

Typical examples of such spaces are the Lebesgue L_p, the sequence l_p and the Sobolev W_p^m spaces for $1 < p < \infty$.

Uniformly smooth Banach spaces enjoy very nice geometrical properties. In 1978, Reich [4] established the following famous inequality:

Theorem R1. ([4]) *Let E be a real uniformly smooth Banach space, then there exists a continuous increasing function $b : R^+ \to R^+$ which satisfies:*
(a) $b(ct) \leq cb(t)$ for all $c \geq 1$,
(b) $b(0) = 0$,
(c) $\|x+y\|^2 \leq \|x\|^2 + 2\langle y, j(x)\rangle + \max\{\|x\|, 1\}\|y\|b(\|y\|)$ for all $x, y \in E$.

Definition 1. Two sequences $\{\lambda_n\}$ and $\{\theta_n\}$ of positive real numbers is called *acceptably paired* if $\{\theta_n\}$ is non-increasing, $\lim_{n\to\infty} \theta_n = 0$ and there exists a strictly increasing sequence $\{n(i)\}_{i=1}^\infty$ of positive integers such that
(1) $\liminf_{i\to\infty} \theta_{n(i)} \sum_{j=n(i)}^{n(i+1)} \lambda_j > 0$
(2) $\lim_{i\to\infty} [\theta_{n(i)} - \theta_{n(i+1)}] \sum_{j=n(i)}^{n(i+1)} \lambda_j = 0$,
(3) $\lim_{i\to\infty} \sum_{j=n(i)}^{n(i+1)} \lambda_j b(\lambda_j) = 0$.

Theorem R2. ([5]) *Let E be a real uniformly smooth Banach space and $A : E \to 2^E$ be m-accretive. If $0 \in R(A)$, then, for each $x \in E$, the strong $\lim_{t\to\infty} J_t(x)$ exists and belongs to $A^{-1}(0)$, where $J_t = (I + tA)^{-1}$ for all $t > 0$.*

For the rest of this paper, let $x^* \in A^{-1}(0)$ be such that $J_t(0) \to x^*$ as $t \to \infty$, which is guaranteed by Theorem R2.

3. Main Results

Theorem 3.1. *Let E be a real uniformly smooth Banach space and $A : D(A) \subseteq E \to 2^E$ be an m-accretive mapping which satisfies a linear growth condition: $\|u\| \leq C(1+\|x\|)$ for some constant $C > 0$ and for all $x \in E$, $u \in Ax$. Let $z \in \overline{D(A)}$ be an arbitrary, but fixed element. Suppose $A^{-1}0 \neq \emptyset$. Let $\{x_n\} \subset D(A)$ be a sequence generated from arbitrary $x_0 \in D(A)$ by*

$$(1) \qquad x_{n+1} \in x_n - \lambda_n(u_n + \theta_n(x_n - z)), \quad \forall u_n \in Ax_n, n \geq 0,$$

where $\{\lambda_n\}$ and $\{\theta_n\}$ are acceptably paired. If $\{\theta_n\}$ is decreasing, $\lim_{n\to\infty}\theta_n = 0$ and $\sum_{n=0}^\infty \lambda_n b(\lambda_n) < \infty$, then $\{x_n\}$ converges strongly to a point $x^* \in A^{-1}(0)$.

Proof. Without loss of generality, we may assume that $z = 0$. Since A is m-accretive, so is $\theta^{-1}A$ and hence $R(I + \theta^{-1}A) = E$ for any $\theta > 0$. Thus, for each i, there exists a unique $y_i \in E$ with $0 \in \theta_i y_i + A y_i$ and, by Theorem R2, $\lim_{\theta_i \to 0} J_{1/\theta_i} = \lim_{t \to \infty} J_t(0) = \lim_{i \to \infty} y_i = x^* \in A^{-1}(0)$ for $t = 1/\theta_i$. For $n \geq i \geq 2$ and $u_{n-1} \in A x_{n-1}$, from (1), it follows that

$$(2) \qquad x_n - y_i = x_{n-1} - y_i - \lambda_{n-1}(u_{n-1} + \theta_{n-1} x_{n-1}).$$

Without loss of generality, we may assume that $x_{n-1} \neq y_i$. Otherwise, replacing $\|x_{n-1} - y_i\|$ by $1 + \|x_{n-1} - y_i\|$. By using (c) of Theorem R1, (1) and (2), we obtain

$$(3) \qquad
\begin{aligned}
\left\| \frac{x_n - y_i}{\|x_{n-1} - y_i\|} \right\|^2 &= \left\| \frac{x_{n-1} - y_i}{\|x_{n-1} - y_i\|} - \lambda_{n-1} \left(\frac{u_{n-1} + \theta_{n-1} x_{n-1}}{\|x_n - y_n\|} \right) \right\|^2 \\
&\leq 1 - \frac{2\lambda_{n-1}}{\|x_{n-1} - y_i\|^2} \langle u_{n-1} + \theta_{n-1} x_{n-1}, j(x_{n-1} - y_i) \rangle \\
&\quad + \lambda_{n-1} \frac{\|u_{n-1} + \theta_{n-1} x_{n-1}\|}{\|x_{n-1} - y_i\|} b \left(\lambda_{n-1} \frac{\|u_{n-1} + \theta_{n-1} x_{n-1}\|}{\|x_{n-1} - y_i\|} \right) \\
&= 1 - \frac{2\lambda_{n-1}}{\|x_{n-1} - y_i\|^2} \langle u_{n-1} + \theta_i x_{n-1}, j(x_{n-1} - y_i) \rangle \\
&\quad + \frac{2\lambda_{n-1}(\theta_i - \theta_{n-1})}{\|x_{n-1} - y_i\|^2} \langle x_{n-1}, j(x_{n-1} - y_i) \rangle \\
&\quad + \lambda_{n-1} \frac{\|u_{n-1} + \theta_{n-1} x_{n-1}\|}{\|x_{n-1} - y_i\|} b \left(\lambda_{n-1} \frac{\|u_{n-1} + \theta_{n-1} x_{n-1}\|}{\|x_{n-1} - y_i\|} \right).
\end{aligned}$$

Since $-\theta_i y_i \in A y_i$, $u_{n-1} \in A x_{n-1}$ and A is accretive, we conclude that

$$\langle u_{n-1} + \theta_i y_i, j(x_{n-1} - y_i) \rangle \geq 0,$$

which gives that

$$(4) \qquad
\begin{aligned}
&\langle u_{n-1} + \theta_i x_{n-1}, j(x_{n-1} - y_i) \rangle \\
&= \langle u_{n-1} + \theta_i y_i, j(x_{n-1} - y_i) \rangle + \theta_i \langle x_{n-1} - y_i, j(x_{n-1} - y_i) \rangle \\
&\geq \theta_i \|x_{n-1} - y_i\|^2.
\end{aligned}$$

Moreover, we have

$$(5) \qquad
\begin{aligned}
|\langle x_{n-1}, j(x_{n-1} - y_i) \rangle| &\leq \|x_{n-1}\| \|x_{n-1} - y_i\| \\
&\leq \frac{1}{2} \|x_{n-1}\|^2 + \frac{1}{2} \|x_{n-1} - y_i\|^2 \\
&\leq \frac{1}{2} (\|x_{n-1} - y_i\| + \|y_i\|)^2 + \frac{1}{2} \|x_{n-1} - y_i\|^2 \\
&\leq \|x_{n-1} - y_i\|^2 + \|y_i\|^2 + \frac{1}{2} \|x_{n-1} - y_i\|^2 \\
&= \frac{3}{2} \|x_{n-1} - y_i\|^2 + \|y_i\|^2.
\end{aligned}$$

Furthermore, using the linear growth condition, we have

$$\frac{\|u_{n-1} + \theta_{n-1}x_{n-1}\|}{\|x_{n-1} - y_i\|} \leq \frac{\|u_{n-1}\| + \|x_{n-1}\|}{\|x_{n-1} - y_i\|}$$

$$\leq \frac{C(1 + \|x_{n-1}\|) + \|x_{n-1}\|}{\|x_{n-1} - y_i\|}$$

(6)

$$\leq \frac{C + (1+C)\|x_{n-1} - y_i + y_i\|}{\|x_{n-1} - y_i\|}$$

$$\leq \frac{C + (1+C)\|x_{n-1} - y_i\| + (1+C)\|y_i\|}{\|x_{n-1} - y_i\|}$$

$$\leq (1+C) + \frac{C + (1+C)\|y_i\|}{\|x_{n-1} - y_i\|}.$$

Using (4)-(6) in (3), we obtain

$$\left\|\frac{x_n - y_i}{\|x_{n-1} - y_i\|}\right\|^2 \leq 1 - 2\lambda_{n-1}\theta_i + 3\lambda_{n-1}(\theta_i - \theta_{n-1}) + \frac{2\lambda_{n-1}(\theta_i - \theta_{n-1})}{\|x_{n-1} - y_i\|^2}\|y_i\|^2$$

(7)

$$+ \left(1 + C + \frac{C + (1+C)\|y_i\|}{\|x_{n-1} - y_i\|}\right)^2 \lambda_{n-1}b(\lambda_{n-1}).$$

Both sides of (7) multiplied by $\|x_{n-1} - y_i\|^2$, it follows that

$$\|x_n - y_i\|^2$$
$$\leq [1 - 2\lambda_{n-1}\theta_i + 3\lambda_{n-1}(\theta_i - \theta_{n-1})]\|x_{n-1} - y_i\|^2 + 2\lambda_{n-1}(\theta_i - \theta_{n-1})\|y_i\|^2$$
$$\quad + ((1+C)\|x_{n-1} - y_i\| + C + (1+C)\|y_i\|)^2\lambda_{n-1}b(\lambda_{n-1})$$
$$\leq [1 - 2\lambda_{n-1}\theta_i + 3\lambda_{n-1}(\theta_i - \theta_{n-1})]\|x_{n-1} - y_i\|^2 + 2\lambda_{n-1}(\theta_i - \theta_{n-1})\|y_i\|^2$$
$$\quad + ((1+C)\|x_{n-1} - y_i\| + C + (1+C)\|y_i\|)^2\lambda_{n-1}b(\lambda_{n-1})$$

(8)

$$\leq [1 - 2\lambda_{n-1}\theta_i + 3\lambda_{n-1}(\theta_i - \theta_{n-1})]\|x_{n-1} - y_i\|^2 + 2\lambda_{n-1}(\theta_i - \theta_{n-1})\|y_i\|^2$$
$$\quad + 2((1+C)^2\|x_{n-1} - y_i\| + 2(C^2 + (1+C)^2\|y_i\|^2))\lambda_{n-1}b(\lambda_{n-1})$$
$$\leq [1 - (2\lambda_{n-1}\theta_i - \lambda_{n-1}(\theta_i - \theta_{n-1}) - 2(1+C)^2\lambda_{n-1}b(\lambda_{n-1}))]\|x_{n-1} - y_i\|^2$$
$$\quad + (2\lambda_{n-1}(\theta_i - \theta_{n-1}) + 4(1+C)^2\lambda_{n-1}b(\lambda_{n-1}))(\|y_i\|^2 + 1)$$
$$= (1 - b_{n-1,i})\|x_{n-1} - y_i\|^2 + a_{n-1,i}(\|y_i\|^2 + 1),$$

where

$$b_{n-1,i} = 2\lambda_{n-1}\theta_i - \lambda_{n-1}(\theta_i - \theta_{n-1}) - 2(1+C)^2\lambda_{n-1}b(\lambda_{n-1})$$

and

$$a_{n-1,i} = 2\lambda_{n-1}(\theta_i - \theta_{n-1}) + 4(1+C)^2\lambda_{n-1}b(\lambda_{n-1}).$$

Applying induction to (8) and noting that $\theta_i - \theta_j \leq \theta_i - \theta_{n-1}$ $(j \leq n-1)$, we deduce

(9)

$$\|x_n - y_i\|^2 \leq \exp\left(-\sum_{j=i}^{n-1} b_{j,i}\right)\|x_i - y_i\|^2 + \sum_{j=i}^{n-1} a_{j,i}(\|y_i\|^2 + 1).$$

Now, taking $i = n(i)$ and $n = n(i+1)$, (9) yields

(10)
$$\left\|x_{n(i+1)} - y_{n(i)}\right\|^2 \leq \exp\left(-\sum_{j=n(i)}^{n(i+1)-1} b_{j,n(i)}\right)\left\|x_{n(i)} - y_{n(i)}\right\|^2$$
$$+ \sum_{j=n(i)}^{n(i+1)-1} a_{j,n(i)}\left(\left\|y_{n(i)}\right\|^2 + 1\right).$$

Thus, by the conditions (1)-(3) of Definition 1 and the fact that

$$\lim_{i\to\infty} \sum_{j=n(i)}^{n(i+1)-1} \lambda_j b(\lambda_j) = 0,$$

there exists $\delta \in (0,1)$ such that

$$\exp\left(-\sum_{j=n(i)}^{n(i+1)-1} b_{j,n(i)}\right) \leq \delta, \quad \varepsilon_{n(i)} := \sum_{j=n(i)}^{n(i+1)-1} a_{j,n(i)} \to 0 \quad (i \to \infty).$$

Hence we have

(11)
$$\left\|x_{n(i+1)} - y_{n(i)}\right\|^2 \leq \delta\left\|x_{n(i)} - y_{n(i)}\right\|^2 + \varepsilon_{n(i)}\left(\left\|y_{n(i)}\right\|^2 + 1\right)$$

and hence

(12)
$$\left\|x_{n(i+1)} - y_{n(i)}\right\| \leq \delta^{1/2}\left\|x_{n(i)} - y_{n(i)}\right\| + \varepsilon_{n(i)}^{1/2}\left(\left\|y_{n(i)}\right\| + 1\right).$$

Since $\lim_{i\to\infty} y_{n(i)} = x^*$, we have

(13)
$$\limsup_{i\to\infty}\left\|x_{n(i+1)} - y_{n(i)}\right\| = \limsup_{i\to\infty}\left\|x_{n(i)} - x^*\right\| = \limsup_{i\to\infty}\left\|x_{n(i)} - y_{n(i)}\right\|.$$

Moreover, by (12) and $\delta < 1$, we have $\limsup_{i\to\infty}\left\|x_{n(i)} - x^*\right\| = 0$, thus $\lim_{i\to\infty} x_{n(i)} = x^*$. Similarly, taking i which satisfies $n(i) \leq n < n(i+1)$, from (8), it follows that

(14)
$$\left\|x_n - y_{n(i)}\right\| \leq \left\|x_{n(i)} - y_{n(i)}\right\| + \varepsilon_{n(i)}^{1/2}\left(\left\|y_{n(i)}\right\| + 1\right),$$

which shows that $\left\|x_n - y_{n(i)}\right\| \to 0$ as $n, i \to \infty$ and hence $x_n \to x^* \in A^{-1}0$ $(n \to \infty)$. This completes the proof. $\qquad\square$

Remark 1. Since a Hilbert space H is 2-uniformly smooth, there exists a constant $d > 0$ such that $b(t) \leq dt$ for all $t \geq 0$. Thereby, if $\{\lambda_n\}$ satisfies the condition (5) of Definition 1 of Bruck [1], then $\{\lambda_n\}$ certainly satisfies the condition (3) of Definition 1.

Remark 2. If E is a q-uniformly smooth space $(q > 1)$, there exists a constant $r > 0$ such that $b(t) \leq rt^{q-1}$ for all $t \geq 0$. Thus, if $\{\lambda_n\}$ satisfies the condition (iii) of Definition 2 of Chidume and Zegeye [2], then $\{\lambda_n\}$ certainly satisfies the condition (3) of Definitin 1.

Remark 3. In Theorem 3.1, the assumption that '$A : D(A) \subseteq E \to 2^E$ is an m-accretive mapping' can be weaken the condition that $A : D(A) \subseteq E \to 2^E$ is an accretive mapping which satisfies the range condition $\overline{D(A)} \subseteq K \subset \bigcap_{r>0} R(I + rA)$, where K is a closed convex subset of E. The conclusions of Theorem 3.1 still valid.

Applying Theorem 3.1 to continuous pseudo-contractions, we have the following convergence theorem:

Theorem 3.2. *Let E be a real uniformly smooth Banach space, K be a nonempty closed convex subset of E and $T : K \to K$ be a continuous pseudo-contraction such that $I - T$ satisfies the linear growth condition: $\|x - Tx\| \leq c(1 + \|x\|)$ for some $c > 0$ and for all $x \in K$. Suppose T has a fixed point in K. Let $\{\lambda_n\}$ and $\{\theta_n\}$ be acceptably paired which satisfy $\lambda_n(1 + \theta_n) \leq 1$ for all $n \geq 1$. Let $x_1 \in K$ and $z \in K$ be arbitrary, but fixed. Then the sequence $\{x_n\}$ defined by*

$$x_{n+1} = (1 - \lambda_n)x_n + \lambda_n T x_n + \lambda_n \theta_n (z - x_n), \quad \forall n \geq 1,$$

is well-defined and converges strongly to a fixed point of T closest to z.

Proof. Since K is convex and $\{x_{n+1}\}$ is a convex combination with $\{x_n\}$, $\{Tx_n\}$ and z, we have $x_{n+1} \in K$ for all $n \geq 1$. Putting $A = I - T$, then $A : K \to E$ is a continuous accretive mapping which satisfies the range condition: $\overline{D(A)} = K \subseteq \bigcap_{r>0} R(I + rA)$. In view of Remark 4, we conclude that $\{x_n\}$ converges strongly to a point $x^* \in A^{-1}(0) = F(T)$. This completes the proof. $\qquad\square$

Remark 4. If E is a uniformly smooth Banach space, K is a nonempty bounded and closed convex subset of E and $T : K \to K$ is a continuous pseudo-contraction, then T has a fixed point in K.

Remark 5. If $\{\theta_n\}$ is decreasing, $\theta_n \to 0$ $(n \to \infty)$ and $\sum_{n=1}^{\infty} \lambda_n \theta_n = \infty$, then there exists a strictly increasing sequence $\{n_{(i)}\}_{i=1}^{\infty}$ of positive integers such that $\sum_{j=n(i)}^{n(i+i)} \lambda_j \theta_j \geq 1$. As $n(i) \leq j \leq n(i+1)$ and $\theta_j \leq \theta_{n(i)}$, we have $\theta_{n(i)} \sum_{j=n(i)}^{n(i+1)} \lambda_j \geq 1$ and hence

$$\liminf_{i \to \infty} \theta_{n(i)} \sum_{j=n(i)}^{n(i+1)} \lambda_j > 0.$$

We note that the condition (1) of Definition 1 is satisfied. Moreover, assume that $\frac{\theta_n - \theta_{n-1}}{\theta_{n+1}} \to 0$ $(n \to \infty)$ and $\theta_{n(i)} \sum_{j=n(i)}^{n(i+1)} \lambda_j$ is bounded. Then we have then

$$
\begin{aligned}
(15) \qquad \lim_{i \to \infty} \left[\theta_{n(i)} - \theta_{n(i+1)}\right] \sum_{j=n(i)}^{n(i+1)} \lambda_j &= \lim_{i \to \infty} \frac{\theta_{n(i)} - \theta_{n(i+1)}}{\theta_{n(i+1)}} \theta_{n(i+1)} \sum_{j=n(i)}^{n(i+1)} \lambda_j \\
&\leq \lim_{i \to \infty} \frac{\theta_{n(i)} - \theta_{n(i+1)}}{\theta_{n(i+1)}} \theta_{n(i)} \sum_{j=n(i)}^{n(i+1)} \lambda_j \\
&= 0
\end{aligned}
$$

and so the condition (2) of Definition 1 is also satisfied.

Finally, suppose that $\sum_{n=1}^{\infty} \lambda_n b(\lambda_n) < \infty$. Then $\lim_{i \to \infty} \sum_{j=n(i)}^{n(i+1)} \lambda_j b(\lambda_j) = 0$ and so the condition (3) of Definition 1 is also satisfied. Consequently, if $\{\lambda_n\}$ and $\{\theta_n\}$ satisfy the following conditions:

(1) $\{\theta_n\}$ is decreasing and $\theta_n \to 0$ $(n \to \infty)$,
(2) $\sum_{n=1}^{\infty} \lambda_n \theta_n = \infty$,
(3) $\frac{\theta_n - \theta_{n-1}}{\theta_{n+1}} \to 0$ $(n \to \infty)$,

(4) $\sum_{n=1}^{\infty} \lambda_n b(\lambda_n) < \infty$,

then $\{\lambda_n\}$ and $\{\theta_n\}$ are acceptably paired.

REFERENCES

1. R. E. Bruck, Jr, *A strongly convergent iterative method for the solution of* $0 \in U(x)$ *for a maximal monotone operator U in Hilbert spaces,* J. Math. Anal. Appl. **48** (1974), 114–126.
2. C. E. Chedume and H. Zegeye, *Global iterative schemes for accretive operators,* J. Math. Anal. Appl. **257** (2001), 364–377.
3. O. Nevanlinna, *Global iteration schemes for monotone operators,* Nonlinear Anal. **3** (1979), 505–514.
4. S. Reich, *An iterative procedure for constructing zeros of accretive sets in Banach spaces,* Nonlinear Anal. **2** (1978), 85–92.
5. S. Reich, *Strong convergence theorems for resolvents of accretive operators in Banach spaces,* J. Math. Anal. Appl. **75** (1980), 287–292.

Nonlinear Functional Analysis and Applications, Volume 1, 219–237

EXISTENCE OF SOLUTIONS FOR NEW SYSTEMS OF GENERALIZED QUASI-VARIATIONAL INCLUSION PROBLEMS IN FC-SPACES

XIE PING DING[1]

ABSTRACT. In this paper, we introduce and study new systems of generalized quasi-variational inclusion problems in FC-spaces which contain many known systems of quasi-variational inclusion problems, systems of quasi-variational disclusion problems and systems of generalized vector quasi-equilibrium problems as very special cases. By applying an existence theorem of maximal elements of set-valued mappings in FC-spaces due to author, we prove some new existence theorems of solutions for new systems of generalized quasi-variational inclusion problems in noncompact FC-spaces.

1. Introduction

Let $\mathbf{R}^n$, $\mathbf{R}^m$ and $\mathbf{R}^p$ be Euclidean spaces, $g : \mathbf{R}^n \times \mathbf{R}^m \to \mathbf{R}^p$ be a single-valued mapping and $Q : \mathbf{R}^n \times \mathbf{R}^m \to 2^{\mathbf{R}^p}$ be a set-valued mapping. In 1979, Robinson [1] studied the following variational inclusion problem: for each $x \in \mathbf{R}^n$, find $\bar{y} \in \mathbf{R}^m$ such that

$$0 \in g(x, \bar{y}) + Q(x, \bar{y}). \tag{VIP}$$

It is well known that the VIP includes a vast fields of variational system with important applications. Since then, various types of variational inclusion problems have been extended, generalized and studied by many authors under different assumptions and underlying spaces, for example, see [2-10] and the references therein. On the other hand, as generalizations of various variational inequality problems, equilibrium problems and variational inclusion problems, many authors introduced and studied various kinds of

Received Januaruy 9, 2008.

2000 *Mathematics Subject Classification.* 49J40, 49J53.

Key words and phrases. Maximal element, system of generalized quasi-variational inclusion problems, Ψ_i-FC-quasiconvex, FC-hull, FC-space.

This project was supported by the NSF of Sichuan Education Department of China (07ZA092) and SZD0406

[1] College of Mathematics and Software Science, Sichuan Normal University, Chengdu, Sichuan 610066, People's Republic of China (*E-mail*: xieping_ding@hotmail.com)

systems of generalized equilibrium problems and generalized variational inclusion problems, for example, see [11-23] and the references therein. Recently, Lin [24] and Lin and Tu [25] studied some systems of generalized quasi-variational inclusion problems with applications in locally convex topological vector spaces. Hai and Khanh [26] studied some systems of set-valued quasivariational inclusion problems in topological vector spaces with applications.

Inspired by this line of research works in this paper, we introduce and study some new systems of generalized quasi-variational inclusion problems in product FC-spaces without convexity structure.

For a nonempty set X, we denote by 2^X the family of all subsets of X. Let I be any index set. For each $i \in I$, let Z_i be a nonempty set, X_i and Y_i be topological spaces. Let $X = \prod_{i \in I} X_i, Y = \prod_{i \in I} Y_i$ and for $x \in X$, $x_i = \pi_i(x)$ be the projection of x onto X_i. For each $i \in I$, let $A_i : X \times Y \to 2^{X_i}$, $T_i, S_i : X \times Y \to 2^{Y_i}$ and $\Phi_i, \Psi_i : Y_i \times X_i \times X \to 2^{Z_i}$ be set-valued mappings. We consider the following systems of generalized quasi-variational inclusions problems (SGQVIP):

(I) Find $(\hat{x}, \hat{y}) \in X \times Y$ such that, for each $i \in I$, $\hat{x}_i \in A_i(\hat{x}, \hat{y})$, $\hat{y}_i \in S_i(\hat{x}, \hat{y})$ and

$$\Phi_i(v_i, u_i, \hat{x}) \subseteq \Psi_i(v_i, \hat{x}_i, \hat{x}), \quad \forall v_i \in T_i(\hat{x}, \hat{y}), \ u_i \in A_i(\hat{x}, \hat{y}). \qquad \text{SGQVIP(I)}$$

(II) Find $(\hat{x}, \hat{y}) \in X \times Y$ such that, for each $i \in I$, $\hat{x}_i \in A_i(\hat{x}, \hat{y})$, $\hat{y}_i \in S_i(\hat{x}, \hat{y})$ and

$$\Phi_i(v_i, u_i, \hat{x}) \not\subseteq \Psi_i(v_i, \hat{x}_i, \hat{x}), \quad \forall v_i \in T_i(\hat{x}, \hat{y}), \ u_i \in A_i(\hat{x}, \hat{y}). \qquad \text{SGQVIP(II)}$$

(III) Find $(\hat{x}, \hat{y}) \in X \times Y$ such that, for each $i \in I$, $\hat{x}_i \in A_i(\hat{x}, \hat{y})$, $\hat{y}_i \in S_i(\hat{x}, \hat{y})$ and

$$\Phi_i(v_i, u_i, \hat{x}) \cap \Psi_i(v_i, \hat{x}_i, \hat{x}) = \emptyset, \quad \forall v_i \in T_i(\hat{x}, \hat{y}), \ u_i \in A_i(\hat{x}, \hat{y}). \qquad \text{SGQVIP(III)}$$

(IV) Find $(\hat{x}, \hat{y}) \in X \times Y$ such that, for each $i \in I$, $\hat{x}_i \in A_i(\hat{x}, \hat{y})$, $\hat{y}_i \in S_i(\hat{x}, \hat{y})$ and

$$\Phi_i(v_i, u_i, \hat{x}) \cap \Psi_i(v_i, \hat{x}_i, \hat{x}) \neq \emptyset, \quad \forall v_i \in T_i(\hat{x}, \hat{y}), \ u_i \in A_i(\hat{x}, \hat{y}). \qquad \text{SGQVIP(IV)}$$

(V) Find $(\hat{x}, \hat{y}) \in X \times Y$ such that, for each $i \in I$, $\hat{x}_i \in A_i(\hat{x}, \hat{y})$, $\hat{y}_i \in S_i(\hat{x}, \hat{y})$ and, for each $u_i \in A_i(\hat{x}, \hat{y})$, there exists $v_i \in T_i(\hat{x}, \hat{y})$ such that

$$\Phi_i(v_i, u_i, \hat{x}) \subseteq \Psi_i(v_i, \hat{x}_i, \hat{x}). \qquad \text{SGQVIP(V)}$$

(VI) Find $(\hat{x}, \hat{y}) \in X \times Y$ such that, for each $i \in I$, $\hat{x}_i \in A_i(\hat{x}, \hat{y})$, $\hat{y}_i \in S_i(\hat{x}, \hat{y})$ and, for each $u_i \in A_i(\hat{x}, \hat{y})$, there exists $v_i \in T_i(\hat{x}, \hat{y})$ such that

$$\Phi_i(v_i, u_i, \hat{x}) \not\subseteq \Psi_i(v_i, \hat{x}_i, \hat{x}). \qquad \text{SGQVIP(VI)}$$

(VII) Find $(\hat{x}, \hat{y}) \in X \times Y$ such that, for each $i \in I$, $\hat{x}_i \in A_i(\hat{x}, \hat{y})$, $\hat{y}_i \in S_i(\hat{x}, \hat{y})$ and, for each $u_i \in A_i(\hat{x}, \hat{y})$, there exists $v_i \in T_i(\hat{x}, \hat{y})$ such that

$$\Phi_i(v_i, u_i, \hat{x}) \cap \Psi_i(v_i, \hat{x}_i, \hat{x}) = \emptyset. \qquad \text{SGQVIP(VII)}$$

(VIII) Find $(\hat{x}, \hat{y}) \in X \times Y$ such that, for each $i \in I$, $\hat{x}_i \in A_i(\hat{x}, \hat{y})$, $\hat{y}_i \in S_i(\hat{x}, \hat{y})$ and, for each $u_i \in A_i(\hat{x}, \hat{y})$, there exists $v_i \in T_i(\hat{x}, \hat{y})$ such that

$$\Phi_i(v_i, u_i, \hat{x}) \cap \Psi_i(v_i, \hat{x}_i, \hat{x}) \neq \emptyset. \qquad \text{SGQVIP(VIII)}$$

If, for each $i \in I$ and $(x, y) \in X \times Y$, $A_i(x, y) = A_i(x)$, $S_i(x, y) = S_i(x)$ and $T_i(x, y) = T_i(x)$, then the SGQVIP(I)-SGQVIP (VIII) reduce to the following problems:

(A) Find $(\hat{x}, \hat{y}) \in X \times Y$ such that, for each $i \in I$, $\hat{x}_i \in A_i(\hat{x})$, $\hat{y}_i \in S_i(\hat{x})$ and

$$\Phi_i(v_i, u_i, \hat{x}) \subseteq \Psi_i(v_i, \hat{x}_i, \hat{x}), \quad \forall v_i \in T_i(\hat{x}), \, u_i \in A_i(\hat{x}). \qquad \text{SGQVIP(A)}$$

(B) Find $(\hat{x}, \hat{y}) \in X \times Y$ such that, for each $i \in I$, $\hat{x}_i \in A_i(\hat{x})$, $\hat{y}_i \in S_i(\hat{x})$ and

$$\Phi_i(v_i, u_i, \hat{x}) \nsubseteq \Psi_i(v_i, \hat{x}_i, \hat{x}), \quad \forall v_i \in T_i(\hat{x}), \, u_i \in A_i(\hat{x}). \qquad \text{SGQVIP(B)}$$

(C) Find $(\hat{x}, \hat{y}) \in X \times Y$ such that, for each $i \in I$, $\hat{x}_i \in A_i(\hat{x})$, $\hat{y}_i \in S_i(\hat{x})$ and

$$\Phi_i(v_i, u_i, \hat{x}) \cap \Psi_i(v_i, \hat{x}_i, \hat{x}) = \emptyset, \quad \forall v_i \in T_i(\hat{x}), \, u_i \in A_i(\hat{x}). \qquad \text{SGQVIP(C)}$$

(D) Find $(\hat{x}, \hat{y}) \in X \times Y$ such that, for each $i \in I$, $\hat{x}_i \in A_i(\hat{x})$, $\hat{y}_i \in S_i(\hat{x})$ and

$$\Phi_i(v_i, u_i, \hat{x}) \cap \Psi_i(v_i, \hat{x}_i, \hat{x}) \neq \emptyset, \quad \forall v_i \in T_i(\hat{x}), \, u_i \in A_i(\hat{x}). \qquad \text{SGQVIP(D)}$$

(E) Find $(\hat{x}, \hat{y}) \in X \times Y$ such that, for each $i \in I$, $\hat{x}_i \in A_i(\hat{x})$, $\hat{y}_i \in S_i(\hat{x})$ and, for each $u_i \in A_i(\hat{x})$, there exists $v_i \in T_i(\hat{x})$ such that

$$\Phi_i(v_i, u_i, \hat{x}) \subseteq \Psi_i(v_i, \hat{x}_i, \hat{x}). \qquad \text{SGQVIP(E)}$$

(F) Find $(\hat{x}, \hat{y}) \in X \times Y$ such that, for each $i \in I$, $\hat{x}_i \in A_i(\hat{x})$, $\hat{y}_i \in S_i(\hat{x})$ and, for each $u_i \in A_i(\hat{x})$, there exists $v_i \in T_i(\hat{x})$ such that

$$\Phi_i(v_i, u_i, \hat{x}) \nsubseteq \Psi_i(v_i, \hat{x}_i, \hat{x}). \qquad \text{SGQVIP(F)}$$

(G) Find $(\hat{x}, \hat{y}) \in X \times Y$ such that, for each $i \in I$, $\hat{x}_i \in A_i(\hat{x})$, $\hat{y}_i \in S_i(\hat{x})$ and, for each $u_i \in A_i(\hat{x})$, there exists $v_i \in T_i(\hat{x})$ such that

$$\Phi_i(v_i, u_i, \hat{x}) \cap \Psi_i(v_i, \hat{x}_i, \hat{x}) = \emptyset. \qquad \text{SGQVIP(G)}$$

(H) Find $(\hat{x}, \hat{y}) \in X \times Y$ such that, for each $i \in I$, $\hat{x}_i \in A_i(\hat{x})$, $\hat{y}_i \in S_i(\hat{x})$ and, for each $u_i \in A_i(\hat{x})$, there exists $v_i \in T_i(\hat{x})$ such that

$$\Phi_i(v_i, u_i, \hat{x}) \cap \Psi_i(v_i, \hat{x}_i, \hat{x}) \neq \emptyset. \qquad \text{SGQVIP(H)}$$

It is easy to see that the SGQVIP(A), SGQVIP(D), SGQVIP(E) and SGQVIP(H) contain, respectively, the (SQVIP2), (SQVIP4) (SQVIP1) and (SQVIP3) introduced and studied by Hai and Khanh [26] as special cases. The SGQVIP(B) SGQVIP(C) SGQVIP(F) and SGQVIP(G) are new.

If, for each $i \in I$, Z_i is a topological vector space and let $\Psi_i(y_i, x_i, x) = \{0\}$ for all $(y_i, x_i, x) \in Y_i \times X_i \times X$, then the SGQVIP(III), SGQVIP(IV), SGQVIP(VII) and SGQVIP(VIII) reduce to the following problems, respectively:

(a) Find $(\hat{x}, \hat{y}) \in X \times Y$ such that, for each $i \in I$, $\hat{x}_i \in A_i(\hat{x}, \hat{y})$, $\hat{y}_i \in S_i(\hat{x}, \hat{y})$ and

$$0 \notin \Phi_i(v_i, u_i, \hat{x}), \quad \forall v_i \in T_i(\hat{x}, \hat{y}), \, u_i \in A_i(\hat{x}, \hat{y}). \qquad \text{(SVDP(a))}$$

(b) Find $(\hat{x}, \hat{y}) \in X \times Y$ such that, for each $i \in I$, $\hat{x}_i \in A_i(\hat{x}, \hat{y})$, $\hat{y}_i \in S_i(\hat{x}, \hat{y})$ and

$$0 \in \Phi_i(v_i, u_i, \hat{x}), \quad \forall v_i \in T_i(\hat{x}, \hat{y}), \, u_i \in A_i(\hat{x}, \hat{y}). \qquad \text{(SVIP(b))}$$

(c) Find $(\hat{x}, \hat{y}) \in X \times Y$ such that, for each $i \in I$, $\hat{x}_i \in A_i(\hat{x}, \hat{y})$, $\hat{y}_i \in S_i(\hat{x}, \hat{y})$ and, for each $u_i \in A_i(\hat{x}, \hat{y})$, there exists $v_i \in T_i(\hat{x}, \hat{y})$ such that

$$0 \notin \Phi_i(v_i, u_i, \hat{x}). \qquad \text{(SVDP(c))}$$

(e) Find $(\hat{x}, \hat{y}) \in X \times Y$ such that, for each $i \in I$, $\hat{x}_i \in A_i(\hat{x}, \hat{y})$, $\hat{y}_i \in S_i(\hat{x}, \hat{y})$ and, for each $u_i \in A_i(\hat{x}, \hat{y})$, there exists $v_i \in T_i(\hat{x}, \hat{y})$ such that

$$0 \in \Phi_i(v_i, u_i, \hat{x}). \qquad \text{(SVIP(e))}$$

It is easy to see that the SVIP(b), SVIP(e), SVDP(a) and SVDP(c) contain, re-
spectinely, the (SVIP1), (SVIP2), SVDP(1) and SVDP(2) introduced and studied by Lin
and Tu [25] as special cases. Hence the SGQVIP(I)-SGQVIP(VIII) include many known
systems of quasi-variational inclusion problems, systems of quasi-variational disclusion
problems and generalized vector quasi-equilibrium problems with wide applications as
very special cases, for example, see [25,26] and the references therein.

In this paper, we introduce the new notions of Ψ_i-FC-quasiconvexity for set-valued
mappings $\Phi_i, \Psi_i : Y_i \times X_i \times X \to 2^{Z_i}$ in FC-space. by using these notions and an existence
theorem of maximal elements for a family of set-valued mappings due to author [27],
some new existence theorems of solutions for the SGQVIP(I)-SGQVIP(VIII) are proved
in noncompact FC-spaces without convexity structure. These results improve, unify and
generalize many known results in recent literature to noncompact FC-spaces without
convexity structure.

2. Preliminaries

For a nonempty set X, we denote by $< X >$ the family of all nonempty finite subsets
of X. Let Δ_n be the standard n-dimensional simplex with vertices $e_0, e_1, \cdots, e_n$. If J
is a nonempty subset of $\{0, 1, \cdots, n\}$, we denote by Δ_J the convex hull of the vertices
$\{e_j : j \in J\}$.

The following notion was introduced by Ben-El-Mechaiekh et al. [28].

Definition 2.1. (X, Γ) is called a *L-convex* space if X is a topological space and
$\Gamma :< X > \to 2^X$ is a mapping such that, for each $N \in < X >$ with $|N| = n + 1$, there
exists a continuous mapping $\varphi_N : \Delta_n \to \Gamma(N)$ satisfying $A \in < N >$ with $|A| = J + 1$
implies $\varphi_N(\Delta_J) \subseteq \Gamma(A)$, where Δ_J is the face of Δ_N corresponding to A.

The following notion of a finitely continuous topological space (in short, FC-space)
was introduced by Ding [29].

Definition 2.2. (X, φ_N) is said to be a *FC-space* if X is a topological space and, for
each $N = \{x_0, \cdots, x_n\} \in < X >$ where some elements in N may be same, there exists a
continuous mapping $\varphi_N : \Delta_n \to X$. A subset D of (X, φ_N) is said to be a *FC-subspace*
of X if, for each $N = \{x_0, \cdots, x_n\} \in < X >$ and $(x_{i_0}, \cdots, x_{i_k}) \subseteq N \cap D$, $\varphi_N(\Delta_k) \subseteq D$,
where $\Delta_k = co(\{e_{i_j} : j = 0, \cdots, k\})$.

Comparing the definitions of L-convex spaces and FC-spaces, it is clear that each
L-convex space must be a FC-space. The following examples show that there exists a
FC-space which is not a L-convex space.

Example 2.1. Let X_1 and X_2 be two nonempty convex subsets of a topological
vector space X with $\text{cl}X_1 \cap \text{cl}X_2 = \emptyset$ and $g : X_2 \to X_1$ be single-valued mapping. Then
$E = X_1 \cup X_2$ is not convex. For each $N = \{x_0, \cdots, x_n\} \in < E >$, define a mapping
$\varphi_N : \Delta_n \to 2^X$ by

$$\varphi_N(\alpha) = \begin{cases} \sum_{i=0}^{n} \alpha_i x_i, & \text{if } N \subset X_1 \text{ or } N \subset X_2, \\ \sum_{i=0}^{j} \alpha_i x_i + \sum_{i=j+1}^{n} \alpha_i g(x_i), & \text{if } N = N_1 \bigcup N_2, \end{cases}$$

for all $\alpha = (\alpha_0, \cdots, \alpha_n) \in \Delta_n$ where $N_1 = \{x_0, \cdots, x_j\} \subseteq X_1$, $N_2 = \{x_{j+1}, \cdots, x_n\} \subseteq X_2$. It is easy to see that φ_N is continuous and hence (E, φ_N) is a FC-space. For any convex subset A of X_1 with $A \neq X_1$ and any subset B of X_2, it is easy to check that the sets A, X_1 and $X_1 \bigcup B$ are all FC-subspaces of E. But the sets X_2, B and $A \bigcup B$ are not FC-subspaces of E. If we define a set-valued mapping $\Gamma :<E> \to 2^E$ by

$$\Gamma(N) = \varphi_N(\Delta_n), \quad \forall N = \{x_0, \cdots, x_n\} \in <E>,$$

then we have that, for each $N = \{x_0, \cdots, x_n\} \in <E>$, $\varphi_N(\Delta_n) \subseteq \Gamma(N)$. But, if $N = N_1 \bigcup N_2$ where $N_1 = \{x_0, \cdots, x_j\} \subseteq X_1$ and $N_2 = \{x_{j+1}, \cdots, x_n\} \subseteq X_2$, then we have $\Gamma(N_2) = \varphi_{N_2}(\Delta_J) \subseteq X_2$ and $\varphi_N(\Delta_J) \subseteq X_1$, where $\Delta_J = \mathrm{co}\{e_k : k = j+1, \cdots, n\}$. Hence we have $\varphi_N(\Delta_J) \not\subseteq \Gamma(N_2)$ and so (E, Γ) is not a L-convex space.

Example 2.2. Let $(X, \|\cdot\|)$ be a strictly convex and reflexive Banach space, X_1 be a nonempty closed convex subset of X and X_2 be a nonempty convex subset of X with $X_1 \cap X_2 = \emptyset$. Then $E = X_1 \cup X_2$ is not convex. For each $N = \{x_0, \cdots, x_n\} \in <E>$, define a mapping $\varphi_N : \Delta_n \to 2^X$ as in Example 2.1, where $g : X_2 \to X_1$ is replaced by the metric projective mapping $P_{X_1} : X_2 \to X_1$. Then (E, φ_N) is a FC-space which is not L-convex spaces.

It is clear that any convex subset of a topological vector space, any H-space introduced by Horvath [30], any G-convex space introduced by Park and Kim [31] and any L-convex spaces introduced by Ben-El-Mechaiekh et al. [28] are all FC-space. Hence it is quite reasonable and valuable to study various nonlinear problems in FC-spaces.

By the definition of FC-subspaces of a FC-space, it is easy to see that, if $\{B_i\}_{i \in I}$ is a family of FC-subspaces of a FC-space (Y, φ_N) and $\cap_{i \in I} B_i \neq \emptyset$, then $\cap_{i \in I} B_i$ is also a FC-subspace of (Y, φ_N), where I is any index set. For a subset A of (Y, φ_N), we can define the FC-hull of A as follows:

$$FC(A) = \cap\{B \subset Y : A \subseteq B \text{ and } B \text{ is } FC - \text{subspace of } Y\}.$$

Clearly, $FC(A)$ is the smallest FC-subspace of Y containing A and each FC-subspace of a FC-space is also a FC-space.

Lemma 2.1. ([27]) *Let (Y, φ_N) be a FC-space and A be a nonempty subset of Y. Then*

$$FC(A) = \bigcup\{FC(N) : N \in <A>\}.$$

Lemma 2.2. ([27]) *Let X be a topological space, (Y, φ_N) be a FC-space and $G : X \to 2^Y$ be such that $G^{-1}(y) = \{x \in X : y \in G(x)\}$ is compactly open in X for each $y \in Y$. Then the mapping $FC(G) : X \to 2^Y$ defined by $FC(G)(x) = FC(G(x))$ for each $x \in X$ satisfies that $(FC(G))^{-1}(y)$ is also compactly open in X for each $y \in Y$.*

Lemma 2.3. ([29]) *Let I be any index set. For each $i \in I$, let $(Y_i, \{\varphi_{N_i}\})$ be a FC-space. Let $Y = \prod_{i \in I} Y_i$ and $\varphi_N = \prod_{i \in I} \varphi_{N_i}$, where N_i is the projection of N onto Y_i. Then $(Y, \{\varphi_N\})$ is also a FC-space.*

The following result is a special case of Corollary 3.3 of Ding [27].

Lemma 2.4. *Let I be any index set. For each $i \in I$, let (X_i, φ_{N_i}) be a FC-space, $X = \prod_{i \in I} X_i$ and K be a compact subset of X. For each $i \in I$, let $G_i : X \to 2^{X_i}$ be such that*

(a) *for each $i \in I$ and $x \in X$, $G_i(x)$ is a FC-subspace of X_i,*

(b) *for each $x \in X$, $\pi_i(x) \notin G_i(x)$ for all $i \in I$,*

(c) *for each $y_i \in X_i$, $G_i^{-1}(y_i)$ is compactly open in X,*

(d) *for each $N_i \in < X_i >$, there exists a nonempty compact FC-subspace L_{N_i} of X_i containing N_i and for each $x \in X \setminus K$, there exists $i \in I$ satisfying $L_{N_i} \cap G_i(x) \neq \emptyset$.*

Then there exists $\hat{x} \in K$ such that $G_i(\hat{x}) = \emptyset$, for each $i \in I$.

3. Existence of Solutions for the SGQVIPs

Throughout this section, unless otherwise specified, we shall fix the following notations and assumptions. Let I be any index set. For each $i \in I$, let (X_i, φ_{N_i}) and (Y_i, φ'_{N_i}) be FC-spaces and Z_i be a nonempty set. Let $X = \prod_{i \in I} X_i$ and $Y = \prod_{i \in I} Y_i$. For each $i \in I$, let $A_i : X \times Y \to 2^{X_i}$, $T_i, S_i : X \times Y \to 2^{Y_i}$ and $\Phi_i, \Psi_i : Y_i \times X_i \times X \to 2^{Z_i}$ be set-valued mappings.

Definition 3.1. For each $i \in I$ and $y \in Y$, Φ_i is said to be:

(1) Ψ_i-*FC-quasiconvex of type* (I) *in the first two arguments if, for each* $N_i = \{u_{i,0}, \cdots, u_{i,n}\} \in < X_i >$ *and* $x \in X$ *with* $x_i \in FC(N_i)$, *there exists* $j \in \{0, \cdots, n\}$ *such that* $\Phi_i(v_i, u_{i,j}, x) \subseteq \Psi_i(v_i, x_i, x)\}$ *for all* $v_i \in T_i(x, y)$,

(2) Ψ_i-*FC-quasiconvex of type* (II) *in the first two arguments if, for each* $N_i = \{u_{i,0}, \cdots, u_{i,n}\} \in < X_i >$ *and* $x \in X$ *with* $x_i \in FC(N_i)$, *there exists* $j \in \{0, \cdots, n\}$ *such that* $\Phi_i(v_i, u_{i,j}, x) \nsubseteq \Psi_i(v_i, x_i, x)\}$ *for all* $v_i \in T_i(x, y)$,

(3) Ψ_i-*FC-FC-quasiconvex of type* (III) *in the first two arguments if, for each* $N_i = \{u_{i,0}, \cdots, u_{i,n}\} \in < X_i >$ *and* $x \in X$ *with* $x_i \in FC(N_i)$, *there exists* $j \in \{0, \cdots, n\}$ *such that* $\psi_i(v_i, u_{i.j}, x) \cap \Psi_i(v_i, x_i, x) = \emptyset$ *for all* $v_i \in T_i(x, y)$,

(4) Ψ_i-*FC-quasiconvex of type* (IV) *in the first two arguments if, for each* $N_i = \{u_{i,0}, \cdots, u_{i,n}\} \in < X_i >$ *and* $x \in X$ *with* $x_i \in FC(N_i)$, *there exists* $j \in \{0, \cdots, n\}$ *such that* $\psi_i(v_i, u_{i.j}, x) \cap \Psi_i(v_i, x_i, x) \neq \emptyset$ *for all* $v_i \in T_i(x, y)$,

Definition 3.2. For each $i \in I$ and $y \in Y$, Φ_i is said to be:

(1) Ψ_i-*FC-weakly quasiconvex of type* (I) *in the first two arguments if, for each* $N_i = \{u_{i,0}, \cdots, u_{i,n}\} \in < X_i >$ *and* $x \in X$ *with* $x_i \in FC(N_i)$, *there exist* $j \in \{0, \cdots, n\}$ *and* $v_i \in T_i(x, y)$ *such that* $\Phi_i(v_i, u_{i,j}, x) \subseteq \Psi_i(v_i, x_i, x)\}$.

(2) Ψ_i-*FC-weakly quasiconvex of type* (II) *in the first two arguments if, for each* $N_i = \{u_{i,0}, \cdots, u_{i,n}\} \in < X_i >$ *and* $x \in X$ *with* $x_i \in FC(N_i)$, *there exist* $j \in \{0, \cdots, n\}$ *and* $v_i \in T_i(x, y)$ *such that* $\Phi_i(v_i, u_{i,j}, x) \nsubseteq \Psi_i(v_i, x_i, x)\}$.

(3) Ψ_i-*FC-weakly quasiconvex of type* (III) *in the first two arguments if, for each* $N_i = \{u_{i,0}, \cdots, u_{i,n}\} \in < X_i >$ *and for each* $x \in X$ *with* $x_i \in FC(N_i)$, *there exist* $j \in \{0, \cdots, n\}$ *and* $v_i \in T_i(x, y)$ *such that* $\psi_i(v_i, u_{i.j}, x) \cap \Psi_i(v_i, x_i, x) = \emptyset$.

(4) Ψ_i-*FC-weakly quasiconvex of type* (IV) *in the first two arguments if, for each* $N_i = \{u_{i,0}, \cdots, u_{i,n}\} \in < X_i >$ *and* $x \in X$ *with* $x_i \in FC(N_i)$, *there exist* $j \in \{0, \cdots, n\}$ *and* $v_i \in T_i(x, y)$ *such that* $\psi_i(v_i, u_{i.j}, x) \cap \Psi_i(v_i, x_i, x) \neq \emptyset$.

Lemma 3.1. *For each $i \in I$, define a set-valued mapping $P_{i,k} : X \times Y \to 2^{X_i}$ $(k = 1, 2, 3, 4)$ by*

$$P_{i,1}(x, y) = \{u_i \in X_i : \Phi(v_i, u_i, x) \nsubseteq \Psi_i(v_i, x_i, x) \text{ for some } v_i \in T_i(x, y)\},$$

$$P_{i,2}(x, y) = \{u_i \in X_i : \Phi(v_i, u_i, x) \subseteq \Psi_i(v_i, x_i, x) \text{ for some } v_i \in T_i(x, y)\},$$

$$P_{i,3}(x, y) = \{u_i \in X_i : \Phi(v_i, u_i, x) \cap \Psi_i(v_i, x_i, x) \neq \emptyset \text{ for some } v_i \in T_i(x, y)\},$$

$$P_{i,4}(x, y) = \{u_i \in X_i : \Phi(v_i, u_i, x) \cap \Psi_i(v_i, x_i, x) = \emptyset \text{ for some } v_i \in T_i(x, y)\}, \text{respectively.}$$

Then, for each $y \in Y$, Φ_i is Ψ_i-FC-quasiconvex of type (I) (resp., type (II), type (III), type (IV)) if and only if, for each $(x, y) \in X \times Y$, $x_i \notin FC(P_{i,1}(x, y))$ (resp., $x_i \notin FC(P_{i,2}(x, y))$, $x_i \notin FC(P_{i,3}(x, y))$, $x_i \notin FC(P_{i,4}(x, y))$).

Proof. We only need to prove the case $k = 1$ since the proof for the cases $k = 2, 3, 4$ is completely similar.

Necessity: Suppose that, for each $y \in Y$, Φ_i is Ψ_i-FC-quasiconvex of type (I). If there exists $(\bar{x}, \bar{y}) \in X \times Y$ such that $\bar{x}_i \in FC(P_{i,1}(\bar{x}, \bar{y}))$, then, by Lemma 2.1, there exists $N_i = \{u_{i,0}, \cdots, u_{i,n}\} \in < P_i(\bar{x}, \bar{y}) >$ such that $\bar{x}_i \in FC(N_i)$. By the definition of $P_{i,1}$, we have that, for each $j \in \{0, \cdots, n\}$, there exists $\bar{v}_{i,j} \in T_i(\bar{x}, \bar{y})$ such that

$$\Phi(\bar{v}_{i,j}, u_{i,j}, \bar{x}) \nsubseteq \Psi_i(\bar{v}_{i,j}, \bar{x}_i, \bar{x})$$

which contradicts that the assumption for each $y \in Y$, Φ_i is Ψ_i-FC-quasiconvex of type (I).

Sufficiency: Suppose that, for each $(x, y) \in X \times Y$, $x_i \notin FC(P_{i,1}(x, y))$. If, for some $\bar{y} \in Y$, Φ_i is not Ψ_i-FC-quasiconvex of type (I), then there exist $N_i = \{u_{i,0}, \cdots, u_{i,n}\} \in < X_i >$ and $\bar{x} \in X$ with $\bar{x}_i \in FC(N_i)$ such that, for each $j \in \{0, \cdots, n\}$, there exists $\bar{v}_{i,j} \in T_i(\bar{x}, \bar{y})$ such that

$$\Phi(\bar{v}_{i,j}, u_{i,j}, \bar{x}) \nsubseteq \Psi_i(\bar{v}_{i,j}, \bar{x}_i, \bar{x}).$$

It follows that $N_i \subseteq P_{i,1}(\bar{x}, \bar{y})$ and hence we have $\bar{x}_i \in FC(N_i) \subseteq FC(P_{i,1}(\bar{x}, \bar{y}))$, which is a contradiction. This completes the proof. $\square$

By using Definition 3.2 and the similar argument as in the proof of Lemma 3.1, we have the following result:

Lemma 3.2. *For each $i \in I$, define a set-valued mapping $P_{i,k} : X \times Y \to 2^{X_i}$, $k = 1, 2, 3, 4$ by*

$$P_{i,1}(x, y) = \{u_i \in X_i : \Phi(v_i, u_i, x) \nsubseteq \Psi_i(v_i, x_i, x), \ \forall v_i \in T_i(x, y)\},$$

$$P_{i,2}(x, y) = \{u_i \in X_i : \Phi(v_i, u_i, x) \subseteq \Psi_i(v_i, x_i, x), \ \forall v_i \in T_i(x, y)\},$$

$$P_{i,3}(x, y) = \{u_i \in X_i : \Phi(v_i, u_i, x) \cap \Psi_i(v_i, x_i, x) \neq \emptyset, \ \forall v_i \in T_i(x, y)\},$$

$$P_{i,4}(x, y) = \{u_i \in X_i : \Phi(v_i, u_i, x) \cap \Psi_i(v_i, x_i, x) = \emptyset, \ \forall v_i \in T_i(x, y)\}, \ \text{respectively.}$$

Then, for each $y \in Y$, Φ_i is Ψ_i-FC-weakly quasiconvex of type (I) (resp., type (II), type (III), type (IV)) if and only if, for each $(x, y) \in X \times Y$, $x_i \notin FC(P_{i,1}(x, y))$ (resp., $x_i \notin FC(P_{i,2}(x, y))$, $x_i \notin FC(P_{i,3}(x, y))$, $x_i \notin FC(P_{i,4}(x, y))$).

Lemma 3.3. ([20]) *Let X and Y be topological spaces and $G : X \to 2^Y$ be a set-valued mapping. Then G is lower semicontinuous in $x \in X$ if and only if, for any $y \in G(x)$ and any net $\{x_\alpha\} \subset X$ satisfying $x_\alpha \to x$, there exists a net $\{y_\alpha\}$ such that $y_\alpha \in G(x_\alpha)$ and $y_\alpha \to y$.*

Lemma 3.4. *For each $i \in I$, let X_i, Y_i and Z_i be topological spaces and $X = \prod_{i \in I} X_i$ and $Y = \prod_{i \in I} Y_i$. For each $i \in I$, let $A_i : X \times Y \to 2^{X_i}$, $T_i, S_i : X \times Y \to 2^{Y_i}$ and $\Phi_i, \Psi_i : Y_i \times X_i \times X \to 2^{Z_i}$ be set-valued mappings such that*

   (a) *T_i is lower semicontinuous on each compact subsets of $X \times Y$,*

   (b) *for each $u_i \in X_i$, $(v_i, x) \mapsto \Phi_i(v_i, u_i, x)$ is lower semicontinuous on each compact subset of $Y_i \times X$,*

   (c) *the mapping Ψ_i is upper semicontinuous on each compact subsets of $Y_i \times X_i \times X$ with closed values.*

Then for each $i \in I$ and $u_i \in X_i$, the set $M_i = \{(x, y) \in X \times Y : \Phi(v_i, u_i, x) \not\subseteq \Psi_i(v_i, x_i, x)$ for some $v_i \in T_i(x, y)\}$ is compactly open in $X \times Y$.

Proof. For each $i \in I$ and $u_i \in X_i$, let $\Omega_i = (X \times Y) \setminus M_i = \{(x, y) \in X \times Y : \Phi(v_i, u_i, x) \subseteq \Psi_i(v_i, x_i, x), \forall v_i \in T_i(x, y)\}$. For any compact subset K of $X \times Y$, if $(x, y) \in \mathrm{cl}_K(\Omega_i \cap K)$, then there exists a net $((x_\lambda, y_\lambda))_{\lambda \in \Lambda} \subseteq \Omega_i \cap K$ such that $(x_\lambda, y_\lambda) \to (x, y) \in K$. Hence we have

$$(3.1) \qquad \Phi_i(v_i, u_i, x_\lambda) \subseteq \Psi_i(v_i, x_{i,\lambda}, x_\lambda), \quad \forall\, v_i \in T_i(x_\lambda, y_\lambda).$$

By (a) and Lemma 3.3, for each $v_i \in T_i(x, y)$, there exists a net $(v_{i,\lambda})_{\lambda \in \Lambda} \subseteq Y_i$ such that $v_{i,\lambda} \in T_i(x_\lambda, y_\lambda)$ and $v_{i,\lambda} \to v_i$. By (3.1), we have

$$(3.2) \qquad \Phi_i(v_{i,\lambda}, u_i, x_\lambda) \subseteq \Psi_i(v_{i,\lambda}, x_{i,\lambda}, x_\lambda), \quad \forall\, \lambda \in \Lambda.$$

Since the mapping $(v_i, x) \mapsto \Phi_i(v_i, u_i, x)$ is lower semicontinuous at (v_i, x), by Lemma 3.3, for each $z_i \in \Phi_i(v_i, u_i, x)$, there exists a net $(z_{i,\lambda})_{\lambda \in \Lambda}$ such that $z_{i,\lambda} \in \Phi_i(v_{i,\lambda}, u_i, x_\lambda)$ and $z_{i,\lambda} \to z_i$. By (3.2), we have

$$(3.3) \qquad z_{i,\lambda} \in \Psi_i(v_{i,\lambda}, x_{i,\lambda}, x_\lambda), \quad \forall\, \lambda \in \Lambda.$$

It follows from the condition (c) and (3.3) that $z_i \in \Psi(v_i, x_i, x)$ and hence

$$\Phi_i(v_i, u_i, x) \subseteq \Psi(v_i, x_i, x).$$

Therefore, $(x, y) \in \Omega_i \cap K$ and Ω_i is compactly closed in $X \times Y$. Hence M_i is compactly open in $X \times Y$. This completes the proof.   □

Lemma 3.5. *For each $i \in I$, let X_i, Y_i and Z_i be topological spaces and $X = \prod_{i \in I} X_i$ and $Y = \prod_{i \in I} Y_i$. For each $i \in I$, let $A_i : X \times Y \to 2^{X_i}$, $T_i, S_i : X \times Y \to 2^{Y_i}$ and $\Phi_i, \Psi_i : Y_i \times X_i \times X \to 2^{Z_i}$ be set-valued mappings such that*

   (a) *T_i is lower semicontinuous on each compact subsets of $X \times Y$,*

   (b) *for each $u_i \in X_i$, $(v_i, x) \mapsto \Phi_i(v_i, u_i, x)$ is upper semicontinuous on each compact subset of $Y_i \times X$ with compact values,*

   (c) *the mapping $(y_i, x_i, x) \mapsto Z_i \setminus \Psi_i(y_i, x_i, x)$ is upper semicontinuous on each compact subsets of $Y_i \times X_i \times X$ with closed values.*

Then, for each $i \in I$ and $u_i \in X_i$, the set $M_i = \{(x, y) \in X \times Y : \Phi(v_i, u_i, x) \subseteq \Psi_i(v_i, x_i, x)$ for some $v_i \in T_i(x, y)\}$ is compactly open in $X \times Y$.

Proof. For each $i \in I$ and $u_i \in X_i$, let $\Omega_i = (X \times Y) \setminus M_i = \{(x,y) \in X \times Y : \Phi(v_i, u_i, x) \nsubseteq \Psi_i(v_i, x_i, x), \forall v_i \in T_i(x,y)\}$. For any compact subset K of $X \times Y$, if $(x,y) \in \mathrm{cl}_K(\Omega_i \cap K)$, then there exists a net $((x_\lambda, y_\lambda))_{\lambda \in \Lambda} \subseteq \Omega_i \cap K$ such that $(x_\lambda, y_\lambda) \to (x,y) \in K$. Hence we have

$$(3.4) \qquad \Phi_i(v_i, u_i, x_\lambda) \nsubseteq \Psi_i(v_i, x_{i,\lambda}, x_\lambda), \quad \forall v_i \in T_i(x_\lambda, y_\lambda).$$

By (a) and Lemma 3.3, for each $v_i \in T_i(x,y)$, there exists a net $(v_{i,\lambda})_{\lambda \in \Lambda} \subseteq Y_i$ such that $v_{i,\lambda} \in T_i(x_\lambda, y_\lambda)$ and $v_{i,\lambda} \to v_i$. By (3.4), we have

$$\Phi_i(v_{i,\lambda}, u_i, x_\lambda) \nsubseteq \Psi_i(v_{i,\lambda}, x_{i,\lambda}, x_\lambda), \quad \forall \lambda \in \Lambda.$$

It follows that, for each $\lambda \in \Lambda$, there exists $z_{i,\lambda} \in \Phi_i(v_{i,\lambda}, u_i, x_\lambda)$ such that $z_{i,\lambda} \in Z_i \setminus \Psi_i(v_{i,\lambda}, x_{i,\lambda}, x_\lambda)$. Let $L = (x_\lambda)_{\lambda \in \Lambda} \bigcup \{x\}$ and $M_i = (v_{i,\lambda})_{\lambda \in \Lambda} \cup \{v_i\}$. Then L and M_i are compact in X and Y_i, respectively. By the condition (b), we know $\Phi_i(M_i, u_i, L)$ is compact in Z_i. Without loss of generality, we can assume that $z_{i,\lambda} \to z_i$ and so we have $z_i \in \Phi_i(v_i, u_i, x)$. By the condition (iii), we have $z_i \in Z_i \setminus \Psi_i(v_i, x_i, x)$. It follows that

$$\Phi_i(v_i, u_i, x) \nsubseteq \Psi(v_i, x_i, x).$$

Therefore, $(x,y) \in \Omega_i \cap K$ and Ω_i is compactly closed in $X \times Y$. Hence M_i is compactly open in $X \times Y$. This completes the proof. $\qquad \square$

By using the similar argument as in the proof of Lemma 3.4 and Lemma 3.5, we have the following results.

Lemma 3.6. *For each $i \in I$, let X_i, Y_i and Z_i be topological spaces and $X = \prod_{i \in I} X_i$ and $Y = \prod_{i \in I} Y_i$. For each $i \in I$, let $A_i : X \times Y \to 2^{X_i}$, $T_i, S_i : X \times Y \to 2^{Y_i}$ and $\Phi_i, \Psi_i : Y_i \times X_i \times X \to 2^{Z_i}$ be set-valued mappings such that*
   (a) *T_i is lower semicontinuous on each compact subsets of $X \times Y$,*
   (b) *for each $u_i \in X_i$, $(v_i, x) \mapsto \Phi_i(v_i, u_i, x)$ is lower semicontinuous on each compact subset of $Y_i \times X$,*
   (c) *the mapping $(y_i, x_i, x) \mapsto Z_i \setminus \Psi_i(y_i, x_i, x)$ is upper semicontinuous on each compact subsets of $Y_i \times X_i \times X$ with closed values.*
Then for each $i \in I$ and $u_i \in X_i$, the set $M_i = \{(x,y) \in X \times Y : \Phi(v_i, u_i, x) \cap \Psi_i(v_i, x_i, x) \neq \emptyset$ for some $v_i \in T_i(x,y)\}$ is compactly open in $X \times Y$.

Lemma 3.7. *For each $i \in I$, let X_i, Y_i and Z_i be topological spaces and $X = \prod_{i \in I} X_i$ and $Y = \prod_{i \in I} Y_i$. For each $i \in I$, let $A_i : X \times Y \to 2^{X_i}$, $T_i, S_i : X \times Y \to 2^{Y_i}$ and $\Phi_i, \Psi_i : Y_i \times X_i \times X \to 2^{Z_i}$ be set-valued mappings such that*
   (a) *T_i is lower semicontinuous on each compact subsets of $X \times Y$,*
   (b) *for each $u_i \in X_i$, $(v_i, x) \mapsto \Phi_i(v_i, u_i, x)$ is upper semicontinuous on each compact subset of $Y_i \times X$ with compact values,*
   (c) *the mapping $(y_i, x_i, x) \mapsto \Psi_i(y_i, x_i, x)$ is upper semicontinuous on each compact subsets of $Y_i \times X_i \times X$ with closed values.*
Then for each $i \in I$ and $u_i \in X_i$, the set $M_i = \{(x,y) \in X \times Y : \Phi(v_i, u_i, x) \cap \Psi_i(v_i, x_i, x) = \emptyset$ for some $v_i \in T_i(x,y)\}$ is compactly open in $X \times Y$.

Lemma 3.8. *For each $i \in I$, let X_i, Y_i and Z_i be topological spaces and $X = \prod_{i \in I} X_i$ and $Y = \prod_{i \in I} Y_i$. For each $i \in I$, let $A_i : X \times Y \to 2^{X_i}$, $T_i, S_i : X \times Y \to 2^{Y_i}$ and $\Phi_i, \Psi_i : Y_i \times X_i \times X \to 2^{Z_i}$ be set-valued mappings such that*

(a) *T_i is upper semicontinuous on each compact subsets of $X \times Y$ with compact values,*

(b) *for each $u_i \in X_i$, $(v_i, x) \mapsto \Phi_i(v_i, u_i, x)$ is lower semicontinuous on each compact subset of $Y_i \times X$,*

(c) *the mapping Ψ_i is upper semicontinuous on each compact subsets of $Y_i \times X_i \times X$ with closed values.*

Then, for each $i \in I$ and $u_i \in X_i$, the set $M_i = \{(x, y) \in X \times Y : \Phi(v_i, u_i, x) \not\subseteq \Psi_i(v_i, x_i, x), \forall v_i \in T_i(x, y)\}$ is compactly open in $X \times Y$.

Proof. For each $i \in I$ and $u_i \in X_i$, let $\Omega_i = (X \times Y) \setminus M_i = \{(x, y) \in X \times Y : \exists v_i \in T_i(x, y) \text{ such that } \Phi_i(v_i, u_i, x) \subseteq \Psi_i(v_i, x_i, x)\}$. For any compact subset K of $X \times Y$, if $(x, y) \in \mathrm{cl}_K(\Omega_i \cap K)$, then there exists a net $(x_\lambda, y_\lambda)_{\lambda \in \Lambda} \subseteq \Omega_i \cap K$ such that $(x_\lambda, y_\lambda) \to (x, y) \in K$. Hence, for each $\lambda \in \Lambda$, there exists $v_{i,\lambda} \in T_i(x_\lambda, y_\lambda)$ such that

$$(3.5) \qquad \Phi_i(v_{i,\lambda}, u_i, x_\lambda) \subseteq \Psi_i(v_{i,\lambda}, x_{i,\lambda}, x_\lambda).$$

Since $\{(x_\lambda, y_\lambda)\}_{\lambda \in \Lambda} \bigcup \{(x, y)\} \subseteq K$, we have $(v_{i,\lambda})_{\lambda \in \Lambda} \subseteq T_i(K)$. By the condition (a), $T_i(K)$ is compact in Y_i, Without loss of generality, we can assume $v_{i,\lambda} \to v_i \in T_i(K)$. It follows from (a) that $T_i|_K$ is a closed mapping and hence $v_i \in T_i(x, y)$. Since the mapping $(v_i, x) \mapsto \Phi_i(v_i, u_i, x)$ is lower semicontinuous at (v_i, x), by Lemma 3.3, for each $z_i \in \Phi_i(v_i, u_i, x)$, there exists net $(z_{i,\lambda})_{\lambda \in \Lambda}$ such that $z_{i,\lambda} \in \Phi_i(v_{i,\lambda}, u_i, x_\lambda)$ and $z_{i,\lambda} \to z_i$. By (3.5), we have

$$(3.6) \qquad z_{i,\lambda} \in \Psi_i(v_{i,\lambda}, x_{i,\lambda}, x_\lambda), \quad \forall \lambda \in \Lambda.$$

It follows from the condition (c) and (3.6) that $z_i \in \Psi(v_i, x_i, x)$ and hence

$$\Phi_i(v_i, u_i, x) \subseteq \Psi(v_i, x_i, x).$$

Therefore, $(x, y) \in \Omega_i \cap K$ and Ω_i is compactly closed in $X \times Y$. Hence M_i is compactly open in $X \times Y$. This completes the proof. $\square$

By using similar argument as in the proof of Lemma 3.8, it is not hard to show that the following results hold.

Lemma 3.9. *For each $i \in I$, let X_i, Y_i and Z_i be topological spaces and $X = \prod_{i \in I} X_i$ and $Y = \prod_{i \in I} Y_i$. For each $i \in I$, let $A_i : X \times Y \to 2^{X_i}$, $T_i, S_i : X \times Y \to 2^{Y_i}$ and $\Phi_i, \Psi_i : Y_i \times X_i \times X \to 2^{Z_i}$ be set-valued mappings such that*

(a) *T_i is upper semicontinuous on each compact subsets of $X \times Y$ with compact values,*

(b) *for each $u_i \in X_i$, $(v_i, x) \mapsto \Phi_i(v_i, u_i, x)$ is upper semicontinuous on each compact subset of $Y_i \times X$ with compact values,*

(c) *the mapping $(y_i, x_i, x) \mapsto Z_i \setminus \Psi_i(y_i, x_i, x)$ is upper semicontinuous on each compact subsets of $Y_i \times X_i \times X$ with closed values.*

Then for each $i \in I$ and $u_i \in X_i$, the set $M_i = \{(x, y) \in X \times Y : \Phi(v_i, u_i, x) \subseteq \Psi_i(v_i, x_i, x), \forall v_i \in T_i(x, y)\}$ is compactly open in $X \times Y$.

Lemma 3.10. *For each $i \in I$, let X_i, Y_i and Z_i be topological spaces and $X = \prod_{i \in I} X_i$ and $Y = \prod_{i \in I} Y_i$. For each $i \in I$, let $A_i : X \times Y \to 2^{X_i}$, $T_i, S_i : X \times Y \to 2^{Y_i}$ and $\Phi_i, \Psi_i : Y_i \times X_i \times X \to 2^{Z_i}$ be set-valued mappings such that*

(a) *T_i is upper semicontinuous on each compact subsets of $X \times Y$ with compact values,*

(b) *for each $u_i \in X_i$, $(v_i, x) \mapsto \Phi_i(v_i, u_i, x)$ is lower semicontinuous on each compact subset of $Y_i \times X$,*

(c) *the mapping $(y_i, x_i, x) \mapsto Z_i \setminus \Psi_i(y_i, x_i, x)$ is upper semicontinuous on each compact subsets of $Y_i \times X_i \times X$ with closed values.*

Then, for each $i \in I$ and $u_i \in X_i$, the set $M_i = \{(x, y) \in X \times Y : \Phi(v_i, u_i, x) \cap \Psi_i(v_i, x_i, x) \neq \emptyset, \ \forall v_i \in T_i(x, y)\}$ is compactly open in $X \times Y$.

Lemma 3.11. *For each $i \in I$, let X_i, Y_i and Z_i be topological spaces and $X = \prod_{i \in I} X_i$ and $Y = \prod_{i \in I} Y_i$. For each $i \in I$, let $A_i : X \times Y \to 2^{X_i}$, $T_i, S_i : X \times Y \to 2^{Y_i}$ and $\Phi_i, \Psi_i : Y_i \times X_i \times X \to 2^{Z_i}$ be set-valued mappings such that*

(a) *T_i is upper semicontinuous on each compact subsets of $X \times Y$ with compact values,*

(b) *for each $u_i \in X_i$, $(v_i, x) \mapsto \Phi_i(v_i, u_i, x)$ is upper semicontinuous on each compact subset of $Y_i \times X$ with compact values,*

(c) *the mapping $(y_i, x_i, x) \mapsto \Psi_i(y_i, x_i, x)$ is upper semicontinuous on each compact subsets of $Y_i \times X_i \times X$ with closed values.*

Then for each $i \in I$ and $u_i \in X_i$, the set $M_i = \{(x, y) \in X \times Y : \Phi(v_i, u_i, x) \cap \Psi_i(v_i, x_i, x) = \emptyset, \forall v_i \in T_i(x, y)\}$ is compactly open in $X \times Y$.

Theorem 3.1. *Suppose that K and H are nonempty compact subsets of X and Y, respectively such that, for each $i \in I$, the following conditions are satisfied:*

(a) *for each $(x, y) \in X \times Y$, $A_i(x, y)$ and $S_i(x, y)$ are both nonempty FC-subspaces of X_i and Y_i, respectively,*

(b) *for each $(u_i, v_i) \in X_i \times Y_i$, $A_i^{-1}(u_i)$, $S_i^{-1}(v_i)$ are compactly open in $X \times Y$ and, for each $i \in I$ and $u_i \in X_i$, the set $M_i = \{(x, y) \in X \times Y : \Phi_i(v_i, u_i, x) \not\subseteq \Psi_i(v_i, x_i, x)$ for some $v_i \in T_i(x, y)\}$ is compactly open in $X \times Y$,*

(c) *for each $y \in Y$, Φ_i is Ψ_i-FC-quasiconvex of type (I) in the first two arguments,*

(d) *the set $W_i = \{(x, y) \in X \times Y : x_i \in A_i(x, y), \ y_i \in S_i(x, y)\}$ is compactly closed in $X \times Y$,*

(e) *for each $N_i \in\ <X_i>$, there exists compact FC-subspace L_{N_i} containing N_i and, for each $M_i \in\ <Y_i>$, there exists compact FC-subspace L_{M_i} of Y_i containing M_i,*

(f) *for each $(x, y) \in X \times Y \setminus K \times H$, there exist $i \in I$, $\bar{u}_i \in A_i(x, y) \cap L_{N_i}$ and $\bar{y}_i \in S_i(x, y) \cap L_{M_i}$ such that $\Phi_i(v_i, u_i, x) \not\subseteq \Psi_i(v_i, x_i, x)$ for some $v_i \in T_i(x, y)$.*

Then there exists $(\hat{x}, \hat{y}) \in K \times H$ such that, for each $i \in I$,

$$\hat{x}_i \in A_i(\hat{x}, \hat{y}), \ \hat{y}_i \in S_i(\hat{x}, \hat{y}) \text{ and } \Phi_i(v_i, u_i, \hat{x}) \subseteq \Psi_i(v_i, \hat{x}_i, \hat{x}), \ \forall v_i \in T_i(\hat{x}, \hat{y}), \ u_i \in A_i(\hat{x}, \hat{y}),$$

i.e., $(\hat{x}, \hat{y})$ is a solution of the SGQVIP(I).

Proof. For each $i \in I$, define a set-valued mapping $P_{i,1} : X \times Y \to 2^{X_i}$ by

$$P_{i,1}(x, y) = \{u_i \in X_i : \Phi_i(v_i, u_i, x) \not\subseteq \Psi_i(v_i.x_i, x) \text{for some } v_i \in T_i(x, y)\}, \ \forall(x, y) \in X \times Y.$$

By (c) and Lemma 3.1, we know that, for each $(x, y) \in X \times Y$,

(3.6) $$x_i \notin FC(P_{i,1}(x, y)).$$

By the condition (b), for each $i \in I$ and $u_i \in X_i$,

$$P_{i,1}^{-1}(u_i) = \{(x,y) \in X \times Y : \Phi_i(v_i, u_i, x) \not\subseteq \Psi_i(v_i, x_i, x) \text{ for some } v_i \in T_i(x,y)\}$$

is compactly open in $X \times Y$. It follows from Lemma 2.2 that $(FC(P_{i,1}))^{-1}(u_i)$ is also compactly open in $X \times Y$ for each $u_i \in X_i$. By Lemma 2.3, for each $i \in I$, $X_i \times Y_i$ is a FC-space and $X \times Y$ is also a FC-space. For each $i \in I$, define a set-valued mapping $G_i : X \times Y \to 2^{X_i \times Y_i}$ by

$$G_i(x,y) = \begin{cases} [A_i(x,y) \cap FC(P_{i,1}(x,y))] \times S_i(x,y), & \text{if } (x,y) \in W_i, \\ A_i(x,y) \times S_i(x,y), & \text{if } (x,y) \notin W_i, \end{cases}$$

By the condition (a), for each $i \in I$ and $(x,y) \in X \times Y$, $G_i(x,y)$ is a FC-subspace of $X_i \times Y_i$. By the definition of W_i and (3.6), for each $i \in I$ and $(x,y) \in X \times Y$, $(x_i, y_i) \notin G_i(x,y)$. For each $i \in I$ and $(u_i, v_i) \in X_i \times Y_i$, we have

$$G_i^{-1}(u_i, v_i) = [A_i^{-1}(u_i)) \cap (FC(P_{i,1}))^{-1}(u_i) \cap S_i^{-1}(v_i)]$$
$$\cup [((X \times Y) \setminus W_i) \cap A_i^{-1}(u_i) \cap S_i^{-1}(v_i)].$$

Since $(FC(P_{i,1}))^{-1}(u_i)$ is compactly open in $X \times Y$ for each $u_i \in X_i$, by the conditions (b) and (d), $G_i^{-1}(u_i, v_i)$ is also compactly open in $X \times Y$. By (e), for each $H_i = N_i \times M_i \in < X_i \times Y_i >$, there exists compact FC-subspace $L_{H_i} = L_{N_i} \times L_{M_i}$ of $X_i \times Y_i$ containing H_i. By (f), for each $(x,y) \in X \times Y \setminus K \times H$, there exists $i \in I$ such that $G_i(x,y) \cap L_{H_i} \neq \emptyset$. All the conditions of Lemma 2.4 are satisfied. By Lemma 2.4, there exists $(\hat{x}, \hat{y}) \in K \times H$ such that $G_i(\hat{x}, \hat{y}) = \emptyset$ for each $i \in I$. If $(\hat{x}, \hat{y}) \notin W_j$ for some $j \in I$, then either $A_j(\hat{x}, \hat{y}) = \emptyset$ or $S_j(\hat{x}, \hat{y}) = \emptyset$, which contradicts the condition (a). Therefore, $(\hat{x}, \hat{y}) \in W_i$ for each $i \in I$. This shows that, for each $i \in I$, $\hat{x}_i \in A_i(\hat{x}, \hat{y})$, $\hat{y}_i \in S_i(\hat{x}, \hat{y})$ and $A_i(\hat{x}, \hat{y}) \cap FC(P_{i,1}(\hat{x}, \hat{y})) = \emptyset$ and hence $A_i(\hat{x}, \hat{y}) \cap P_{i,1}(\hat{x}, \hat{y}) = \emptyset$. Therefore, for each $i \in I$,

$$\hat{x}_i \in A_i(\hat{x}, \hat{y}), \ \hat{y}_i \in S_i(\hat{x}, \hat{y}) \text{ and } \Phi_i(v_i, u_i, \hat{x}) \subset \Psi_i(v_i, \hat{x}_i, \hat{x}), \ \forall v_i \in T_i(\hat{x}, \hat{y}), \ u_i \in A_i(\hat{x}, \hat{y}),$$

i.e., $(\hat{x}, \hat{y})$ is a solution of the SGQVIP(I). This completes the proof. $\qquad\square$

Remark 3.1. Theorem 3.1 improves and generalizes Theorem 3.2 of Hai and Khanh [26] in the following ways:

(1) the mathematical model of SGQVIP(I) is more general than the mathematical model of (SQIP2) in [26];

(2) for each $i \in I$, X_i and Y_i may be FC-spaces without convexity structure and Z_i may be any nonempty set;

(3) the condition (c) of Theorem 3.1 is weaker than the condition (i_2) in Theorem 3.2 in [26].

By applying Lemma 2.4, Lemma 3.1 and similar argument as in the proof of Theorem 3.1, we can obtain the following results:

Theorem 3.2. *Suppose that K and H are nonempty compact subsets of X and Y, respectively, such that, for each $i \in I$, the following conditions are satisfied:*

(a) for each $(x,y) \in X \times Y$, $A_i(x,y)$ and $S_i(x,y)$ are both nonempty FC-subspaces of X_i and Y_i respectively,

(b) *for each* $(u_i, v_i) \in X_i \times Y_i$, $A_i^{-1}(u_i)$, $S_i^{-1}(v_i)$ *are compactly open in* $X \times Y$ *and, for each* $u_i \in X_i$, *the set* $\{(x,y) \in X \times Y : \Phi_i(v_i, u_i, x) \subseteq \Psi_i(v_i.x_i, x)$ *for some* $v_i \in T_i(x,y)\}$ *is compactly open in* $X \times Y$,

(c) *for each* $y \in Y$, Φ_i *is* Ψ_i-*FC-quasiconvex of type* (II) *in the first two arguments*,

(d) *the set* $W_i = \{(x,y) \in X \times Y : x_i \in A_i(x,y), \ y_i \in S_i(x,y)\}$ *is compactly closed in* $X \times Y$,

(e) *for each* $N_i \in \langle X_i \rangle$, *there exists compact FC-subspace* L_{N_i} *containing* N_i *and, for each* $M_i \in \langle Y_i \rangle$, *there exists compact FC-subspace* L_{M_i} *of* Y_i *containing* M_i,

(f) *for each* $(x,y) \in X \times Y \setminus K \times H$, *there exist* $i \in I$, $\bar{u}_i \in A_i(x,y) \cap L_{N_i}$ *and* $\bar{y}_i \in S_i(x,y) \cap L_{M_i}$ *such that* $\Phi_i(v_i, u_i, x) \subseteq \Psi_i(v_i, x_i, x)$ *for some* $v_i \in T_i(x,y)$.
Then there exists $(\hat{x}, \hat{y}) \in K \times H$ *such that, for each* $i \in I$,

$$\hat{x}_i \in A_i(\hat{x}, \hat{y}), \ \hat{y}_i \in S_i(\hat{x}, \hat{y}) \text{ and } \Phi_i(v_i, u_i, \hat{x}) \not\subseteq \Psi_i(v_i, \hat{x}_i, \hat{x}), \ \forall v_i \in T_i(\hat{x}, \hat{y}), u_i \in A_i(\hat{x}, \hat{y}),$$

i.e., $(\hat{x}, \hat{y})$ *is a solution of the* SGQVIP(II).

Theorem 3.3. *Suppose that* K *and* H *are nonempty compact subsets of* X *and* Y, *respectively, such that, for each* $i \in I$, *the following conditions are satisfied:*

(a) *for each* $(x,y) \in X \times Y$, $A_i(x,y)$ *and* $S_i(x,y)$ *are both nonempty FC-subspaces of* X_i *and* Y_i *respectively,*

(b) *for each* $(u_i, v_i) \in X_i \times Y_i$, $A_i^{-1}(u_i)$, $S_i^{-1}(v_i)$ *are compactly open in* $X \times Y$ *and, for each* $u_i \in X_i$, *the set* $\{(x,y) \in X \times Y : \Phi_i(v_i, u_i, x) \cap \Psi_i(v_i.x_i, x) \neq \emptyset$ *for some* $v_i \in T_i(x,y)\}$ *is compactly open in* $X \times Y$,

(c) *for each* $y \in Y$, Φ_i *is* Ψ_i-*FC-quasiconvex of type* (III) *in the first two arguments,*

(d) *the set* $W_i = \{(x,y) \in X \times Y : x_i \in A_i(x,y), \ y_i \in S_i(x,y)\}$ *is compactly closed in* $X \times Y$,

(e) *for each* $N_i \in \langle X_i \rangle$, *there exists compact FC-subspace* L_{N_i} *containing* N_i *and, for each* $M_i \in \langle Y_i \rangle$, *there exists compact FC-subspace* L_{M_i} *of* Y_i *containing* M_i,

(f) *for each* $(x,y) \in X \times Y \setminus K \times H$, *there exist* $i \in I$, $\bar{u}_i \in A_i(x,y) \cap L_{N_i}$ *and* $\bar{y}_i \in S_i(x,y) \cap L_{M_i}$ *such that* $\Phi_i(v_i, u_i, x) \cap \Psi_i(v_i, x_i, x) \neq \emptyset$ *for some* $v_i \in T_i(x,y)$.
Then there exists $(\hat{x}, \hat{y}) \in K \times H$ *such that, for each* $i \in I$,

$$\hat{x}_i \in A_i(\hat{x}, \hat{y}), \ \hat{y}_i \in S_i(\hat{x}, \hat{y}) \text{ and } \Phi_i(v_i, u_i, \hat{x}) \cap \Psi_i(v_i, \hat{x}_i, \hat{x}) = \emptyset, \ \forall v_i \in T_i(\hat{x}, \hat{y}), u_i \in A_i(\hat{x}, \hat{y}),$$

i.e., $(\hat{x}, \hat{y})$ *is a solution of the* SGQVIP(III).

Remark 3.2. Theorem 3.1 improves and generalizes Theorem 3.4 of Hai and Khanh [26] in the following ways:

(1) the mathematical model of SGQVIP(I) is more general than the mathematical model of (SQIP2) in [26];

(2) for each $i \in I$, X_i and Y_i may be FC-spaces without convexity structure and Z_i may be any nonempty set;

(3) the condition (c) of Theorem 3.3 is weaker than the condition (i_4) in Theorem 3.4 in [26].

Theorem 3.4. *Suppose that* K *and* H *are nonempty compact subsets of* X *and* Y, *respectively, such that, for each* $i \in I$, *the following conditions are satisfied:*

(a) *for each* $(x,y) \in X \times Y$, $A_i(x,y)$ *and* $S_i(x,y)$ *are both nonempty FC-subspaces of* X_i *and* Y_i, *respectively,*

(b) *for each* $(u_i, v_i) \in X_i \times Y_i$, $A_i^{-1}(u_i)$, $S_i^{-1}(v_i)$ *are compactly open in* $X \times Y$ *and, for each* $u_i \in X_i$, *the set* $\{(x,y) \in X \times Y : \Phi_i(v_i, u_i, x) \cap \Psi(v_i.x_i, x) = \emptyset$ *for some* $v_i \in T_i(x,y)\}$ *is compactly open in* $X \times Y$,

(c) *for each* $y \in Y$, Φ_i *is* Ψ_i-*FC-quasiconvex of type* (IV) *in the first two arguments*,

(d) *the set* $W_i = \{(x,y) \in X \times Y : x_i \in A_i(x,y), y_i \in S_i(x,y)\}$ *is compactly closed in* $X \times Y$,

(e) *for each* $N_i \in < X_i >$, *there exists compact FC-subspace* L_{N_i} *containing* N_i *and, for each* $M_i \in < Y_i >$, *there exists compact FC-subspace* L_{M_i} *of* Y_i *containing* M_i,

(f) *for each* $(x,y) \in X \times Y \setminus K \times H$, *there exist* $i \in I$, $\bar{u}_i \in A_i(x,y) \cap L_{N_i}$ *and* $\bar{y}_i \in S_i(x,y) \cap L_{M_i}$ *such that* $\Phi_i(v_i, u_i, x) \cap \Psi_i(v_i, x_i, x) = \emptyset$ *for some* $v_i \in T_i(x,y)$. *Then there exists* $(\hat{x}, \hat{y}) \in K \times H$ *such that for each* $i \in I$,

$$\hat{x}_i \in A_i(\hat{x}, \hat{y}),\ \hat{y}_i \in S_i(\hat{x}, \hat{y})\ and\ \Phi_i(v_i, u_i, \hat{x}) \cap \Psi_i(v_i, \hat{x}_i, \hat{x}) \neq \emptyset,\ \forall v_i \in T_i(\hat{x}, \hat{y}), u_i \in A_i(\hat{x}, \hat{y}),$$

i.e., $(\hat{x}, \hat{y})$ *is a solution of the* SGQVIP(IV).

If, for each $i \in I$, Z_i is a topological space, by applying Theorems 3.1-3.4 and Lemmas 3.4-3.7, we have the following results:

Corollary 3.1. *If the condition that, for each* $i \in I$ *and* $u_i \in X_i$, *the set* $M_i = \{(x,y) \in X \times Y : \Phi_i(v_i, u_i, x) \not\subseteq \Psi_i(v_i, x_i, x)$ *for some* $v_i \in T_i(x,y)\}$ *is compactly open in* $X \times Y$ *in* (b) *of Theorem 3.1 is replaced by the conditions* (a)-(c) *of Lemma 3.4, Then the conclusion of Theorem 3.1 holds.*

Corollary 3.2. *If the condition that, for each* $i \in I$ *and* $u_i \in X_i$, *the set* $M_i = \{(x,y) \in X \times Y : \Phi_i(v_i, u_i, x) \subseteq \Psi_i(v_i, x_i, x)$ *for some* $v_i \in T_i(x,y)\}$ *is compactly open in* $X \times Y$ *in* (b) *of Theorem 3.2 is replaced by the conditions* (a)-(c) *of Lemma 3.5. Then the conclusion of Theorem 3.2 holds.*

Corollary 3.3. *If the condition that, for each* $i \in I$ *and* $u_i \in X_i$, *the set* $M_i = \{(x,y) \in X \times Y : \Phi_i(v_i, u_i, x) \cap \Psi_i(v_i, x_i, x) \neq \emptyset$ *for some* $v_i \in T_i(x,y)\}$ *is compactly open in* $X \times Y$ *in* (b) *of Theorem 3.3 is replaced by the conditions* (a)-(c) *of Lemma 3.6. Then the conclusion of Theorem 3.3 holds.*

Corollary 3.4. *If the condition that, for each* $i \in I$ *and* $u_i \in X_i$, *the set* $M_i = \{(x,y) \in X \times Y : \Phi_i(v_i, u_i, x) \cap \Psi_i(v_i, x_i, x) = \emptyset$ *for some* $v_i \in T_i(x,y)\}$ *is compactly open in* $X \times Y$ *in* (b) *of Theorem 3.4 is replaced by the conditions* (a)-(c) *of Lemma 3.7. Then the conclusion of Theorem 3.4 holds.*

Theorem 3.5. *Suppose that* K *and* H *are nonempty compact subsets of* X *and* Y, *respectively, such that, for each* $i \in I$, *the following conditions are satisfied:*

(a) *for each* $(x,y) \in X \times Y$, $A_i(x,y)$ *and* $S_i(x,y)$ *are both nonempty FC-subspaces of* X_i *and* Y_i, *respectively*,

(b) *for each* $(u_i, v_i) \in X_i \times Y_i$, $A_i^{-1}(u_i)$, $S_i^{-1}(v_i)$ *are compactly open in* $X \times Y$ *and, for each* $i \in I$ *and* $u_i \in X_i$, *the set* $M_i = \{(x,y) \in X \times Y : \Phi_i(v_i, u_i, x) \not\subseteq \Psi_i(v_i, x_i, x), \forall v_i \in T_i(x,y)\}$ *is compactly open in* $X \times Y$,

(c) *for each* $y \in Y$, Φ_i *is* Ψ_i-*FC-weakly quasiconvex of type* (I) *in the first two arguments*,

(d) *the set* $W_i = \{(x,y) \in X \times Y : x_i \in A_i(x,y),\ y_i \in S_i(x,y)\}$ *is compactly closed in* $X \times Y$,

(e) *for each* $N_i \in < X_i >$, *there exists compact FC-subspace* L_{N_i} *containing* N_i *and, for each* $M_i \in < Y_i >$, *there exists compact FC-subspace* L_{M_i} *of* Y_i *containing* M_i,

(f) *for each* $(x,y) \in X \times Y \setminus K \times H$, *there exist* $i \in I$, $\bar{u}_i \in A_i(x,y) \cap L_{N_i}$ *and* $\bar{y}_i \in S_i(x,y) \cap L_{M_i}$ *such that* $\Phi_i(v_i, u_i, x) \not\subseteq \Psi_i(v_i, x_i, x)$ *for all* $v_i \in T_i(x,y)$.

Then there exists $(\hat{x}, \hat{y}) \in K \times H$ *such that, for each* $i \in I$, $\hat{x}_i \in A_i(\hat{x}, \hat{y})$, $\hat{y}_i \in S_i(\hat{x}, \hat{y})$ *and, for each* $u_i \in A_i(\hat{x}, \hat{y})$, *there exists* $v_i \in T_i(\hat{x}, \hat{y})$ *such that,*

$$\Phi_i(v_i, u_i, \hat{x}) \subseteq \Psi_i(v_i, \hat{x}_i, \hat{x}),$$

i.e., $(\hat{x}, \hat{y})$ *is a solution of the* SGQVIP(V).

Proof. For each $i \in I$, define a set-valued mapping $P_{i,1} : X \times Y \to 2^{X_i}$ by

$$P_{i,1}(x,y) = \{u_i \in X_i : \Phi_i(v_i, u_i, x) \not\subseteq \Psi_i(v_i.x_i, x),\ \forall v_i \in T_i(x,y)\},\ \forall (x,y) \in X \times Y\}.$$

By (c) and Lemma 3.2, we know that, for each $(x,y) \in X \times Y$,

(3.7) $$x_i \notin FC(P_{i,1}(x,y)).$$

By the condition (b), for each $i \in I$ and $u_i \in X_i$,

$$P_{i,1}^{-1}(u_i) = \{(x,y) \in X \times Y : \Phi_i(v_i, u_i, x) \not\subseteq \Psi_i(v_i, x_i, x),\ \forall v_i \in T_i(x,y)\}$$

is compactly open in $X \times Y$. It follows from Lemma 2.2 that $(FC(P_{i,1}))^{-1}(u_i)$ is also compactly open in $X \times Y$ for each $u_i \in X_i$. By Lemma 2.3, for each $i \in I$, $X_i \times Y_i$ is a FC-space and $X \times Y$ is also a FC-space. For each $i \in I$, define a set-valued mapping $G_i : X \times Y \to 2^{X_i \times Y_i}$ by

$$G_i(x,y) = \begin{cases} [A_i(x,y) \cap FC(P_{i,1}(x,y))] \times S_i(x,y), & \text{if } (x,y) \in W_i, \\ A_i(x,y) \times S_i(x,y), & \text{if } (x,y) \notin W_i, \end{cases}$$

By using similar argument as in the proof of Theorem 3.1, it is easy to prove that there exists $(\hat{x}, \hat{y}) \in K \times H$ such that $G_i(\hat{x}, \hat{y}) = \emptyset$ for each $i \in I$. If $(\hat{x}, \hat{y}) \notin W_j$ for some $j \in I$, then either $A_j(\hat{x}, \hat{y}) = \emptyset$ or $S_j(\hat{x}, \hat{y}) = \emptyset$, which contradicts the condition (i). Therefore, $(\hat{x}, \hat{y}) \in W_i$ for each $i \in I$. This shows that for each $i \in I$, $\hat{x}_i \in A_i(\hat{x}, \hat{y})$, $\hat{y}_i \in S_i(\hat{x}, \hat{y})$ and $A_i(\hat{x}, \hat{y}) \cap FC(P_{i,1}(\hat{x}, \hat{y})) = \emptyset$ and hance $A_i(\hat{x}, \hat{y}) \cap P_{i,1}(\hat{x}, \hat{y}) = \emptyset$. Therefore, for each $i \in I$, $\hat{x}_i \in A_i(\hat{x}, \hat{y})$, $\hat{y}_i \in S_i(\hat{x}, \hat{y})$ and, for each $u_i \in A_i(\hat{x}, \hat{y})$, there exists $v_i \in T_i(\hat{x}, \hat{y})$ such that

$$\Phi_i(v_i, u_i, \hat{x}) \subseteq \Psi(v_i, \hat{x}_i, \hat{x}),$$

i.e., $(\hat{x}, \hat{y})$ is a solution of the SGQVIP(V). This completes the proof. $\square$

Remark 3.2. Theorem 3.5 improves and generalizes Theorem 3.1 of Hai and Khanh [26] in the following ways:

(1) the mathematical model of SGQVIP (I) is more general than the mathematical model of (SQIP2) in [26];

(2) for each $i \in I$, X_i and Y_i may be FC-spaces without convexity structure and Z_i may be any nonempty set;

(3) the condition (c) of Theorem 3.5 is weaker than the condition (a) in Theorem 3.1 in [26].

By using similar argument as in the proof of Theorem 3.5, it is not hard to prove the following results.

Theorem 3.6. *Suppose that K and H are nonempty compact subsets of X and Y, respectively, such that, for each $i \in I$, the following conditions are satisfied:*

(a) for each $(x, y) \in X \times Y$, $A_i(x, y)$ and $S_i(x, y)$ are both nonempty FC-subspaces of X_i and Y_i, respectively,

(b) for each $(u_i, v_i) \in X_i \times Y_i$, $A_i^{-1}(u_i)$, $S_i^{-1}(v_i)$ are compactly open in $X \times Y$ and for each $u_i \in X_i$, the set $\{(x, y) \in X \times Y : \Phi_i(v_i, u_i, x) \subseteq \Psi_i(v_i.x_i, x), \forall v_i \in T_i(x, y)\}$ is compactly open in $X \times Y$,

(c) for each $y \in Y$, Φ_i is Ψ_i-FC-weakly quasiconvex of type (II) in the first two arguments,

(d) the set $W_i = \{(x, y) \in X \times Y : x_i \in A_i(x, y), y_i \in S_i(x, y)\}$ is compactly closed in $X \times Y$,

(e) for each $N_i \in < X_i >$, there exists compact FC-subspace L_{N_i} containing N_i and, for each $M_i \in < Y_i >$, there exists compact FC-subspace L_{M_i} of Y_i containing M_i,

(f) for each $(x, y) \in X \times Y \setminus K \times H$, there exist $i \in I$, $\bar{u}_i \in A_i(x, y) \cap L_{N_i}$ and $\bar{y}_i \in S_i(x, y) \cap L_{M_i}$ such that $\Phi_i(v_i, u_i, x) \subseteq \Psi_i(v_i, x_i, x)$ for all $v_i \in T_i(x, y)$.

Then there exists $(\hat{x}, \hat{y}) \in K \times H$ such that, for each $i \in I$, $\hat{x}_i \in A_i(\hat{x}, \hat{y})$, $\hat{y}_i \in S_i(\hat{x}, \hat{y})$ and, for each $u_i \in A_i(\hat{x}, \hat{y})$, there exists $v_i \in T_i(\hat{x}, \hat{y})$ such that

$$\Phi_i(v_i, u_i, \hat{x}) \nsubseteq \Psi_i(v_i, \hat{x}_i, \hat{x}),$$

i.e., $(\hat{x}, \hat{y})$ is a solution of the SGQVIP(VI).

Theorem 3.7. *Suppose that K and H are nonempty compact subsets of X and Y, respectively, such that, for each $i \in I$, the following conditions are satisfied:*

(a for each $(x, y) \in X \times Y$, $A_i(x, y)$ and $S_i(x, y)$ are both nonempty FC-subspaces of X_i and Y_i, respectively,

(b) for each $(u_i, v_i) \in X_i \times Y_i$, $A_i^{-1}(u_i)$, $S_i^{-1}(v_i)$ are compactly open in $X \times Y$ and, for each $u_i \in X_i$, the set $\{(x, y) \in X \times Y : \Phi_i(v_i, u_i, x) \cap \Psi_i(v_i.x_i, x) \neq \emptyset, \forall v_i \in T_i(x, y)\}$ is compactly open in $X \times Y$,

(c) for each $y \in Y$, Φ_i is Ψ_i-FC-weakly quasiconvex of type (III) in the first two arguments,

(d) the set $W_i = \{(x, y) \in X \times Y : x_i \in A_i(x, y), y_i \in S_i(x, y)\}$ is compactly closed in $X \times Y$,

(e) for each $N_i \in < X_i >$, there exists compact FC-subspace L_{N_i} containing N_i and, for each $M_i \in < Y_i >$, there exists compact FC-subspace L_{M_i} of Y_i containing M_i,

(f) for each $(x, y) \in X \times Y \setminus K \times H$, there exist $i \in I$, $\bar{u}_i \in A_i(x, y) \cap L_{N_i}$ and $\bar{y}_i \in S_i(x, y) \cap L_{M_i}$ such that $\Phi_i(v_i, u_i, x) \cap \Psi_i(v_i, x_i, x) \neq \emptyset$ for all $v_i \in T_i(x, y)$.

Then there exists $(\hat{x}, \hat{y}) \in K \times H$ such that, for each $i \in I$, $\hat{x}_i \in A_i(\hat{x}, \hat{y})$, $\hat{y}_i \in S_i(\hat{x}, \hat{y})$ and, for each $u_i \in A_i(\hat{x}, \hat{y})$, there exists $v_i \in T_i(\hat{x}, \hat{y})$ such that

$$\Phi_i(v_i, u_i, \hat{x}) \cap \Psi_i(v_i, \hat{x}_i, \hat{x}) = \emptyset,$$

i.e., $(\hat{x}, \hat{y})$ is a solution of the SGQVIP(VII).

Remark 3.3. Theorem 3.7 improves and generalizes Theorem 3.3 of Hai and Khanh [26] in the following ways:

(1) the mathematical model of SGQVIP(I) is more general than the mathematical model of (SQIP2) in [26];

(2) for each $i \in I$, X_i and Y_i may be FC-spaces without convexity structure and Z_i may be any topological spaces;

(3) the condition (c) of Theorem 3.5 is weaker than the condition (i_3) in Theorem 3.3 in [26].

Theorem 3.8. *Suppose that K and H are nonempty compact subsets of X and Y, respectively, such that, for each $i \in I$, the following conditions are satisfied:*

(a) for each $(x,y) \in X \times Y$, $A_i(x,y)$ and $S_i(x,y)$ are both nonempty FC-subspaces of X_i and Y_i, respectively,

(b) for each $(u_i, v_i) \in X_i \times Y_i$, $A_i^{-1}(u_i)$, $S_i^{-1}(v_i)$ are compactly open in $X \times Y$ and, for each $u_i \in X_i$, the set $\{(x,y) \in X \times Y : \Phi_i(v_i, u_i, x) \cap \Psi(v_i.x_i, x) = \emptyset, \ \forall v_i \in T_i(x,y)\}$ is compactly open in $X \times Y$,

(c) for each $y \in Y$, Φ_i is Ψ_i-FC-weakly quasiconvex of type (IV) in the first two arguments,

(d) the set $W_i = \{(x,y) \in X \times Y : x_i \in A_i(x,y), y_i \in S_i(x,y)\}$ is compactly closed in $X \times Y$,

(e) for each $N_i \in < X_i >$, there exists compact FC-subspace L_{N_i} containing N_i and, for each $M_i \in < Y_i >$, there exists compact FC-subspace L_{M_i} of Y_i containing M_i,

(f) for each $(x,y) \in X \times Y \setminus K \times H$, there exist $i \in I$, $\bar{u}_i \in A_i(x,y) \cap L_{N_i}$ and $\bar{y}_i \in S_i(x,y) \cap L_{M_i}$ such that $\Phi_i(v_i, u_i, x) \cap \Psi_i(v_i, x_i, x) = \emptyset$ for all $v_i \in T_i(x,y)$.

Then there exists $(\hat{x}, \hat{y}) \in K \times H$ such that, for each $i \in I$, $\hat{x}_i \in A_i(\hat{x}, \hat{y})$, $\hat{y}_i \in S_i(\hat{x}, \hat{y})$ and, for each $u_i \in A_i(\hat{x}, \hat{y})$, there exists $v_i \in T_i(\hat{x}, \hat{y})$ such that

$$\Phi_i(v_i, u_i, \hat{x}) \cap \Psi_i(v_i, \hat{x}_i, \hat{x}) \neq \emptyset,$$

i.e., $(\hat{x}, \hat{y})$ is a solution of the SGQVIP(VIII).

If, for each $i \in I$, Z_i is a topological space, by applying Theorems 3.5-3.8 and Lemmas 3.8-3.11, we have the following results:

Corollary 3.5. *If the condition that, for each $i \in I$ and $u_i \in X_i$, the set $M_i = \{(x,y) \in X \times Y : \Phi_i(v_i, u_i, x) \not\subseteq \Psi_i(v_i, x_i, x), \forall v_i \in T_i(x,y)\}$ is compactly open in $X \times Y$ in (b) of Theorem 3.5 is replaced by the conditions (a)-(c) of Lemma 3.8. Then the conclusion of Theorem 3.5 holds.*

Corollary 3.6. *If the condition that, for each $i \in I$ and $u_i \in X_i$, the set $M_i = \{(x,y) \in X \times Y : \Phi_i(v_i, u_i, x) \subseteq \Psi_i(v_i, x_i, x), \forall v_i \in T_i(x,y)\}$ is compactly open in $X \times Y$ in (b) of Theorem 3.6 is replaced by the conditions (a)-(c) of Lemma 3.9. Then the conclusion of Theorem 3.6 holds.*

Corollary 3.7. *If the condition that, for each $i \in I$ and $u_i \in X_i$, the set $M_i = \{(x,y) \in X \times Y : \Phi_i(v_i, u_i, x) \cap \Psi_i(v_i, x_i, x) \neq \emptyset, \forall v_i \in T_i(x,y)\}$ is compactly open in $X \times Y$ in (b) of Theorem 3.7 is replaced by the conditions (a)-(c) of Lemma 3.10. Then the conclusion of Theorem 3.7 holds.*

Corollary 3.8. *If the condition that, for each $i \in I$ and $u_i \in X_i$, the set $M_i = \{(x,y) \in X \times Y : \Phi_i(v_i, u_i, x) \cap \Psi_i(v_i, x_i, x) = \emptyset, \forall v_i \in T_i(x,y)\}$ is compactly open in*

$X \times Y$ in (b) of Theorem 3.8 is replaced by the conditions (a)-(c) of Lemma 3.11. Then the conclusion of Theorem 3.8 holds.

REFERENCES

1. S. M. Robinson, *Generalized equation and their solutions,* Math. Program. Study **10** (1979), 128–141.

2. A. Hassouni and A. Moudafi, *A perturbed algorithm for variational inclusions,* J. Math. Anal. Appl. **185(3)** (1994), 706–721.

3. S. Adly, *Perturbed algorithms and sensitivity analysis for a general class of variational inclusions,* J. Math. Anal. Appl. **201(3)** (1996), 609–630.

4. X. P. Ding, *Perturbed proximal point algorithm for generalized quasivariational inclusions,* J. Math. Anal. Appl. **210(1)** (1997), 88–101.

5. X. P. Ding, *Perturbed Ishikawa type iterative algorithm for generalized quasivariational inclusions,* Appl. Math. Comput. **141** (2003), 359–373.

6. Y. P. Fang and N. J. Huang, *H-monotone operator and resolvent operator technique for variational inclusions,* Appl. Math. Comput. **145** (2003), 795–803.

7. X. P. Ding, *Sensitivity analysis for generalized nonlinear implicit quasi-variational inclusions,* Appl. Math. Lett. **17(2)** (2004), 225–235.

8. X. P. Ding and J. C. Yao, *Existence and algorithm of solutions for mixed quasi-variational-like inclusions in Banach spaces,* Comput. Math. Appl. **49(5-6)** (2005), 857–869.

9. X. P. Ding, *Predictor-Corrector iterative algorithms for solving generalized mixed quasi-variational-like inclusion,* J. Comput. Appl. Math. **182(1)** (2005), 1–12.

10. X. P. Ding, *Parametric completely generalized mixed implicit quasi-variational inclusions involving h-maximal monotone mappings,* J. Comput. Appl. Math. **182(2)** (2005), 252–269.

11. Q. H. Ansari and J. C. Yao, *A fixed point theorem and its applications to a system of cariational inequalities,* Bull. Aust. Math. Soc. **59** (1999), 433o-442.

12. Q. H. Ansari, S. Schaible and J. C. Yao, *Systems of vector equilibrium problems and its applications,* J. Optim. Theory Appl. **107** (2000), 547–557.

13. Q. H. Ansari, S. Schaible and J. C. Yao, *The system of generalized vector equilibrium problems with applications,* J. Global Optim. **22** (2002), 3–16.

14. L.J. Lin and Y.H. Liu, *Existence theorems for systems of generalized vector quasi-equilibrium problems and optimization problems,* J. Optim. Theory Appl. **130** (2006), 461–475.

15. L. J. Lin, *System of Generalized Vector Quasi-Equilibrium Problems with Applications to Fixed Point Theorems for a Family of Nonexpansive Multivalued Mappings,* J. Global Optim. **34** (2006), 15–32.

16. N. X. Hai and P. Q. Khanh, *System of multivalued quasiequilibrium problems,* Adv. Nomlinear Var. Inequal. **9** (2006), 97–108.

17. X. P. Ding, *Generalized game and system of generalized vector quasi-equilibrium problems in G-convex spaces,* Acta Math. Scientia **26(4)** (2006), 506–515.

18. X. P. Ding, *System of generalized vector quasi-equilibrium problems in locally FC-spaces,* Acta Math. Sinica **22(5)** (2006), 1529–1538.

19. X. P. Ding, *System of generalized vector quasi-equilibrium problems on product FC-spaces,* Acta Math. Scientia **27(3)** (2007), 522–534.

20. X. P. Ding, *Maximal elements of G_{KKM}-majorized mappings in product FC-spaces and applications (II),* Nonlinear Anal. **67(12)** (2007), 3411–3423.

21. X. P. Ding, *The generalized game and system of generalized vector quasi-equilibrium problems in locally FC-uniform spaces,* Nonlinear Anal., doi:10.1016/j.na.2006.12.03.

22. X. P. Ding, *Maximal elements and generalized games involving condensing mappings in locally FC-uniform spaces and applications (I),* Appl. Math. Mech. **28(12)** (2007), 1561–1568.

23. X. P. Ding, *Maximal elements and generalized games involving condensing mappings in locally FC-uniform spaces and applications (II),* Appl. Math. Mech. **28(12)** (2007), 1569–1580.

24. L. J. Lin, *Systems of generalized quasivariational inclusions problems with applications to variational analysis and optimization problems,* J. Global Optim. **38(1)** (2007), 21–39.

25. L. J. Lin and C. I. Tu, *The studies of systems of variational inclusions problems and applications,* Nonlinear Anal. 2007, doi:10.1016/j.na.2007.07.041.

26. N. X. Hai and P.Q. Khanh, *Systems of set-valued quasivariational inclusion problems,* J. Optim. Theory Appl. 2007, DOI 10.1007/s10957-007-9222-0.

27. X. P. Ding, *Maximal elements of G_{KKM}-majorized mappings in product FC-spaces and applications (I),* Nonlinear Anal. **67(3)** (2007), 963–973.

28. H. Ben-El-Mechaiekh, S. Chebbi, M. Flornzano and J. V. Llinares, *Abstract convexity and fixed points,* J. Math. Anal. Appl. **222** (1998), 138–150.

29. X. P. Ding, *Maximal element theorems in product FC-spaces and generalized games,* J. Math. Anal. Appl. **305(1)** (2005), 29–42.

30. C. D. Horvath, *Contractibility and generalized convexity,* J. Math. Anal. Appl. **156** (1991), 341–357.

31. S. Park and H. Kim, *Foundations of the KKM theory on generalized convex spaces,* J. Math. Anal. Appl. **209** (1997), 551–571.

Nonlinear Functional Analysis and Applications, Volume 1, 239–246

COMMON FIXED POINTS FOR TWO PAIRS OF
COMPATIBLE MAPPINGS IN 2-METRIC SPACES

QINHUA WU[1], LILI WANG[2], HYEONG KUG KIM[3] AND SHIN MIN KANG[4,*]

ABSTRACT. In this paper, two common fixed point theorems for two pairs of compatible mappings of type (P) in a complete 2-metric space are established. Our results improve and generalize several previously known results given by some authors.

1. Introduction

In 1963, Gähler [2] introduced and investigated the concept and properties of a 2-metric space, respectively. In particular, he established that a 2-metric d is a continuous function of any one of its three arguments but it need not be continuous in two arguments. If it is continuous in any two arguments, it is continuous in all three arguments. A 2-metric d which is continuous in all of its arguments will be called continuous.

In 1975, Iséki [4], for the first time, developed a fixed point theorem in 2-metric spaces. Afterwards a quite number of authors ([1, 3-12]) obtained various existence and uniqueness results of fixed points, common fixed points and coincidence points for certain classes of contractive and expansive type mappings in 2-metric spaces.

Motivated and inspired by the results in [1, 3-12], in this paper, we show the existence and uniqueness of common fixed points for two pairs of compatible mappings of type (P) in a complete 2-metric space. Our results extend, improve and unify the corresponding results in [6] and [7].

Received February 12, 2008.   * Corresponding author.

2000 *Mathematics Subject Classification*. 54H25.

Key words and phrases. Compatible mappings of type (P), common fixed point, 2-metric space.

[1] Department of Basic Science Subjects, Huaihai Institute of Technology, Lianyungang, Jiangsu 222005, People's Republic of China (*E-mail*: wuqinhua575@sina.com)

[2] Department of Mathematics, Liaoning Normal University, Dalian, Liaoning 116029, People's Republic of China (*E-mail*: lili_wang@yahoo.cn)

[3] Department of Mathematics, Gyeongsang National University, Jinju 660-701, Korea (*E-mail*: hyeong4@hanmail.net)

[4] Department of Mathematics and the Research Institute of Natural Science, Gyeongsang National University, Jinju 660-701, Korea *E-mail*: smkang@gnu.ac.kr)

2. Preliminaries

Throughout this paper, ω and $\mathbb{N}$ denote the sets of nonnegative and positive integers, respectively. Let $\mathbb{R}^+ = [0, +\infty)$. We consider the family Φ of all functions $\phi : (\mathbb{R}^+)^{10} \to \mathbb{R}^+$ with the following properties:

$(P1)$ ϕ is non-decreasing and upper semicontinuous in each variables;

$(P2)$ $\phi^*(t) = \max\{\phi(t, t, t, 2t, 0, t, t, t, t, t), \phi(t, t, t, 0, 2t, t, t, t, t, t),$
$\phi(t, 0, 0, t, t, t, 0, t, 0, t)\} < t$ for each $t > 0$.

Definition 2.1. A sequence $\{x_n\}_{n \in \mathbb{N}}$ in a 2-metric space (X, d) is said to be *convergent* to a point $x \in X$ if $\lim_{n \to \infty} d(x_n, x, a) = 0$ for all $a \in X$. The point x is called the limit of the sequence $\{x_n\}_{n \in \mathbb{N}}$ in X.

Definition 2.2. A sequence $\{x_n\}_{n \in \mathbb{N}}$ in a 2-metric space (X, d) is said to be a *Cauchy sequence* if $\lim_{m,n \to \infty} d(x_m, x_n, a) = 0$ for all $a \in X$.

Definition 2.3. A 2-metric space (X, d) is said to be *complete* if every Cauchy sequence in X is convergent.

Note that, in a 2-metric space (X, d), a convergent sequence need not be a Cauchy sequence, but every convergent sequence is a Cauchy sequence when the 2-metric d is continuous on X ([21]).

Definition 2.4. Let f and g be mappings from a 2-metric space (X, d) into itself. f and g are said to be *compatible* if

$$\lim_{n \to \infty} d(fgx_n, gfx_n, a) = 0, \quad \forall a \in X,$$

whenever $\{x_n\}_{n \in \mathbb{N}} \subseteq X$ such that $\lim_{n \to \infty} fx_n = \lim_{n \to \infty} fx_n = t$ for some $t \in X$; f and g are said to be *compatible of type* (P) if

$$\lim_{n \to \infty} d(ffx_n, ggx_n, a) = 0, \quad \forall a \in X,$$

wherever $\{x_n\}_{n \in \mathbb{N}} \subseteq X$ such that $\lim_{n \to \infty} fx_n = \lim_{n \to \infty} gx_n = t$ for some $t \in X$.

Definition 2.5. A mapping f from a 2-metric space (X, d) into itself is said to be *continuous at* $x \in X$ if for every sequence $\{x_n\}_{n \in \mathbb{N}} \subset X$ such that $\lim_{n \to \infty} d(x_n, x, a) = 0$ for all $a \in X$, $\lim_{n \to \infty} d(fx_n, fx, a) = 0$. f is called *continuous in* X if it is so at all points of X.

Lemma 2.1. *For every* $t > 0$, $c(t) < t$ *if and only if* $\lim_{n \to \infty} c^n(t) = 0$, *where* c^n *denotes the n-times composition of c.*

Lemma 2.2. ([11]) *Let* f *and* g *be compatible mappings of type* (P) *from a 2-metric spaces* (X, d) *into itself. If* $ft = gt$ *for some* $t \in X$, *then* $fgt = ggt = gft = fft$.

3. Common Fixed Point Theorems

Now we provide the existence and uniqueness of common fixed points for two pairs of compatible mappings of type (P) in complete 2-metric spaces.

Theorem 3.1. *Let* (X, d) *be a complete 2-metric space with* d *continuous in* X *and let* f, g, h, t *be mappings from* (X, d) *into itself such that*

(C1) *the pairs f, h and g, t are compatible of type (P);*
(C2) *one of f, g, h and t is continuous, $f(X) \subseteq t(X)$ and $g(X) \subseteq h(X)$;*
(C3) *there exists $\phi \in \Phi$ such that*

$$d(fx, gy, a)$$

$$\leq \phi\Big(d(hx, ty, a), d(hx, fx, a), d(ty, gy, a), d(hx, gy, a), d(ty, fx, a),$$

$$\frac{1}{2}\big[d(hx, ty, a) + d(fx, gy, a)\big], \frac{d(hx, fx, a)d(ty, gy, a)}{1 + d(fx, gy, a)},$$

$$\frac{d(hx, gy, a)d(ty, fx, a)}{1 + d(fx, gy, a)}, \frac{d(hx, fx, a)d(ty, gy, a)}{1 + d(hx, ty, a)},$$

$$\frac{d(hx, gy, a)d(ty, fx, a)}{1 + d(hx, ty, a)}\Big), \quad \forall x, y, a \in X.$$

Then f, g, h and t have a unique common fixed point in X.

Proof. Let x_0 be an arbitrary point in X. It follows from $f(X) \subseteq t(X)$ and $g(X) \subseteq h(X)$ that there exist sequences $\{x_n\}_{n \in \omega}$ and $\{y_n\}_{n \in \mathbb{N}}$ in X such that $y_{2n+1} = tx_{2n+1} = fx_{2n}$, $y_{2n+2} = hx_{2n+2} = gx_{2n+1}$ for all $n \in \omega$. Define $d_n(a) = d(y_n, y_{n+1}, a)$ for any $a \in X$ and $n \in \mathbb{N}$. We claim that

$$(3.1) \qquad \lim_{n \to \infty} d_n(a) = 0, \quad \forall a \in X.$$

Assume that $d_{2n}(y_{2n+2}) > 0$ for some $n \in \mathbb{N}$. Indeed, by (C3) we gain that

$$d_{2n}(y_{2n+2}) = d(fx_{2n}, gx_{2n+1}, y_{2n})$$

$$\leq \phi\Big(0, 0, d_{2n}(y_{2n+2}), 0, 0, \frac{1}{2}d_{2n}(y_{2n+2}), 0, 0, 0, 0\Big)$$

$$\leq \phi^*(d_{2n}(y_{2n+2})) < d_{2n}(y_{2n+2}),$$

which is absurd. Hence $d_{2n}(y_{2n+2}) = 0$ for all $n \in \mathbb{N}$. Similarly, we have $d_{2n-1}(y_{2n+1}) = 0$ for all $n \in \mathbb{N}$. Hence $d_n(y_{n+2}) = 0$ for all $n \in \mathbb{N}$. Notice that, for all $a \in X$ and $n \in \mathbb{N}$,

$$(3.2) \qquad d(y_n, y_{n+2}, a) \leq d_n(y_{n+2}) + d_n(a) + d_{n+1}(a) = d_n(a) + d_{n+1}(a).$$

In light of (C3) and (3.2), we have

$$d_{2n+1}(a)$$

$$= d(fx_{2n}, gx_{2n+1}, a)$$

$$\leq \phi\Big(d_{2n}(a), d_{2n}(a), d_{2n+1}(a), d(y_{2n}, y_{2n+2}, a), 0, \frac{1}{2}\big[d_{2n}(a) + d_{2n+1}(a)\big],$$

$$\frac{d_{2n}(a)d_{2n+1}(a)}{1 + d_{2n+1}(a)}, 0, \frac{d_{2n}(a)d_{2n+1}(a)}{1 + d_{2n}(a)}, 0\Big)$$

$$\leq \phi\Big(d_{2n}(a), d_{2n}(a), d_{2n+1}(a), d_{2n}(a) + d_{2n+1}(a), 0, \frac{1}{2}\big[d_{2n}(a) + d_{2n+1}(a)\big],$$

$$\frac{d_{2n}(a)d_{2n+1}(a)}{1 + d_{2n+1}(a)}, 0, \frac{d_{2n}(a)d_{2n+1}(a)}{1 + d_{2n}(a)}, 0\Big), \quad \forall a \in X, n \in \mathbb{N}.$$

Suppose that $d_{2n+1}(a) > d_{2n}(a)$ for some $a \in X$ and $n \in \mathbb{N}$ in the above inequalities, it is easy to verify that

$$d_{2n+1}(a) \leq \phi\big(d_{2n+1}(a), d_{2n+1}(a), d_{2n+1}(a), 2d_{2n+1}(a), 0,$$
$$d_{2n+1}(a), d_{2n+1}(a), d_{2n+1}(a), d_{2n+1}(a), d_{2n+1}(a)\big)$$
$$< d_{2n+1}(a),$$

which is a contradiction. Hence we get that $d_{2n+1}(a) \leq d_{2n}(a)$ for any $a \in X$ and $n \in \mathbb{N}$ and so $d_{2n+1}(a) \leq \phi^*(d_{2n}(a))$ for any $a \in X$ and $n \in \mathbb{N}$.

Similarly, by (C3) and (3.2), we also have $d_{2n}(a) \leq \phi^*(d_{2n-1}(a))$ for any $a \in X$ and $n \in \mathbb{N}$.

Consequently, $d_{n+1}(a) \leq \phi^*(d_n(a))$ for all $a \in X$ and $n \in \mathbb{N}$. It follows that

$$(3.3) \qquad d_{n+1}(a) \leq \phi^*(d_n(a)) \leq \phi^{*2}(d_{n-1}(a)) \leq \cdots \leq \phi^{*n}(d_1(a)), \quad \forall a \in X, n \in \mathbb{N}.$$

It follows from (3.3) and Lemma 2.1 that (3.1) holds.

Let n, m be in $\mathbb{N}$. (3.3) leads to $0 = d_n(y_m)$ for $n \geq m$. Note that, for $n < m$,

$$d_n(y_m) = d(y_n, y_{n+1}, y_m)$$
$$\leq d(y_n, y_{n+1}, y_{m-1}) + d(y_n, y_{m-1}, y_m) + d(y_{m-1}, y_{n+1}, y_m)$$
$$= d_n(y_{m-1}) + d_{m-1}(y_n) + d_{m-1}(y_{n+1})$$
$$\leq d_n(y_{m-1}) + d_n(y_n) + d_n(y_{n+1})$$
$$= d_n(y_{m-1}) \leq d_n(y_{m-2}) \leq \cdots \leq d_n(y_{n+1}) = 0.$$

Thus, for all $n, m \in \mathbb{N}$, $d_n(y_m) = 0$.

Next, we show that, for all $i, j, k \in \mathbb{N}$,

$$(3.4) \qquad\qquad\qquad d(y_i, y_j, y_k) = 0.$$

Without loss of generality, we may assume that $i \leq j$. It follows that

$$d(y_i, y_j, y_k) \leq d(y_i, y_j, y_{i+1}) + d(y_i, y_{i+1}, y_k) + d(y_{i+1}, y_j, y_k)$$
$$= d_i(y_j) + d_i(y_k) + d(y_{i+1}, y_j, y_k)$$
$$= d(y_{i+1}, y_j, y_k) \leq d(y_{i+2}, y_j, y_k)$$
$$\leq \cdots \leq d(y_{j-1}, y_j, y_k) = d_{j-1}(y_k) = 0.$$

Therefore, (3.4) holds.

In order to prove that $\{y_n\}_{n \in \mathbb{N}}$ is a Cauchy sequence, it is sufficient to prove that $\{y_{2n}\}_{n \in \mathbb{N}}$ is a Cauchy sequence. Clearly, $\{y_{2n}\}_{n \in \mathbb{N}}$ is a Cauchy sequence. Otherwise, for any given $\epsilon > 0$ and $a \in X$ such that, for each even integer $2k$, there exist even integers $2m(k)$ and $2n(k)$ with

$$2m(k) > 2n(k) > 2k \quad \text{and} \quad d(y_{2m(k)}, y_{2n(k)}, a) \geq \epsilon.$$

Further, let $2m(k)$ denote the least even integer exceeding $2n(k)$ satisfying

$$(3.5) \qquad d(y_{2m(k)-2}, y_{2n(k)}, a) \leq \epsilon \quad \text{and} \quad d(y_{2m(k)}, y_{2n(k)}, a) > \epsilon.$$

On account of (3.4) and (3.5), we see that

$$
\begin{aligned}
\epsilon &< d(y_{2m(k)}, y_{2n(k)}, a) \\
&\leq d(y_{2m(k)-2}, y_{2n(k)}, a) + d(y_{2m(k)}, y_{2m(k)-2}, a) + d(y_{2m(k)}, y_{2n(k)}, y_{2m(k)-2}) \\
&\leq \epsilon + d(y_{2m(k)}, y_{2m(k)-2}, y_{2m(k)-1}) + d(y_{2m(k)}, y_{2m(k)-1}, a) \\
&\quad + d(y_{2m(k)-1}, y_{2m(k)-2}, a) \\
&= \epsilon + d_{2m(k)-1}(a) + d_{2m(k)-2}(a),
\end{aligned}
$$

which means that

$$
(3.6) \qquad \lim_{k \to \infty} d(y_{2m(k)}, y_{2n(k)}, a) = \epsilon, \quad \forall a \in X.
$$

Notice that, for any $k \in \mathbb{N}$ and $a \in X$,

$$
|d(y_{2n(k)}, y_{2m(k)-1}, a) - d(y_{2n(k)}, y_{2m(k)}, a)| \leq d_{2m(k)-1}(a) + d_{2m(k)-1}(y_{2n(k)}),
$$

$$
|d(y_{2n(k)+1}, y_{2m(k)}, a) - d(y_{2n(k)}, y_{2m(k)}, a)| \leq d_{2n(k)}(a) + d_{2n(k)}(y_{2m(k)}),
$$

$$
|d(y_{2n(k)+1}, y_{2m(k)-1}, a) - d(y_{2n(k)}, y_{2m(k)-1}, a)| \leq d_{2n(k)}(a) + d_{2n(k)}(y_{2m(k)-1}).
$$

It is easy to check that

$$
(3.7) \qquad
\begin{aligned}
\lim_{k \to \infty} d(y_{2n(k)}, y_{2m(k)-1}, a) &= \lim_{k \to \infty} d(y_{2n(k)+1}, y_{2m(k)}, a) \\
&= \lim_{k \to \infty} d(y_{2n(k)+1}, y_{2m(k)-1}, a) = \epsilon, \quad \forall a \in X.
\end{aligned}
$$

In terms of (C3) and (3.4), we infer that, for all $a \in X$ and $k \in \mathbb{N}$,

$$
\begin{aligned}
d(&y_{2n(k)+1}, y_{2m(k)}, a) \\
&= d(fx_{2n(k)}, gx_{2m(k)-1}, a) \\
&\leq \phi\Big(d(y_{2n(k)}, y_{2m(k)-1}, a), d_{2n(k)}(a), d_{2m(k)-1}(a), d(y_{2n(k)}, y_{2m(k)}, a), \\
&\qquad d(y_{2m(k)-1}, y_{2n(k)+1}, a), \tfrac{1}{2}\big[d(y_{2n(k)}, y_{2m(k)-1}, a) + d(y_{2n(k)+1}, y_{2m(k)}, a)\big], \\
&\qquad \frac{d_{2n(k)}(a) d_{2m(k)-1}(a)}{1 + d(y_{2n(k)+1}, y_{2m(k)}, a)}, \frac{d(y_{2n(k)}, y_{2m(k)}, a) d(y_{2m(k)-1}, y_{2n(k)+1}, a)}{1 + d(y_{2n(k)+1}, y_{2m(k)}, a)}, \\
&\qquad \frac{d_{2n(k)}(a) d_{2m(k)-1}(a)}{1 + d(y_{2n(k)}, y_{2m(k)-1}, a)}, \frac{d(y_{2m(k)}, y_{2m(k)}, a) d(y_{2m(k)-1}, y_{2n(k)+1}, a)}{1 + d(y_{2n(k)}, y_{2m(k)-1}, a)} \Big).
\end{aligned}
$$

Taking the limit as $k \to \infty$, by (3.1), (3.6) and (3.7), we immediately conclude that

$$
\begin{aligned}
\epsilon &\leq \phi\Big(\epsilon, 0, 0, \epsilon, \epsilon, \epsilon, 0, \frac{\epsilon^2}{1+\epsilon}, 0, \frac{\epsilon^2}{1+\epsilon}\Big) \\
&\leq \phi^*(\epsilon) < \epsilon,
\end{aligned}
$$

which is impossible and therefore $\{y_n\}_{n \in \mathbb{N}}$ is a Cauchy sequence in X. By completeness of (X, d), we conclude that $\{y_n\}_{n \in \mathbb{N}}$ converges to a point $u \in X$. Thus fx_{2n}, hx_{2n}, gx_{2n+1}, and $tx_{2n+1} \to u$ as $n \to \infty$.

244 Qinhua Wu, Lili Wang, Hyeong Kug Kim and Shin Min Kang

Now, suppose that f is continuous. It follows from (C1) and Lemma 2.2 that $fhx_{2n}, hhx_{2n} \to fu$ as $n \to \infty$. In light of (C3), for all $a \in X$ and $n \in \omega$, we have

$$d(fhx_{2n}, gx_{2n+1}, a)$$

$$\leq \phi\Big(d(hhx_{2n}, tx_{2n+1}, a), d(hhx_{2n}, fhx_{2n}, a), d(tx_{2n+1}, gx_{2n+1}, a), d(hhx_{2n}, gx_{2n+1}, a),$$

$$d(tx_{2n+1}, fhx_{2n}, a), \frac{1}{2}\big[d(hhx_{2n}, tx_{2n+1}, a) + d(fhx_{2n}, gx_{2n+1}, a)\big],$$

$$\frac{d(hhx_{2n}, fhx_{2n}, a)d(tx_{2n+1}, gx_{2n+1}, a)}{1 + d(fhx_{2n}, gx_{2n+1}, a)}, \frac{d(hhx_{2n}, gx_{2n+1}, a)d(tx_{2n+1}, fhx_{2n}, a)}{1 + d(fhx_{2n}, gx_{2n+1}, a)},$$

$$\frac{d(hhx_{2n}, fhx_{2n}, a)d(tx_{2n+1}, gx_{2n+1}, a)}{1 + d(hhx_{2n}, tx_{2n+1}, a)}, \frac{d(hhx_{2n}, gx_{2n+1}, a)d(tx_{2n+1}, fhx_{2n}, a)}{1 + d(hhx_{2n}, tx_{2n+1}, a)}\Big).$$

Letting $n \to \infty$, we get

$$d(fu, u, a) \leq \phi\Big(d(fu, u, a), 0, 0, d(fu, u, a), d(fu, u, a),$$

$$0, \frac{d^2(fu, u, a)}{1 + dfu, u, a)}, 0, \frac{d^2(fu, u, a)}{1 + dfu, u, a)}\Big)$$

$$\leq \phi^*(d(fu, u, a)), \quad \forall a \in X,$$

which implies that $fu = u$. It follows from (C2) that there exists a point $v \in X$ with $u = tv$. From (C3), we know that

$$d(fx_{2n}, gv, a)$$

$$\leq \phi\Big(d(hx_{2n}, tv, a), d(hx_{2n}, fx_{2n}, a), d(tv, gv, a), d(hx_{2n}, gv, a), d(tv, fx_{2n}, a),$$

$$\frac{1}{2}\big[d(hx_{2n}, tv, a) + d(fx_{2n}, gv, a)\big], \frac{d(hx_{2n}, fx_{2n}, a)d(tv, gv, a)}{1 + d(fx_{2n}, gv, a)},$$

$$\frac{d(hx_{2n}, gv, a)d(tv, fx_{2n}, a)}{1 + d(fx_{2n}, gv, a)}, \frac{d(hx_{2n}, fx_{2n}, a)d(tv, gv, a)}{1 + d(hx_{2n}, tv, a)},$$

$$\frac{d(hx_{2n}, gv, a)d(tv, fx_{2n}, a)}{1 + d(hx_{2n}, tv, a)}\Big), \quad \forall a \in X, n \in \mathbb{N}.$$

As $n \to \infty$ in the above inequality, we obtain

$$d(u, gv, a) \leq \phi\Big(0, 0, d(u, gv, a), d(u, gv, a), 0, \frac{1}{2}d(u, gv, a), 0, 0, 0, 0\Big)$$

$$\leq \phi^*(d(u, gv, a)), \quad \forall a \in X,$$

which means that $u = gv = tv = fu$. It follows from Lemma 2.2 that $gu = gtv = tgv = tu$. Using (C3), we have

$$d(fx_{2n}, gu, a)$$

$$\leq \phi\Big(d(hx_{2n}, tu, a), d(hx_{2n}, fx_{2n}, a), d(tu, gu, a), d(hx_{2n}, gu, a), d(tu, fx_{2n}, a),$$

$$\frac{1}{2}\big[d(hx_{2n}, tu, a) + d(fx_{2n}, gu, a)\big], \frac{d(hx_{2n}, fx_{2n}, a)d(tu, gu, a)}{1 + d(fx_{2n}, gu, a)},$$

$$\frac{d(hx_{2n}, gu, a)d(tu, fx_{2n}, a)}{1 + d(fx_{2n}, gu, a)}, \frac{d(hx_{2n}, fx_{2n}, a)d(tu, gu, a)}{1 + d(hx_{2n}, tu, a)},$$

$$\frac{d(hx_{2n}, gu, a)d(tu, fx_{2n}, a)}{1 + d(hx_{2n}, tu, a)}\Big), \quad \forall a \in X, n \in \mathbb{N}.$$

Taking the limit as $n \to \infty$ in the above inequality, we conclude that

$$d(u, gu, a) \le \phi\Big(d(u, gu, a), 0, 0, d(u, gu, a), d(u, gu, a), d(u, gu, a),$$

$$0, \frac{d^2(u, gu, a)}{1 + d(u, gu, a)}, 0, \frac{d^2(u, gu, a)}{1 + d(u, gu, a)}\Big)$$

$$\le \phi^*(d(u, gu, a)), \quad \forall a \in X,$$

which gives that $u = gu$. (C2) means that there is $w \in X$ with $u = hw$. In terms of (C3), we know that

$$d(fw, u, a) = d(fw, gu, a)$$

$$\le \phi\Big(0, d(u, fw, a), 0, 0, d(u, fw, a), \frac{1}{2}d(u, fw, a), 0, 0, 0, 0\Big)$$

$$\le \phi^*(d(u, fw, a)), \quad \forall a \in X,$$

which implies that $fw = u$. Hence $fw = hw$. Lemma 2.2 ensures that $u = fu = fhw = hfw = hu$. That is, f, g, h and t have a common fixed point $u \in X$. It is easy to show that u is a unique common fixed point of f, g, h and t.

Similarly, we can complete the proof when g or h or t is continuous. This completes the proof.   □

Clearly, we have the following result.

Theorem 3.2. *Let (X, d) be a complete 2-metric space with d continuous in X and let f, h, t be mappings from (X, d) into itself such that*
(C4) *the pairs f, h and f, t are compatible of type (P);*
(C5) *one of f, h and t is continuous and $f(X) \subseteq t(X) \cap h(X)$;*
(C6) *there exists $\phi \in \Phi$ such that*

$$d(fx, fy, a)$$

$$\le \phi\Big(d(hx, ty, a), d(hx, fx, a), d(ty, fy, a), d(hx, fy, a), d(ty, fx, a),$$

$$\frac{1}{2}\big[d(hx, ty, a) + d(fx, fy, a)\big], \frac{d(hx, fx, a)d(ty, fy, a)}{1 + d(fx, fy, a)},$$

$$\frac{d(hx, fy, a)d(ty, fx, a)}{1 + d(fx, fy, a)}, \frac{d(hx, fx, a)d(ty, gy, a)}{1 + d(hx, ty, a)},$$

$$\frac{d(hx, fy, a)d(ty, fx, a)}{1 + d(hx, ty, a)}\Big), \quad \forall x, y, a \in X.$$

Then f, h and t have a unique common fixed point in X.

Remark 3.1. Theorem 3.1 extends and unifies Theorem 2 of Khan-Fisher [6] and Theorem 1 of Kubiak [7].

REFERENCES

1. R. P. Dubey, *Some fixed point theorems on expansion mappings in 2-metric spaces*, Pure Appl. Math. Sci. **32** (1990), 33-37.

2. S. Gähler, *2-metrische Räume und ihr topologische Struktur*, Math. Nachr. **26** (1963), 115–148.

3. M. Imdad, M. S. Khan and M. D. Khan, *A common fixed point theorem in 2-metric spaces*, Math. Japon. **36** (1991), 907–914

4. K. Iséki, *Fixed point theorems in 2-metric spaces*, Math. Seminar Notes, Kobe Univ. **3** (1975), 133–136.

5. K. Iséki, P. L. Sharma and B. K. Sharma, *Contraction type mappings on 2-metric spaces*, Math. Japon. **21** (1976), 67–70.

6. M. S. Khan, *On fixed point theorems in 2-metric spaces*, Publ. Inst. Math. (Beograd) (N.S.) **41** (1980), 107–112.

7. M. S. Khan and M. Swaleh, *Results concerning fixed pints in 2-metric spaces*, Math. Japon. **29** (1984), 519–525.

8. Z. Liu, *Compatible mappings and fixed points*, Acta Sci. Math. **65** (1999), 371–383.

9. Z. Liu and F. R. Zhang, *Characterizations of common fixed points in 2-metric spaces*, Rostock Math. Kolloq. **55** (2001), 49–64.

10. Z. Liu, F. R. Zhang and J. F. Mao, *Common fixed points for compatible mappings of type (A)*, Bull. Malaysian Math. Soc. **22** (1999), 67–86.

11. H. K. Pathak, S. S. Chang and Y. J. Cho, *Fixed point theorems for compatible mappings of type (P)*, Indian J. Math. **36** (1994), 151–166.

12. D. Tan, Z. Liu and J. K. Kim, *Common fixed point for compatible mappings of type (P) in 2-metric spaces*, Nonlinear Funct. Anal. Appl. **8** (2003), 215–232.

Nonlinear Functional Analysis and Applications, Volume 1, 247–257

GENERALIZED CONTRACTION MAPPING PRINCIPLE IN MENGER PROBABILISTIC METRIC SPACES

Shih-sen Chang[1], Yeol Je Cho[2,*] and Jong Kyu Kim[3]

ABSTRACT. In this paper, some new fixed point theorems for nonlinear contractive type and nonlinear compatible type mapping in complete Menger probabilistic metric spaces are proved.

1. Introduction and Preliminaries

Menger introduced the notion of a probabilistic metric space in 1942 and since then the theory of probabilistic metric spaces has developed in many directions ([3, 9]). The idea of Menger was to use distribution functions instead of non-negative real numbers as values of the metric. The notion of a probabilistic metric space corresponds to situations when we do not know exactly the distance between two points, but we know probabilities of possible values of this distance. A probabilistic generalization of metric spaces appears to be interest in the investigation of physical quantities and physiological thresholds. It is also of fundamental importance in probabilistic functional analysis.

The purpose of this paper is to prove some existence theorems of fixed points for nonlinear contractive type and nonlinear compatible type mappings in complete Menger probabilistic metric spaces. In the sequel, we shall adopt the usual terminology, notation and conventions of the theory of probabilistic metric as in [1-5, 9].

Throughout this paper, let $\mathbb{R}$ be the set of all real numbers and $\mathbb{R}^+$ be the set of all nonnegative real numbers. A mapping $\mathcal{F} : \mathbb{R} \to \mathbb{R}^+$ is called a *distribution function*

Received November 9, 2009.   * Corresponding author.

2000 *Mathematics Subject Classification.* 54E40, 54E35, 54H25.

Key words and phrases. Probabilistic metric spaces, completeness, nonlinear contraction, compatible mapping, fixed point theorem.

[1] Department of Mathematics, Yibin University, Yibin, Sichuan 644007, People's Republic of China (*E-mail:* changss@yahoo.cn)

[2] Department of Mathematics Education and the Research Institute of Natural Sciences, Gyeongsang National University, Jinju 660-701, Korea (*E-mail:* yjcho@gnu.ac.kr)

[3] Department of Mathematics Education, Kyungnam University, Masan 631-701, Korea (*E-mail:* jongkyuk@kyungnam.ac.kr)

(briefly, d.f.) if it is left-continuous and non-decreasing with

$$\inf_{x\in\mathbb{R}} \mathcal{F}(x) = 0, \quad \sup_{x\in\mathbb{R}} \mathcal{F}(x) = 1.$$

In the sequel, we denote by Δ^+ the set of all distribution functions on $\mathbb{R}$. The space Δ^+ is partially ordered by the usual point-wise ordering of functions, i.e., $\mathcal{F} \leq \mathcal{G}$ if and only if $\mathcal{F}(t) \leq \mathcal{G}(t)$ for all $t \in \mathbb{R}$. The maximal element for Δ^+ in this order is the d.f. given by

$$H(t) = \begin{cases} 0, & t \leq 0, \\ 1, & t > 0. \end{cases}$$

Definition 1.1. ([3, 9]) A mapping $T : [0,1] \times [0,1] \to [0,1]$ is called a *continuous t-norm* if T satisfies the following conditions:

(a) T is commutative and associative;
(b) T is continuous;
(c) $T(a,1) = a$ for all $a \in [0,1]$;
(d) $T(a,b) \leq T(c,d)$ whenever $a \leq c$, $b \leq d$ for all $a,b,c,d \in [0,1]$.

Two typical examples of continuous t-norm are $T(a,b) = ab$ and $T(a,b) = \min\{a,b\}$.

Now t-norms are recursively defined by $T^1 = T$ and

$$T^n(x_1, x_2, \cdots, x_{n+1}) = T(T^{n-1}(x_1, x_2, \cdots, x_n), x_{n+1})$$

for all $n \geq 2$ and $x_i \in [0,1]$ $(i = 1, 2, \cdots, n+1)$.

Definition 1.2. A *Menger Probabilistic Metric space* (briefly, Menger PM-space) is a triple $(X, \mathcal{F}, T)$, where X is a non-empty set, T is a continuous t-norm and $\mathcal{F}$ is a mapping from $X \times X$ into Δ^+ satisfying the following conditions (in the sequel, we use $F_{x,y}$ to denote $\mathcal{F}(x,y)$): for all $x, y, z \in X$,

(PM1) $F_{x,y}(t) = H(t)$ for all $t > 0$ if and only if $x = y$;
(PM2) $F_{x,y}(t) = F_{y,x}(t)$;
(PM3) $F_{x,z}(t+s) \geq T(F_{x,y}(t), F_{y,z}(s))$ for all $x, y, z \in X$ and $t, s \geq 0$.

Definition 1.3. Let $(X, \mathcal{F}, T)$ be a Menger PM-space.

(1) A sequence $\{x_n\}$ in X is said to be *convergent* to $x \in X$ if, for every $\epsilon > 0$ and $\lambda > 0$, there exists a positive integer N such that $F_{x_n,x}(\epsilon) > 1 - \lambda$ whenever $n \geq N$.

(2) A sequence $\{x_n\}$ in X is called a *Cauchy sequence* if, for every $\epsilon > 0$ and $\lambda > 0$, there exists a positive integer N such that $F_{x_n,x_m}(\epsilon) > 1 - \lambda$ whenever n, $m \geq N$.

(3) A Menger PM-space $(X, \mathcal{F}, T)$ is said to be *complete* if every Cauchy sequence in X is convergent to a point in X.

Definition 1.4. Let $(X, \mathcal{F}, T)$ be a Menger PM space and $p \in X$ be a given point.
(1) For any given $\epsilon > 0$ and $\lambda > 0$ the set

$$Np(\epsilon, \lambda) = \{q \in X : F_{p,q}(\epsilon) > 1 - \lambda\}$$

is called the strong (ϵ, λ)-*neighborhood* of p.

(2) The strong neighborhood system for X is the union $\bigcup_{p\in X} M_p$, where $M_p = \{N_p(\epsilon, \lambda) : \epsilon > 0, \lambda > 0\}$.

Remark 1. It should be pointed out that the strong neighborhood system determines a Hausdorff topology σ on X ([3, 9]).

Definition 1.5. ([3]) Let X be a nonempty set and $\{d_\alpha : \alpha \in (0,1)\}$ be a family of mappings from $X \times X$ into $\mathbb{R}^+$. The ordered pair $(X, d_\alpha : \alpha \in (0,1))$ is called a *generating space of quasi-metric family*, where $\{d_\alpha : \alpha \in (0,1)\}$ is called the *family of quasi-metric on X* if it satisfies the following conditions:

(QM-1) $d_\alpha(x,y) = 0$ for all $\alpha \in (0,1)$ if and only if $x = y$;
(QM-2) $d_\alpha(x,y) = d_\alpha(y,x)$ for all $\alpha \in (0,1)$ and $x, y \in X$;
(QM-3) for any given $\alpha \in (0,1)$, there exists $\mu \in (0,\alpha]$ such that

$$d_\alpha(x,y) \le d_\mu(x,z) + d_\mu(z,y), \quad \forall x, y, z \in X$$

(QM-4) for any give $x, y \in X$, the function $\alpha \mapsto d_\alpha(x,y)$ is nonincreasing and left-continuous.

Lemma 1.1. *Let $(X, \mathcal{F}, T)$ be a Menger PM-space with a t-norm T satisfying the following conditions:*

$$(1.1) \qquad\qquad \sup_{t<1} T(t,t) = 1.$$

For any given $\lambda \in (0,1)$, define a mapping $E_{\lambda,F}(x,y) : X \times X \to R^+$ as follows:

$$(1.2) \qquad\qquad E_{\lambda,F}(x,y) = \inf\{t > 0 : F_{x,y} > 1 - \lambda\}.$$

Then we have the following:

(1) $(X, E_{\lambda,F} : \lambda \in (0,1))$ *is a generating space of the quasi-metric family* $\{E_{\lambda,F} : \lambda \in (0,1)\}$.

(2) *The topology induced by the quasi-metric family* $\{E_{\lambda,F} : \lambda \in (0,1)\}$ *on X coincides with the (ϵ, λ)-topology on X.*

Proof. (1) From Definition 1.5, it is easy to see that the family $\{E_{\lambda,F} : \lambda \in (0,1)\}$ satisfies the conditions (QM-1) and (QM-2).

Next, we prove that $E_{\lambda,F}$ is left-continuous in $\lambda \in (0,1)$. In fact, for any given $\lambda_1 \in (0,1)$ and $\epsilon > 0$, by the definition of $E_{\lambda,F}$, there exists $t_1 > 0$ such that $t_1 < E_{\lambda_1,F}(x,y)+\epsilon$ and $F_{x,y}(t_1) > 1 - \lambda_1$. Letting $\delta = F_{x,y}(t_1) - (1 - \lambda_1) > 0$ and $\lambda \in (\lambda_1 - \delta, \lambda_1)$, we have

$$1 - \lambda_1 < 1 - \lambda < 1 - (\lambda_1 - \delta) = F_{x,y}(t_1).$$

This implies that

$$t_1 \in \{t > 0 : F_{x,y}(t) > 1 - \lambda\}.$$

Hence we have

$$E_{\lambda_1,F}(x,y) \le E_{\lambda,F}(x,y) = \inf\{t > 0 : F_{x,y} > 1 - \lambda\} \le t_1 < E_{\lambda_1,F}(x,y) + \epsilon,$$

which shows $\lambda \mapsto E_{\lambda,F}$ is left-continuous.

Now, we prove that, for any given $(x,y) \in X \times X$, $E_{\lambda,F}$ is nonincreasing in $\lambda \in (0,1)$. In fact, for any $\lambda_1, \lambda_2 \in (0,1)$ with $\lambda_1 < \lambda_2$, we have

$$\{t > 0 : F_{x,y}(t) > 1 - \lambda_1\} \subset \{t > 0 : F_{x,y}(t) > 1 - \lambda_2\}.$$

Hence it follows that

$$E_{\lambda_2,F} = \inf\{t > 0 : F_{x,y}(t) > 1 - \lambda_2\}$$
$$\leq E_{\lambda_1,F} = \inf\{t > 0 : F_{x,y}(t) > 1 - \lambda_1\}$$

and so the condition (QM-4) is proved.

Finally, we prove that $\{E_{\lambda,F} : \lambda \in (0,1)\}$ also satisfies the condition (QM-3).

In fact, for any given $x, y \in X$ and $\lambda \in (0,1)$, by the condition (1.1), there exists $\mu \in (0, \lambda]$ such that

$$T^n(1 - \mu, 1 - \mu, \cdots, 1 - \mu) > 1 - \lambda.$$

Letting $E_{\mu,F}(x, x_1) = \delta_1$, $E_{\mu,F}(x_1, x_2) = \delta_2, \cdots$, $E_{\mu,F}(x_n, y) = \delta_{n+1}$, then, from (1.2), we have that that, for any $\epsilon > 0$,

$$F_{x,x_1}(\delta_1 + \epsilon) > 1 - \mu, \ F_{x_1,x_2}(\delta_2 + \epsilon) > 1 - \mu, \ \cdots F_{x_n,y}(\delta_{n+1} + \epsilon) > 1 - \mu$$

and so

$$F_{x,y}(\delta_1 + \delta_2 + \cdots + \delta_{n+1} + (n+1)\epsilon)$$
$$\geq T^n(F_{x,x_1}(\delta_1 + \epsilon), \ F_{x_1,x_2}(\delta_2 + \epsilon), \ \cdots, F_{x_n,y}(\delta_{n+1} + \epsilon))$$
$$\geq T^n(1 - \mu, \ 1 - \mu, \ \cdots, \ 1 - \mu) > 1 - \lambda.$$

This implies that

$$E_{\lambda,F} = \inf\{t > 0 : F_{x,y}(t) > 1 - \lambda\} \leq \delta_1 + \delta_2 + \cdots + \delta_{n+1} + (n+1)\epsilon.$$

By the arbitrariness of $\epsilon > 0$, we have

(1.3) $$E_{\lambda,F} \leq E_{\mu,F}(x, x_1) + E_{\mu,F}(x_1, x_2) + \cdots + E_{\mu,F}(x_n, y)$$

for any $x, y, x_1, x_2, \cdots, x_n \in X$. Especially, if $n = 1$, then the condition (QM-3) is proved. The conclusion (1) is proved.

Now. we prove the conclusion (2). For the purpose, it is sufficient to prove that, for any given $\epsilon > 0$ and $\lambda \in (0,1)$,

$$E_{\lambda,F}(x, y) < \epsilon \quad \Longleftrightarrow \quad F_{x,y}(\epsilon) > 1 - \lambda.$$

In fact, if $E_{\lambda,F}(x, y) < \epsilon$, then it follow from (1.2) that $F_{x,y}(\epsilon) > 1 - \lambda$.

Conversely, if $F_{x,y}(\epsilon) > 1 - \lambda$, since $F_{x,y}$ is a left-continuous distribution function, there exists $\mu > 0$ such that $F_{x,y}(\epsilon - \mu) > 1 - \lambda$ and so $E_{\lambda,F}(x, y) \leq \epsilon - \mu < \epsilon$. This completes the proof. ∎

Remark 2. From Lemma 1.1, it is easy to see that a sequence $\{x_n\}$ in a Menger PM-space $(X, \mathcal{F}, T)$ is convergent in the (ϵ, λ)-topology σ if and only if $E_{\lambda,F}(x_n, x) \to 0$ for any $\lambda \in (0,1)$. Also, a sequence $\{x_n\}$ in a Menger PM-space $(X, \mathcal{F}, T)$ is a Cauchy sequence in the (ϵ, λ)-topology if and only if $E_{\lambda,F}(x_n, x_m) \to 0$ for any $\lambda \in (0,1)$ as $n, m \to \infty$.

Definition 1.6. A function $\phi : [0, \infty) \to [0, \infty)$ is said to *satisfy the condition* (Φ) if it is non-decreasing and $\sum_{n=1}^{\infty} \phi^n(t) < \infty$ for all $t > 0$, where $\phi^n(t)$ denotes the n-th iterative function of $\phi(t)$.

Remark 3. ([11]) If $\phi : [0, \infty) \to [0, \infty)$ satisfies the condition (Φ), then $\phi(t) < t$ for any $t > 0$. Further, if $t \leq \phi(t)$, then $t = 0$.

Lemma 1.2. ([8]) Let $(X, \mathcal{F}, T)$ be a Menger PM-space. Suppose that the function $\phi : [0, \infty) \to [0, \infty)$ is onto and strictly increasing. Then

$$\inf\{\phi^n(t) > 0, F_{x,y}(t) > 1 - \lambda\} \leq \phi^n(\inf\{t > 0 : F_{x,y}(t) > 1 - \lambda\})$$

for any $x, y \in X$, $\lambda \in (0, 1)$ and $n \geq 1$.

Lemma 1.3. Let $(X, \mathcal{F}, T)$ be a Menger PM space and $\{x_n\}$ be a sequence in X such that

(1.4) $$F_{x_n, x_{n+1}}(\phi^n(t)) \geq F_{x_0, x_1}(t), \quad \forall t > 0, \ n \geq 1,$$

where $\phi : [0, \infty) \to [0, \infty)$ is onto, strictly increasing and satisfies the condition (Φ). If

$$E_F(x_0, x_1) := \sup_{\lambda \in (0,1)} \{E_{\lambda, F}(x_0, x_1)\} < \infty,$$

then $\{x_n\}$ is a Cauchy sequence in X.

Proof. For any $\lambda \in (0, 1)$, it follows from Lemma 1.2 and the condition (1.4) that

$$\begin{aligned}
E_{\lambda, F}(x_n, x_{n+1}) &= \inf\{t > 0, F_{x_n, x_{n+1}}(t)) > 1 - \lambda\} \\
&= \inf\{\phi^n(\phi^n)^{-1}(t)) > 0, F_{x_n, x_{n+1}}(\phi^n(\phi^n)^{-1}(t)) > 1 - \lambda\} \\
&\leq \inf\{\phi^n(\phi^n)^{-1}(t)) > 0, F_{x_0, x_1}(\phi^n)^{-1}(t)) > 1 - \lambda\} \\
(1.5) \qquad &\leq \phi^n(\inf\{(\phi^n)^{-1}(t) > 0, F_{x_0, x_1}(\phi^n)^{-1}(t)) > 1 - \lambda\}) \\
&= \phi^n(\inf\{t > 0, F_{x_0, x_1}(t) > 1 - \lambda\}) \\
&= \phi^n(E_{\lambda, F}(x_0.x_1)) \\
&\leq \phi^n(E_F(x_0, x_1)).
\end{aligned}$$

For any given positive integers m, n with $m > n$ and $\lambda \in (0, 1)$, by Lemma 1.1, there exists $\mu \in (0, \lambda]$ such that

$$\begin{aligned}
E_{\lambda, F}(x_n, x_m) &\leq E_{\mu, F}(x_n, x_{n+1}) + E_{\mu, F}(x_{n+1}, x_{n+2}) + \cdots + E_{\mu, F}(x_{m-1}, x_m) \\
&\leq \sum_{j=n}^{m-1} \phi^j(E_F(x_0, x_1)) \to 0 \quad (n, m \to \infty).
\end{aligned}$$

Therefore, by Lemma 1.1, $\{x_n\}$ is a Cauchy sequence in X. This completes the proof. $\square$

2. Generalized Contraction Mapping Principle in Menger PM-spaces

Theorem 2.1. Let $(X, \mathcal{F}, T)$ be a complete Menger PM-space, $\{A_i : i = 1, 2, \cdots\}$ be a sequence of mappings from X into itself such that, for any two mappings A_i, A_j $(i \neq j)$, $x, y \in X$ and $t > 0$,

(2.1) $$F_{A_i x, A_j y}(\phi(t)) \geq \min\{F_{x,y}(t), \ F_{x, A_i x}(t), \ F_{y, A_j y}(t),$$

where $\phi : [0, \infty) \to [0, \infty)$ is onto, strictly increasing, continuous and satisfies the condition (Φ). If there exists $x_0 \in X$ such that

(2.2) $$E_F(x_0, A_1 x_0) := \sup_{\lambda \in (0,1)} E_{\lambda, F}(x_0, A_1 x_0) < \infty,$$

then $\{A_i\}$ has a unique common fixed point x^* in X. Further, the sequence $\{x_n\}$ defined by

$$(2.3) \qquad\qquad x_n = A_n x_{n-1}, \quad \forall n \geq 1,$$

converges to x^* in the (ϵ, λ)-topology of X.

Proof. It follows from (2.1) and (2.3) that

$$
\begin{aligned}
(2.4) \quad F_{x_1,x_2}(\phi(t)) &= F_{A_1 x_0, A_2 x_1}(\phi(t)) \\
&\geq \min\{F_{x_0,x_1}(t), F_{x_0,x_1}(t), F_{x_1,x_2}(t)\} \\
&= \min\{F_{x_0,x_1}(t), F_{x_1,x_2}(t)\}, \quad \forall t > 0
\end{aligned}
$$

If $F_{x_1,x_2}(t) < F_{x_0,x_1}(t)$, then, from (2.4), it follows that

$$F_{x_1,x_2}(\phi(t)) \geq F_{x_1,x_2}(t), \quad \forall t > 0.$$

By induction, we can prove that, for any positive integer n,

$$F_{x_1,x_2}(\phi^n(t)) \geq F_{x_1,x_2}(t), \quad \forall t > 0.$$

Letting $n \to \infty$, we have $F_{x_1,x_2}(t) = 0$ for all $t > 0$. This contradicts that $F_{x_1,x_2}(t)$ is a distribution function. Hence we have

$$F_{x_1,x_2}(\phi(t)) \geq F_{x_0,x_1}(t), \quad \forall t > 0.$$

Similarly, we have

$$
\begin{aligned}
F_{x_2,x_3}(\phi^2(t)) &= F_{A_2 x_1, A_3 x_2}(\phi^2(t)) \\
&\geq \min\{F_{x_1,x_2}(\phi(t)), F_{x_1,x_2}(\phi(t)), F_{x_2,x_3}(\phi(t))\} \\
&= F_{x_1,x_2}(\phi(t)) \\
&\geq F_{x_0,x_1}(t), \quad \forall t > 0.
\end{aligned}
$$

By induction, we can prove that

$$(2.5) \qquad F_{x_n,x_{n+1}}(\phi^n(t)) \geq F_{x_0,x_1}(t), \quad \forall n \geq 1, t > 0.$$

It follows from the condition (2.2) and Lemma 1.3 that $\{x_n\}$ is a Cauchy sequence in X. Since X is complete, there exists a point $x^* \in X$ such that $x_n \to x^*$.

Next, we prove that x^* is the unique common fixed point of $\{A_i\}$. In fact, for any given positive integers i, n with $n > i$ and $\lambda \in (0,1)$, by Lemma 1.2, it follows that

$$
\begin{aligned}
E_{\lambda,F}(x_n, T_i x^*) &= \inf\{t > 0, F_{T_n x_{n-1}, T_i x^*}(t) > 1 - \lambda\} \\
&\leq \inf\{t > 0, \min\{F_{x_{n-1},x^*}(\phi^{-1}(t)), F_{x_{n-1},x_n}(\phi^{-1}(t)), \\
&\qquad F_{x^*, T_i(x^*)}(\phi^{-1}(t))\} > 1 - \lambda\} \\
&= \inf\{\phi(\phi^{-1})(t) > 0, \min\{F_{x_{n-1},x^*}(\phi^{-1}(t)), F_{x_{n-1},x_n}(\phi^{-1}(t)), \\
&\qquad F_{x^*, T_i(x^*)}(\phi^{-1}(t))\} > 1 - \lambda\} \\
&\leq \phi(\inf\{t > 0, \min\{F_{x_{n-1},x^*}(t), F_{x_{n-1},x_n}(t), F_{x^*, T_i(x^*)}(t)\} > 1 - \lambda\}) \\
&\leq \phi(\max\{E_{\lambda,F}(x_{n-1}, x^*), E_{\lambda,F}(x_{n-1}, x_n), E_{\lambda,F}(x^*, T_i x^*)\}).
\end{aligned}
$$

By virtue of the continuity of ϕ, we have

$$
\begin{aligned}
E_{\lambda,F}(x^*, T_i x^*) &= \lim_{n\to\infty} E_{\lambda,F}(x_n, T_i x^*) \\
&\leq \lim_{n\to\infty} \phi(\max\{E_{\lambda,F}(x_{n-1}, x^*), E_{\lambda,F}(x_{n-1}, x_n), E_{\lambda,F}(x^*, T_i x^*)\}) \\
&= \phi(E_{\lambda,F}(x^*, T_i x^*)).
\end{aligned}
$$

From Remark 3, it follows that $E_{\lambda,F}(x^*, T_i x^*) = 0$, i.e., $x^* = T_i x^*$ for all $i \geq 1$.

Next, we prove that x^* is the unique common fixed point of $\{A_i\}$ in X. In fact, if $y \in X$ is also common fixed point of $\{A_i\}$, then, for any $\lambda \in (0,1)$ and any positive integers i, j wuth $i \neq j$,

$$
\begin{aligned}
E_{\lambda,F}(x^*, y) &= E_{\lambda,F}(T_i x^*, T_j y) \\
&= \inf\{t > 0, F_{T_i x^*, T_j y}(t) > 1 - \lambda\} \\
&\leq \inf\{t > 0, \min\{F_{x^*, y}(\phi^{-1}(t)), F_{x^*, x^*}(\phi^{-1}(t)), F_{y,y}(\phi^{-1}(t))\} > 1 - \lambda\} \\
&= \inf\{\phi(\phi^{-1})(t) > 0, F_{x^*, y}(\phi^{-1}(t)) > 1 - \lambda\} \\
&\leq \phi(\inf\{(\phi)^{-1}(t) > 0, F_{x^*, y}(\phi^{-1}(t)) > 1 - \lambda\}) \\
&= \phi(\inf\{t > 0, F_{x^*, y}(t) > 1 - \lambda\}) \\
&= \phi(E_{\lambda,F}(x^*, y)),
\end{aligned}
$$

i.e., $E_{\lambda,F}(x^*, y) = 0$ by Remark 3 and so $x^* = y$. This completes the proof. $\quad\square$

3. Fixed Point Theorems for Compatible Mappings

Definition 3.1. Let $(X, \mathcal{F}, T)$ be a Menger PM-space and f, S be two mappings from X into itself. f and S are called *compatible* if $F_{Sfx_n, fSx_n}(t) \to H(t)$ for any $t > 0$ whenever $\{x_n\}$ is a sequence in X such that $\{fx_n\}$ and $\{Sx_n\}$ converge, in the (ϵ, λ)-topology, to x for some $x \in X$ as $n \to \infty$.

Remark 4. It should be point out that the concept of compatible mappings was introduced by Jungck [6] in metric space. The concept of compatible mappings introduced here is a generalization of Jungck [6].

Teorem 3.1. *Let $(X, \mathcal{F}, T)$ be a complete Menger PM-space and $f, g, S, G : X \to X$ be mappings satisfying the following conditions:*
(a) $S(X) \subset g(X)$, $G(X) \subset f(X)$;
(b) $F_{Sx, Ty}(\phi(t)) \geq \min\{F_{f(x), g(y)}(t), F_{f(x), S(x)}(t), F_{g(y), G(y)}(t)\}$ for all $x, y \in X$, $t > 0$,
where the function $\phi : [0, \infty) \to [0, \infty)$ is onto, strictly increasing, continuous and satisfies the condition (Φ). If f or g is continuous, the pairs S, f and G, g both are compatible and there exists an $x_0 \in X$ such that

$$
\text{(3.1)} \qquad E_F(z_0, z_1) := \sup_{\lambda \in (0,1)} E_{\lambda,F}(z_0), z_1) < \infty,
$$

where $z_0 = S(x_0)$, $z_1 = G(x_1)$ and $g(x_1) = S(x_0)$, then the pairs S, f and G, g have a unique common fixed point z in X.

Proof. By the condition (a), there exists $x_1 \in X$ such that $Sx_0 = g(x_1) = z_0$. For this point x_1, there exists $x_2 \in X$ such that $G(x_1) = f(x_2) = z_1$. Inductively, we can construct sequences $\{x_n\}$ and $\{z_n\}$ as follows:

$$(3.2) \qquad \begin{cases} g(x_{2n+1}) = S(x_{2n}) = z_{2n}, \\ f(x_{2n+2}) = G(x_{2n+1}) = z_{2n+1}, \quad \forall n \geq 0. \end{cases}$$

It follows from the condition (b) that, for any $t > 0$,

$$(3.3) \qquad \begin{aligned} F_{z_{2n},z_{2n+1}}(\phi(t)) &= F_{S(x_{2n}),G(x_{2n+1})}(\phi(t)) \\ &\geq \min\{F_{f(x_{2n}),g(x_{2n+1})}(t), \; F_{f(x_{2n}),Sx_{2n}}(t), F_{g(x_{2n+1}),G(x_{2n+1})}(t)\} \\ &= \min\{F_{z_{2n-1},z_{2n}}(t), F_{z_{2n},z_{2n+1}}(t)\}. \end{aligned}$$

If $F_{z_{2n-1},z_{2n}}(t) \geq F_{z_{2n},z_{2n+1}}(t)$, then, from (3.3), we have $F_{z_{2n},z_{2n+1}}(\phi(t)) \geq F_{z_{2n},z_{2n+1}}(t)$ for any $t > 0$ and so

$$F_{z_{2n},z_{2n+1}}(\phi^m(t)) \geq F_{z_{2n},z_{2n+1}}(t), \quad \forall t > 0, \, m \geq 1.$$

Since ϕ satisfies the condition (Φ), we have

$$0 = \lim_{m \to \infty} F_{z_{2n},z_{2n+1}}(\phi^m(t)) \geq F_{z_{2n},z_{2n+1}}(t), \quad \forall t > 0,$$

which contradicts the fact that $F_{z_{2n},z_{2n+1}}(t)$ is a distribution function. Therefore

$$F_{z_{2n},z_{2n+1}}(\phi(t)) \geq F_{z_{2n-1},z_{2n}}(t), \quad \forall t > 0.$$

Similarly, we can prove that

$$F_{z_{2n+1},z_{2n+2}}(\phi(t)) \geq F_{z_{2n},z_{2n+1}}(t), \quad \forall t > 0.$$

These show that, for any positive integer $m \geq 1$,

$$F_{z_m,z_{m+1}}(\phi(t)) \geq F_{z_{m-1},z_m}(t), \quad \forall t > 0,$$

i.e.,

$$(3.4) \qquad F_{z_m,z_{m+1}}(t) \geq F_{z_{m-1},z_m}(\phi^{-1}(t)), \quad \forall t > 0.$$

On the other hand, it follows from Lemma 1.2 that, for any $\lambda \in (0,1)$,

$$\begin{aligned} E_{\lambda,F}(z_m, z_{m+1}) &= \inf\{t > 0, F_{z_m,z_{m+1}}(t) > 1 - \lambda\} \\ &\leq \inf\{\phi(\phi)^{-1}(t) > 0, F_{z_{m-1},z_m}(\phi^{-1}(t)) > 1 - \lambda\} \\ &\leq \phi(\inf\{\phi)^{-1}(t) > 0, F_{z_{m-1},z_m}(\phi^{-1}(t)) > 1 - \lambda\}) \\ &= \phi(\inf\{t > 0, F_{z_{m-1},z_m}(t) > 1 - \lambda\}) \\ &= \phi(E_{\lambda,F}(z_{m-1}, z_m)). \end{aligned}$$

By induction, we can prove that

$$(3.5) \qquad E_{\lambda,F}(z_m, z_{m+1}) \leq \phi(E_{\lambda,F}(z_{m-1}, z_m)) \leq \cdots \leq \phi^m(E_{\lambda,F}(z_0, z_1)).$$

By Lemma 1.1, for any given $\lambda \in (0,1)$ and for any positive integers m, n wuth $m > n$, there exists $\mu \in (0,\mu]$ such that

$$E_{\lambda,F}(z_m, z_n) \leq E_{\mu,F}(z_m, z_{m-1}) + E_{\mu,F}(z_{m-1}, z_{m-2}) + \cdots + E_{\mu,F}(z_{n-1}, z_n)$$

$$\leq \sum_{j=n}^{m-1} \phi^j(E_{\mu,F}(z_0, z_1))$$

$$\leq \sum_{j=n}^{m-1} \phi^j(E_F(z_0, z_1)) \to 0 \quad (n, m \to \infty).$$

This implies that $\{z_n\}$ is a Cauchy sequence in X. Without loss of generality, we can assume that $z_n \to z^* \in X$. Therefore, we have

(3.6) $$\begin{cases} f(x_{2n}) = z_{2n-1} \to z^*, & g(x_{2n+1}) = z_{2n} \to z^*, \\ S(x_{2n}) = z_{2n} \to z^*, & G(x_{2n+1}) = z_{2n+1} \to z^*. \end{cases}$$

By the assumption, without loss of generality, we can assume that f is continuous. Then $f^2(x_{2n}) \to f(z^*)$ and $fS(x_{2n}) \to f(z^*)$. Since S and f are compatible, we have

(3.7) $$F_{Sf(x_{2n}), fS(x_{2n})}(t) \to H(t), \quad \forall t > 0,$$

and so

$$F_{Sf(x_{2n}), fz^*}(t) \geq T(F_{Sf(x_{2n}), fS(x_{2n})}(t - \phi(t)), \ F_{fS(x_{2n}), fz^*}(\phi(t)) \to H(t), \quad \forall t > 0,$$

which shows that

(3.8) $$Sfx_{2n} \to f(z^*) \quad (n \to \infty).$$

Again, for any positive integer $n \geq 1$ and $\lambda \in (0,1)$, it follows from Lemma 1.2 that

$$E_{\lambda,F}(Sf(x_{2n}), G(x_{2n+1}))$$
$$= \inf\{t > 0, F_{Sf(x_{2n}), Gx_{2n+1}}(t) > 1 - \lambda\}$$
$$\leq \inf\{\phi(\phi)^{-1}(t) > 0, \min\{F_{f^2(x_{2n}), g(x_{2n+1})}(\phi^{-1}(t)), \ F_{f^2(x_{2n}), Sf(x_{2n})}(\phi^{-1}(t)),$$
$$F_{g(x_{2n+1}), G(x_{2n+1})}(\phi^{-1}(t))\}\}$$
$$\leq \phi(\max\{E_{\lambda,F}(f^2(x_{2n}), g(x_{2n+1})), \ E_{\lambda,F}(f^2(x_{2n}), Sf(x_{2n})),$$
$$E_{\lambda,F}(g(x_{2n+1}), G(x_{2n+1}))\}).$$

Therefore, we have

$$\lim_{n\to\infty} E_{\lambda,F}(Sf(x_{2n}), Gx_{2n+1}) = E_{\lambda,F}(fz^*, z^*) \leq \phi(E_{\lambda,F}(f(z^*), z^*)).$$

By Remark 3, $E_{\lambda,F}(fz^*, z^*) = 0$, i.e., $z^* = fz^*$.

Similarly, we can prove that

$$E_{\lambda,F}(Sz^*, G(x_{2n+1})$$
$$\leq \phi(\max\{E_{\lambda,F}(fz^*), g(x_{2n+1})), \ E_{\lambda,F}(fz^*, Sz^*), E_{\lambda,F}(g(x_{2n+1}), G(x_{2n+1}))\}).$$

Hence we have

$$
\begin{aligned}
&E_{\lambda,F}(Sz^*, z^*) \\
&= \lim_{n\to\infty} E_{\lambda,F}(Sz^*, G(x_{2n+1})) \\
&\leq \lim_{n\to\infty} \phi(\max\{E_{\lambda,F}(fz^*, g(x_{2n+1})), E_{\lambda,F}(fz^*, Sz^*), E_{\lambda,F}(g(x_{2n+1}), G(x_{2n+1}))\}) \\
&= \phi(\max\{0, E_{\lambda,F}(Sz^*, z^*), 0\}) \\
&= \phi(E_{\lambda,F}(Sz^*, z^*))
\end{aligned}
$$

and so $E_{\lambda,F}(Sz^*, z^*) = 0$, i.e., $Sz^* = z^*$. Select $z \in X$ such that $g(z) = z^* = S(z^*)$. Then $G(g(z)) = G(z^*)$ and, for any $\lambda \in (0,1)$,

$$
\begin{aligned}
E_{\lambda,F}(z^*, G(z)) &= E_{\lambda,F}(Sz^*, G(z)) \\
&= \inf\{t > 0, F_{Sz^*,G(z)}(t) > 1 - \lambda\} \\
&\leq \inf\{\phi(\phi)^{-1}(t) > 0, \min\{F_{f(z^*),g(z)}(\phi^{-1}(t)), F_{f(z^*),S(z^*)}(\phi^{-1}(t)), \\
&\qquad F_{g(z),G(z)}(\phi^{-1}(t))\} > 1 - \lambda\} \\
&\leq \phi(\max\{E_{\lambda,F}(f(z^*), g(z)), E_{\lambda,F}(f(z^*), S(z^*)), E_{\lambda,F}(g(z), G(z))\}) \\
&= \phi(\max\{E_{\lambda,F}(z^*, z^*)), \ E_{\lambda,F}(z^*, S(z^*)), E_{\lambda,F}(z^*, G(z))\}) \\
&= \phi(\max\{0, 0, E_{\lambda,F}(z^*, G(z))\}).
\end{aligned}
$$

This implies that $z^* = G(z)$ and so $g(G(z)) = g(z^*)$. Since the pair G, g are compatible and $E_{\lambda,F}(G(z), g(z)) = E_{\lambda,F}(z^*, z^*) = 0$, we get

$$
E_{\lambda,F}(Gz^*, g(z^*)) = E_{\lambda,F}(G(g(z)), g(G(z))) = 0.
$$

This shows that $Gz^* = g(z^*)$. Again, since, for any $\lambda \in (0,1)$,

$$
\begin{aligned}
E_{\lambda,F}(z^*, G(z^*)) &= E_{\lambda,F}(S(z^*), G(z^*)) \\
&= \inf\{t > 0, F_{S(z^*),G(z^*)}(t) > 1 - \lambda\} \\
&\leq \inf\{\phi(\phi)^{-1}(t) > 0, \min\{F_{f(z^*),g(z^*)}(\phi^{-1}(t)), \ F_{f(z^*),S(z^*)}(\phi^{-1}(t)), \\
&\qquad F_{g(z^*),G(z^*)}(\phi^{-1}(t))\} > 1 - \lambda\} \\
&\leq \phi(\inf\{\phi^{-1}(t) > 0, \min\{F_{z^*,G(z^*)}(\phi^{-1}(t)), F_{z^*,z^*}(\phi^{-1}(t)), \\
&\qquad F_{G(z^*),G(z^*)}(\phi^{-1}(t))\}) > 1 - \lambda\} \\
&= \phi(\max\{E_{\lambda,F}(z^*, G(z^*)), 0, 0\}) \\
&= \phi(E_{\lambda,F}(z^*, G(z^*)).
\end{aligned}
$$

This implies that $E_{\lambda,F}(z^*, G(z^*)) = 0$, i.e., $z^* = G(z^*)$ and so z^* is a common fixed point of the pairs S, f and G, g in X.

If $y \in X$ is another common fixed point of the pairs S, f and G, g, then we have

$$
\begin{aligned}
E_{\lambda,F}(y, z^*) &= E_{\lambda,F}(S(y), G(z^*)) \\
&= \inf\{t > 0, F_{S(y),G(z^*)}(t) > 1 - \lambda\} \\
&\leq \inf\{t > 0, \min\{F_{y,z^*}(\phi^{-1}(t)),\ F_{f(y),S(y)}(\phi^{-1}(t)), \\
&\qquad F_{g(z^*),G(z^*)}(\phi^{-1}(t))\} > 1 - \lambda\} \\
&\leq \phi(\max\{E_{\lambda,F}(y, z^*), 0, 0\}) \\
&= \phi(E_{\lambda,F}(y, z^*)),
\end{aligned}
$$

i.e., $y = z^*$. This completes the proof. □

Remark 5. If the mappings $S, G : X \to X$ given in Theorem 2.1 are multi-valued, we can also prove that the conclusion of Theorem 2.1 still holds.

REFERENCES

1. C. Alsina, B. Schweizer and A. Sklar, *Continuity properties of probabilistic norms,* J. Math. Anal. Appl. **208** (1997), 446–452.
2. A. Bharucha-Reid, *Fixed point theorems in probabilistic analysis,* Bull. Amer. Math. Soc. **82** (1976), 641–657.
3. S. S. Chang, Y. J. Cho and S. M. Kang, *Nonlinear Operator Theory in Probabilistic Metric Spaces,* Nova Science Publishers Inc., New York, 2001.
4. S. S. Chang, B. S. Lee, Y.J. Cho and Y. Q. Chen, *Generalized contraction mapping principle and differential equations in probabilistic metric spaces,* Proc. Amer. Math. Soc. **124(8)** (1996), 641–657.
5. O. Hadžic and E. Pap, *Fixed Point Theory in PM Spaces,* Kluwer Academic Publishers, Dordrecht, 2001.
6. G. Jungck, *Commuting maps and fixed points,* Amer. Math. Monthly **83** (1976), 261–263.
7. M. A. Khamsi and V. Y. Kreinovich, *Fixed point theorems for dissipative mappings in complete probabilistic metric spaces,* Math. Japon. **44** (1996), 513–520.
8. D. O'Regan and R. Saadati, *Nonlinear contraction theorems in probabilistic spaces,* Appl. Math. Comput. **195** (2008), 86–93.
9. B. Schweizer and A. Sklar, *Probabilistic Metric Spaces,* Elsevier, North-Holland, New York, 1983.
10. B. Singh and S. Jain, *A fixed point theorem in Menger space through weak compatibility,* J. Math. Anal. Appl. **301** (2005), 439–448.
11. S. S. Zhang, *Fixed Point Theory and Applications,* Chongqing Publishing Press, Chongqing, 1984 (in Chinese).